Planning and Design of Small Hydro Power Projects

B. N. Asthana

Planning and Design of Small Hydro Power Projects

CP

Springer

B. N. Asthana
Irrigation Department
UP, India

Visiting Faculty & Emeritus Fellow at
WRDM
IIT, Roorkee
Roorkee, India

Drongo Advisory Services Pvt. Ltd.
New Delhi, India

ISBN 978-3-031-96270-7 ISBN 978-3-031-96271-4 (eBook)
https://doi.org/10.1007/978-3-031-96271-4

Jointly published with Capital Publishing Company

This Springer imprint is published by the registered company Springer Nature Switzerland AG
The registered company address is: Gewerbestrasse 11, 6330 Cham, Switzerland

Foreword

It is my privileged pleasure to introduce the Book on *Small Hydro* authored by Dr. B. N. Asthana, a renowned expert and Consultant in Water Resources. Dr. Asthana has not only a vast professional experience in Uttar Pradesh Irrigation Department (UPID) for nearly four decades but also a rich teaching experience of about 15 years as Visiting Faculty at the Department of WRDM at IIT Roorkee, post his illustrious professional Career in UPID. I find that he has authored the book which is quite comprehensive in covering all the aspects that a student or emerging professional in small hydro development would seek to learn and lean upon.

I had a good interaction with Dr. Asthana for several decades when I held senior positions in Central Water Commission and later, as Secretary General of the International Commission on Irrigation and Drainage, a global organisation of its own kind of esteem in Water Management. I recall the comprehensive publication on Sediment Management in Water Resources Projects, covering Reservoir Sedimentation, which was brought out by the Central Board of Irrigation and Power (CBIP) about a decade back under the guidance and mentoring of a Technical Committee that I was privileged to chair. Dr. Asthana's contribution in respect of various chapters of this publication, as the author, was impressive.

His mature Consultancy advice in respect of several Water Resources Development Projects including small hydro projects, not only in India but also abroad, (like Afghanistan, Salma Dam) was crucial for Consultancy Organisation like WAPCOS (India) Ltd. who engaged Dr. Asthana to provide a high calibre technical consultancy in its contracts with Clients. Dr. Asthana has also authored a book on *Dam Engineering.*

As I go through the contents of the chapters of the book, I find that Dr. Asthana has left no stones unturned in covering the entire gamut of engineering planning, design and construction aspects on Small Hydro. He has not hesitated to draw the rich treasure that is available on the subject from the Department of Hydro and Alternate Energy with its base in IIT Roorkee, India, and various other organisations, where necessary.

Given the new thrust to find a substitute to large hydro that had associated environmental and issues related to resettlement of affected population on a sizeable scale, this work is timely. All agencies, like Government and the Public as well as Private Sector (IPPs) are attempting to launch as many small hydro projects for Capacity addition to the National Energy Grid. In the emerging scenario and special impetus for Green Energy that relies largely on Solar, Wind, etc., stabilising the grid is assuming importance with as much hydro, as is feasible to ensure dependable energy supply.

I hope that the book serves the intent with which it had been authored and would remain as a valuable addition to the students and practising young professionals on the subject.

I wish the book a great success.

Sept 11, 2024

Bangalore, India

Er. Mukuteswara Gopalakrishnan
FNAE
Formerly Secretary General,
International Commission on Irrigation
and Drainage, New Delhi
Formerly Member, Central Water
Commission, Government of India,
New Delhi
Presently Adviser to Various
Government Organizations and PSUs
of Water Resources and Hydro Power
Development Projects Including Tehri
Hydro Uttarakhand

Preface

The technique of generating electrical energy using water as an input was first developed through a small capacity generating equipment in USA in 1882. This technology was soon adopted in developing small hydro projects by many countries. India commissioned first small hydro project in 1897. Thereafter, rapid advances in engineering, manufacturing and construction techniques made it possible to construct large hydro power plants along with dam projects. India also started constructing large power projects after independence.

The experience of large multipurpose projects made it necessary to find an alternative to large hydro projects that had associated environmental and re-habitational issues of sizeable magnitude. Hence, a new thrust was given to small hydro projects as an alternative to hydro power generation and many such projects were launched by both public and private agencies. It also fitted well with the concept of Green Energy generation.

The experience with the construction of small hydro projects necessitated the need for developing a suitable technology for planning and design of such projects. The engineers engaged in such projects had also to be trained for the purpose. Several institutions like CBI&P, AHEC, IIT Roorkee came forward to work in this direction. Attempts were made to transfer the technical advances made in this subject to the field engineers through manuals, seminars, workshops and conferences at national and international levels which were held at regular intervals. Some institutes started short terms courses for the students as well as the field engineers.

The author had the opportunity to be associated with these activities. He also taught the subject at graduate level at IIT Roorkee for more than 10 years after his retirement as Chief Engineer from Irrigation Department, Uttar Pradesh (India), where he had been associated with planning and design of both large and small hydro projects in various capacities. He also worked as consultant to a number of small hydro projects. The author in this period experienced that it is difficult for the students and field engineers to assimilate the advances in the technology that are mostly available in the Journals, Seminar and Conference proceedings and the manuals.

With above experience the author has attempted this book and has tried to give the basic concepts, principles and methodology of planning and design of different types of small hydro projects (SHPs). The book is in two parts. In the first part, (Chaps. 2 to 9) deal with the layout of all the three types of SHPs along with the necessary structures required with each type, the type of electromechanical equipment and their selection criterion, size of the equipments and their impact on the layout and the size of power house, planning procedures and field investigations, hydrological and power studies and measures of sediment management. The chapters in the second part of the book deal with the design aspects of diversion structures, storage dams, intake structures, water conductors, cross drainage works, desilting structures, surge tanks, forebay, penstocks, powerhouse and hydro mechanical works. It is hoped that the book will be useful for students and professionals.

The author gratefully acknowledges the support and cooperation of the family, friends and colleagues which he received during the preparation of this text. The author is especially grateful to his brother Shri Vinod Asthana IRTS (Retd.) and Director, Drongo Advisory Services along with his office for the support during the compilation of this book. The author further gratefully acknowledges the efforts put in by Shri D. K. Agarwal, Engineer-in-Chief (Retd.), Irrigation Department (Uttar Pradesh) and Consultant AHEC, IIT Roorkee and Prof. Gopal Chauhan (Retd.) from Department of WRD&M, IIT Roorkee, in meticulously reviewing the entire text and offering valuable suggestions and comments. The author is grateful to Er. M. Gopal Krishnan (Former Secretary General ICID, New Delhi) for writing the Foreword for this book. Author also acknowledges the efforts put in by Shri Yashpal of IIT Roorkee and Shri Surender Singh Auto-CAD Operator in offering all the secretarial help in compiling the text of this book.

Author has made efforts to individually acknowledge the source of data and information used in the book, but special mention must be made of the organisations like CBI&P, BIS, AHEC-IIT Roorkee and USBR whose publications are made use of in the book.

Author would also like to gratefully acknowledge his debt to all those public and private organisations and the Institutes with whom he worked on planning and design of several small hydro projects and utilised the engineering experience gained there in authoring the book.

It is hoped that the book will be useful to the students and engineers. The chapters in first part deal with the layout of all the three type of SHPs along with the necessary structures required with each type, type of electro-mechanical equipments and their selection criteria, size of equipments and their impact on layout and size of powerhouse, planning procedures and field investigations, hydrological and power studies and measures of sediment management.

Chapters of second part deal with the design aspects of diversions structures, storage dams, intake structures, water conductors, cross drainage works, desilting structures, surge tanks, forebay, penstocks, power house and hydro-mechanical equipments.

Noida, India B. N. Asthana

Contents

Abbreviations

AHEC	Alternate Hydro Energy Centre IIT, Roorkee
BIS	Bureau of Indian Standards Manak Bhawan Delhi
CBI&P	Central Board of Irrigation and Power, New Delhi
cm	Centimetre
Cumec	Cubic Metre/Sec
CWC	Central Water Commission, New Delhi
CWPRS	Central Water Power Research Station, Pune
DPR	Detailed Project Report
F.S.L.	Full Supply Level
FRL	Full Reservoir Level
GPS	Global Positioning System
GRP	Glass Fibre Reinforced Plastics
GW	Gega Watt
HDPE	High Density Polyethylene
HEP	Hydroelectric Project
HFL	High Flood Level
I.S.	Indian Standards Code
IIT	Indian Institute of Technology
IMD	Indian Metrological Department
IRI	Irrigation Research Institute, Roorkee
Kg/Cm^2	Kilogram Per Centimetre Square
Kw	Kilowatt
Kwh	Kilo Watt Hour
LDPE	Low-Density Polyethylene
M	Metre
m^3	Cubic Metre
MCM	Million Cubic Metre
MDDL	Minimum Drawdown Level
mm	Millimetre
Mm^3	Million Metre Cube
MU	Million Units

MW	Mega Watt
MWL	Maximum Water Level
NIH	National Institute of Hydrology, Roorkee
PMF	Probable Maximum Flood
PVC	Poly Venyl Chloride
RCC	Reinforced Cement Concrete
ROR	Run of River
SHP	Small Hydro Project
T/m^2	Tonnes Per Metre Square
TWL	Tail Water Level
UP	Uttar Pradesh, India
WRD&M	Water Resources Development and Management, IIT, Roorkee

Chapter 1
Introduction

1.1 General

Small hydro projects (SHPs), as the name suggests, are basically hydro power projects with small generating capacity. In India, hydropower projects with an installed capacity equal to or less than 25 MW are termed small hydro projects. These are further classified as below

•	5 MW to 25 MW (with unit capacity not more than 5 MW)	– Small hydro projects
•	101 kW to 2 MW	– Mini hydro projects
•	Up to 100 kW	– Micro hydro projects

Since an SHP is a hydro power project, the principle of power generation, the requirement of generating equipment, i.e. turbine and generator and other components of the project of an SHP are similar to that of a major hydro project. However, the technical specifications for the planning and design of a major hydro project are not very rigidly followed in the case of an SHP. Attempts are also made to use locally available materials for the construction of an SHP to reduce the cost and time of construction of the project.

1.2 Hydro Power

The electrical power generated by using water as a resource is known as hydropower. Water is a natural yearly renewable resource available on earth through a hydrological cycle in the form of rain and snow and is used for various other purposes besides hydro power generation. Water sustains life on earth, so its sources have been harnessed by man since times immemorial. The use of water for hydropower generation is recent, about 150 years old. Hydro power generation is directly proportional to the product of flow (discharge) and the drop or head. Hence, wherever the flow and the drop, natural

B. N. Asthana, *Planning and Design of Small Hydro Power Projects*,
https://doi.org/10.1007/978-3-031-96271-4_1

or man made is available, hydropower can be generated. Thus, the potential/kinetic energy of water is converted into electrical energy. To achieve this process of energy conversion, water is made to act under the water head on a wheel (turbine). The wheel moves which in turn rotates a conductor (rotor) in a magnetic field (generator) to produce electrical energy. Hydropower generation is a non-consumptive use of water. Thus the water coming out of this process of generating hydropower can be put to other consumptive and non-consumptive uses. It can be said that hydro power generation is a by-product of a water resource development project. In this way, the input cost for hydropower generation is practically NIL.

It is also well recognised that hydropower generation is a non-polluting and eco-friendly source of energy having minimal impact on the environment as compared to thermal or nuclear. Thermal generation uses fossil fuels whose availability is limited and contributes to climate change. The input cost of fuel required for thermal and nuclear power generation is affected by inflation and keeps on increasing with time. Hence, it is seen that in the long run hydropower generation is cheaper than thermal and nuclear power.

Hydro power stations have an inherent ability to quickly start or stop generation (within 2 to 3 min) and to adjust load variation on the grid. These are ideally suited for meeting peak demand on the grid. Gas power station is an alternative to hydropower to meet peak demand but it is costlier. Hence hydropower generation is necessary for optimal operation of a power grid.

Besides the above advantages, there are some aspects that are against hydro power projects. These are

- It is capital intensive, i.e. the initial cost per MW is higher than a thermal plant.
- The construction time required to complete the project is more than thermal plants.
- Hydrological and geological risks are involved in hydro projects which many times increase cost as well as construction time.
- Dam projects have adverse environmental impacts and involve the rehabilitation of a large population.
- It involves working in remote and undeveloped areas. This often causes problems in engaging the required manpower.

The impact of the above adverse aspects of a major hydro project is to a great extent reduced in small hydro projects.

1.3 Historical Development of Hydropower

Hydropower generation dates back to 1882 when a hydro plant with a small capacity (a few kWs) was commissioned at Wisconsin (USA). Soon after other countries such as Sweden and Japan started hydropower generation. The hydropower plant harnessing Niagara Falls, which was started in 1901, attracted international attention to the concept of hydropower generation and its transmission to long distances. Since then it has been a long journey of technological innovations that have made it now possible to develop a single unit of 800 MW with turbine efficiency of 93–94% and generators of 98% efficiency. Long-distance power transmission with a voltage of 800 KVA has also been possible. The Itaipu project in Brazil and the recently commissioned Three Gorges project in China are the examples of world's largest hydro power projects with 800 MW units.

In India, the first hydropower plant with a capacity of 200 kW at Darjeeling dates back to the year 1897. The first hydropower plant of a significant capacity of 4.5 MW was commissioned at Sivasamudram in Karnataka in 1902 and it is still in operation. A small plant was constructed at Pathri (near Hardwar) utilising a fall on the Ganga Canal in 1912. Thereafter, eight other falls on Ganga Canal were harnessed to develop hydropower till 1936 and many of them are still in operation after renovation. Pathri power plant was rebuilt in 1954 for a capacity of 21 MW by combining a number of canal falls. The pace of power development up to 1947 (the year when India gained independence) was rather slow. The total installed capacity by that time was 1362 MW out of which 508 MW was hydro power (through small capacity plants). Soon after independence, the need for more power was felt to accelerate the pace of economic and industrial development in the country. The development of hydropower through multipurpose water resource projects was given priority by the Government to meet part of the power demand. A large number of such major hydro projects were constructed on several major rivers and the first among these was Bhakra on river Sutlej.

Unlike India, China after independence in 1948 started hydro power development on a large scale with small hydro projects. It was taken up as a socially, economically and environmentally sustainable development programme for poverty alleviation. Under this programme the local social groups had to manage funds, construction and operation of the SHPs and the distribution of power locally through isolated grids. Thus, this programme was carried out as a social movement for poverty alleviation. In a short period about 50,000 SHPs with installed capacity of about 16,000 MW were developed. Thereafter the major hydro power projects were taken up in China.

1.4 Hydro Power Development Policy in India

The hydropower development in India was started in 1897 with a small capacity plant and this development through SHPs continued till 1947. After independence hydropower was developed through major multipurpose water resources projects to meet increasing power requirements. This kind of development continued till the 1980s. At this stage, the share of hydro power was about 40% of the total installed capacity of the country.

The hydropower projects being capital intensive, the Government felt the difficulty in financing these projects. The policy planners also raised issues like the rehabilitation of large populations and adverse environmental impacts of reservoirs against the construction of major hydro projects. Thus, the policy being pursued by the Government after independence received a setback and the pace of hydropower development was significantly reduced. The two major projects Tehri and Sardar Sarovar dams which were under construction were completed in the first decade of the twenty-first century. A few run-of-river power projects have also been constructed in the last two decades. At this stage, the Government revived the policy of developing small hydro projects in the private sector. The response of the private sector was initially not very encouraging in taking up such hydropower projects. In order to meet the fast-growing demand of power the Government pursued the policy of developing nuclear power plants and super and mega thermal plants.

Since the start of twenty-first century, the issues of global warming and climate change have become burning issues and these affected the energy policies of many countries including India. In the new policy, the Government has emphasised more on harnessing renewable sources of energy such as wind, solar and biomass. It includes small hydro projects also. India is pursuing vigorously, the development of solar energy as a major component of the policy of harnessing renewable sources of energy. The current solar power capacity is 23 GW and the programme is to raise it to 100 GW by 2022 which will be about 3/4th of the total capacity of renewable energy. The Government of India has also planned that the share of renewable energy in total installed capacity by 2030 shall be 40% which at present is about 21%. Thus, it is evident that in view of the threats of disaster due to climate change, India is committed to reducing thermal power generation and increasing power generation through renewable energy sources. The renewable energy sources and their contribution to carbon dioxide emission are given below

•	Wind	7–g/kWh
•	Geothermal	7–9 g/kWh
•	Small Hydro	9–g/kWh
•	Biomass	17–27 g/kWh
•	Solar Thermal Electric	26–38 g/kWh
•	Solar photo voltaic	99–167 g/kWh

As compared with the above figures, the CO_2 emission from a coal-based thermal power generation is 955 g/kwh. Thus, to arrest the rate of global warming it is desirable to reduce coal-based thermal power generation and boost energy generation through renewable sources.

Perusing the above policy, the share of hydropower generation at present stands at about 11% of the total power generation. There is need to increase it for proper grid operation. Thus all efforts are required to harness hydropower both through SHPs and other types of hydro projects. The Govt. is thus presently favouring construction of SHPs and other major hydro projects specially pumped storage schemes wherever these are technologically, economically and environmentally viable.

1.5 Potential of Hydro Power Development Through SHP

As per an estimate, the small hydro power potential of the world is assessed as 1,80,000 MW (180 GW) out of which about 30,000 MW has been harnessed. In India, the potential of SHP is assessed as about 15,000 MW out of which only a few thousand have been harnessed. So there is a vast scope for developing SHPs. These can be located on canals using canal falls and small streams for power generation. An SHP on a natural stream utilises the flow and slope of the stream as head for generation. The head can be created by constructing a dam. In some cases, these two types of heads may be combined for power generation.

1.6 Advantages of SHP

As stated above SHP has advantages over major hydro as well as other sources of energy. These advantages are summarised below

- It is non-polluting and eco-friendly.
- It can use local materials and labour for construction.
- Its construction time and cost are less.
- It generally does not create rehabilitation and environmental problems.
- It requires a small infrastructure.
- It can be built in remote areas and supply power for consumption locally reducing the cost of long distance transmission.
- It requires small size civil structures such as river diversion structures, water conductors and power houses, which are simple to design and construct.
- It may use small generating units of standard size manufactured by local firms without involving long distance transport.
- It is simple to operate.
- It can be operated for a few hours to meet daily peak load requirements, with a relatively small upstream storage facility.
- It requires minimal maintenance.
- Local social groups may own, build and operate an SHP for local consumption of electricity in isolated localities.

1.7 Types of SHP Development

The power generation of an SHP is the product of discharge and head so the small hydro power development is of the following two types:

(i) Large discharge and small head: This type of project generally utilises the small canal falls. A minimum canal drop of 1.05 m has been utilsed at Ganzhutan (China) and 1.4 m at Kakroi, Haryana (India) to generate 3 × 100 kW with 610

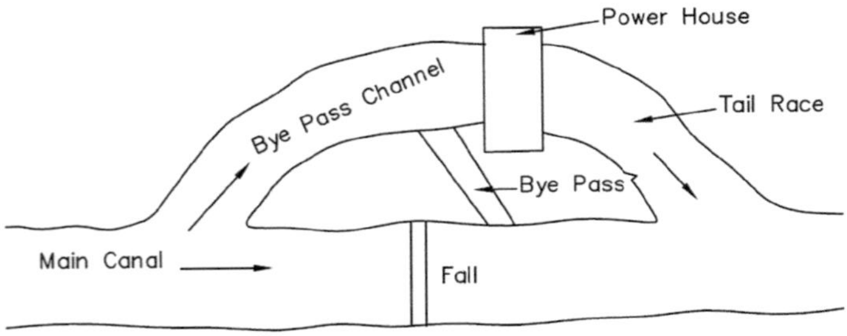

Fig. 1.1 Showing SHP on a canal fall

cumec discharge. However, manufacturers generally do not recommend turbines to be used below a head of 3.0 m. A typical sketch showing the layout of this type of SHP is given in Fig. 1.1. Such SHPs are operational only when the canal is in operation.

This type of development is also possible on rivers where steep gradients and rapids exist naturally in a small reach of the river. A low diversion weir is made to utilise the river drop for power generation. In these cases, if the pond level created by the diversion structure is above the high flood level, the power house is operational round the year and if the pond level is below the high flood level then there will be no head available for power generation during floods. Such development is feasible on a river that carries either no or only a small quantity of silt during floods. A large number of such developments are made in Europe for navigation and power generation purposes. The typical sketch showing the layout of such development is given in Fig. 1.2, which also shows that the power house may be located on both ends of the weir.

(ii) Low discharge and high head: Small streams which are normally the tributaries of a main river are characterised by low discharge but with high gradients. These provide ideal locations for developing such types of small hydro projects. Such tributaries in the Himalayas are generally perennial and the power house is operational throughout the year. A typical layout sketch of such a scheme is shown in Fig. 1.3.

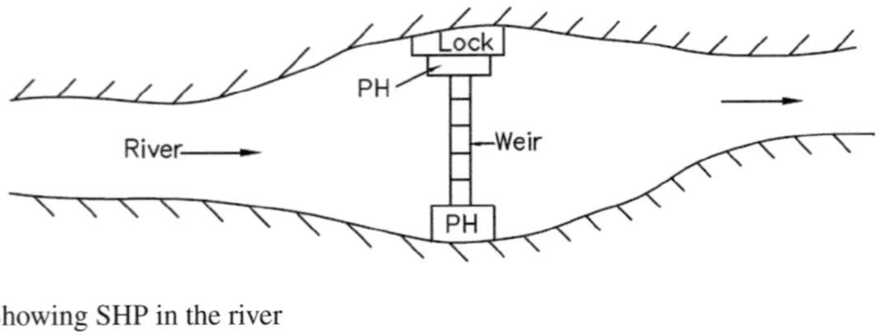

Fig. 1.2 Showing SHP in the river

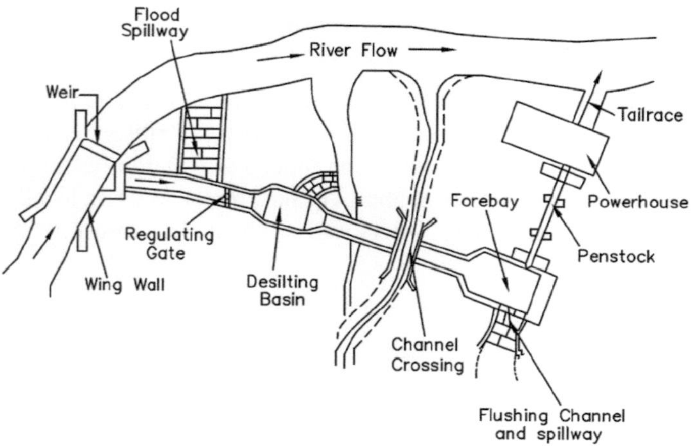

Fig. 1.3 Typical layout of small hydropower RoR scheme

Normally the tributaries of Southern Indian rivers do not carry any discharge during non-monsoon season. Hence, SHPs on such tributaries are built at the toe of a storage dam which stores the excess river flow during monsoon and the same is used to run the SHP during non-monsoon period. The flow released from the SHP is generally used for other purposes such as irrigation or meeting domestic water requirements. A typical sketch of such type of development is shown in Fig. 1.4.

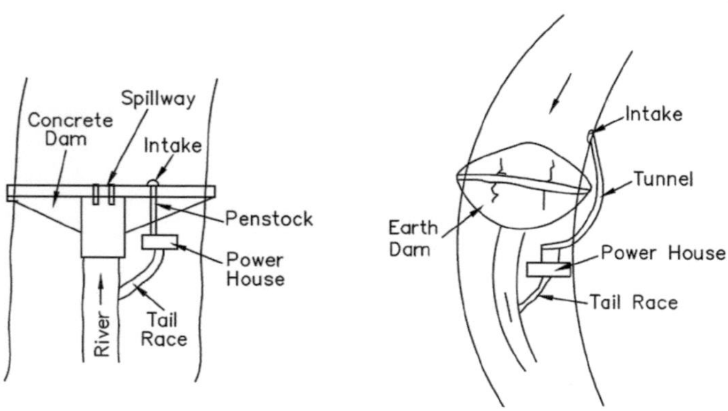

(a) Concrete Dam with Power House (b) Earth Dam and Power House

Fig. 1.4 Showing SHP below a dam

1.8 Challenges in SHP Development

There is a need to develop the SHPs up to the extent of available potential as part of the renewable energy programme. But the challenge at present is to reduce both the project cost as well as the construction time. The solution lies in training and involving the local population in all phases of the project, in using locally available construction material and in using power locally without a long transmission network.

Chapter 2
Layout of Small Hydro Projects

2.1 General

The layout of a SHP depends on the type of SHP and the structures required for each type. The most common types of SHPs are as follows:

(i) Run-of-river type.
(ii) In-stream type.
(iii) Dam-based type.
(iv) Canal fall type.

The first three types of SHPs use natural river stream flows for power generation and the fourth type uses canal flows. The salient features of the layout of the above types of SHPs and the required structures are given in the following paras but the actual layout of each SHP is site-specific and depends on the topography, geology and hydrology of the site.

2.2 Run-of-River-Type Schemes (RoR)

2.2.1 Layout of RoR Scheme

The typical RoR scheme diverts the stream flows, by constructing a diversion structure across the stream and an intake on its side, into a water conductor system that carries water to the power house. The water when discharged from the power house after power generation meets the same stream again or any other natural water course. Such schemes are more suited to perennial streams having steep gradients. The river bed drop available between the diversion site and the power house location is utilised as the head for power generation. These are generally high-head and low-discharge

B. N. Asthana, *Planning and Design of Small Hydro Power Projects*,
https://doi.org/10.1007/978-3-031-96271-4_2

schemes. A typical layout of a RoR scheme is shown in Fig. 1.3. A RoR scheme generally comprises the following civil structures:

- Diversion structure.
- Intake structure.
- Desilting arrangement.
- Water conductor system.
- Forebay/balancing reservoir.
- Surge tank.
- Penstock.
- Power house.
- Tail race.

The type and layout of these structures are dependent on the topography and geology of the site. These are described in the following paras.

2.2.2 Diversion Structure

These are of several types: (i) raised crest, (ii) trench weir, (iii) barrage and (iv) other types.

(i) Raised Crest Type: These are of small height (say upto 5–6 m). The crest is made of boulders, stones, concrete, soil, etc. The crest height is fixed in such a way that it is adequate to raise the non-monsoon water level to the extent required to feed the water conductor. The boulder bars are made of wire crates filled with boulders. These are made water tight by providing concrete facing. Where wood and bamboo are locally available, crib dams are made of balli and bamboo crates filled with boulders. Water tightness is provided by wood planking and soil. Where boulder bars or crib dams are not feasible, concrete/masonry weirs are provided. All these raised crests are suitable for rocky foundations. These are shown in Fig. 2.1.

(ii) Trench Weir: A trench weir (also known as Tyrolean or bottom intake) is a simple rectangular or trapezoidal trench, which is made up of stone masonry or concrete, is constructed across the stream and covered with a trash rack at the top. It is generally made in full width of the stream below the river bed. The trash rack is made sloping in the flow direction so that boulders may roll down. The trench is also provided with a bed slope in the direction towards the intake well which is constructed at one of the banks of the stream. The layout of a trench weir along with the intake well is shown in Fig. 2.2.

    Trench weirs have been extensively used on small hill streams for SHPs. These can be constructed on any type of foundation and are techno-economically viable. Since no part of this weir is above the river bed, it is not liable to damage due to rolling boulders. It does not alter the regime of the stream.

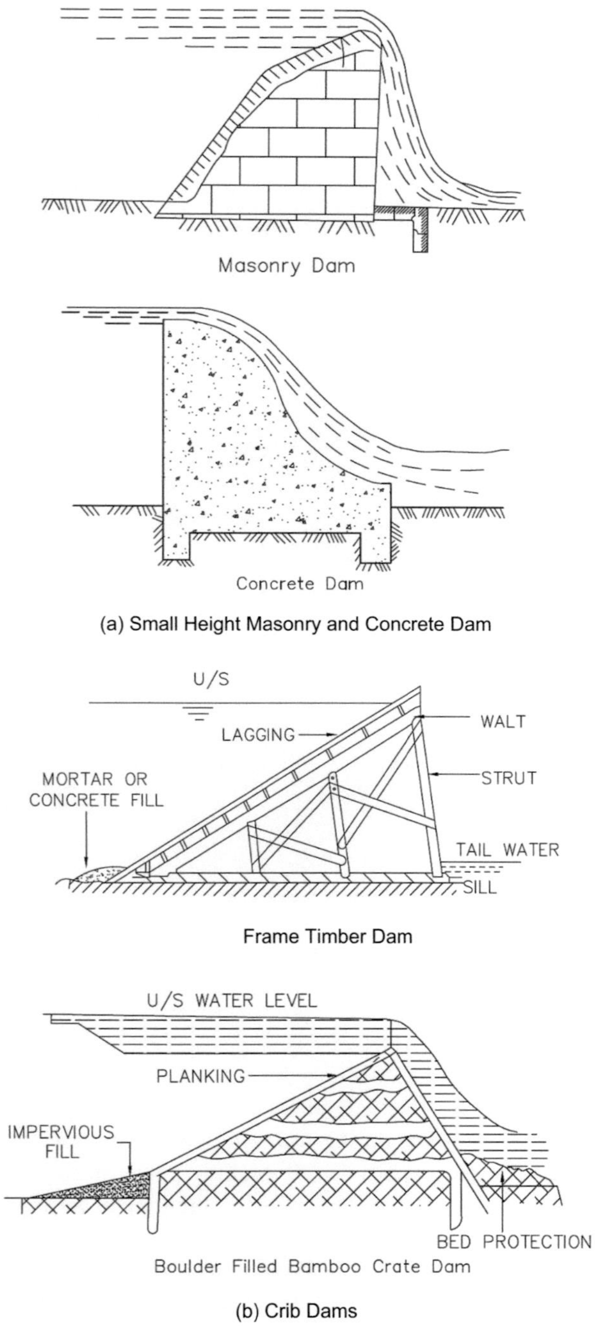

Fig. 2.1 Raised crest-type diversion structures

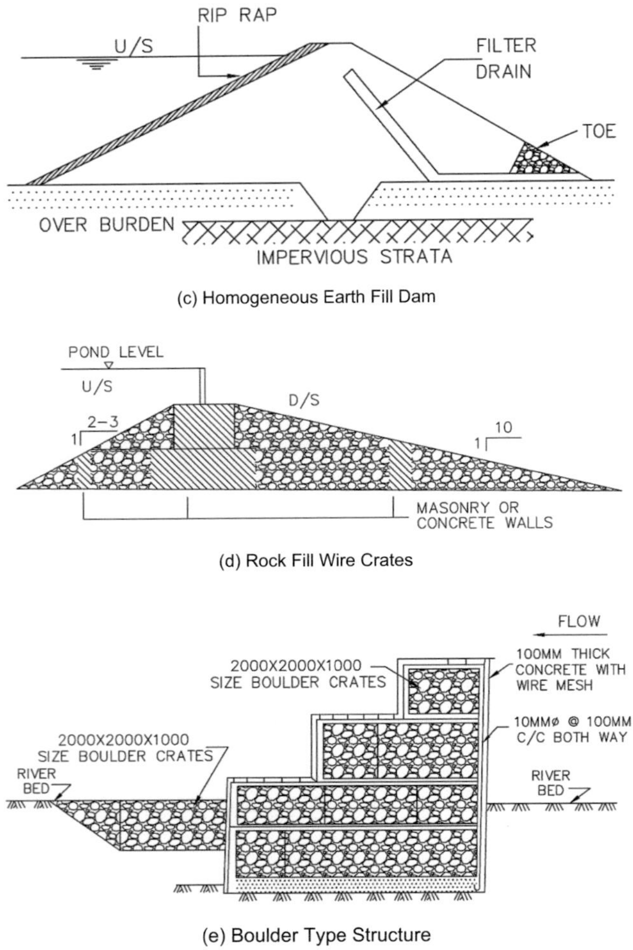

Fig. 2.1 (continued)

(iii) Barrage: The weirs described above divert the flow which occurs in the stream at any point in time. So, the SHP with the weir generates power to full capacity only during monsoon but during non-monsoon when stream flows are low the power generation is quite less. If peaking power during non-monsoon is required to be generated, instead of a weir a high-gated barrage is constructed which may store stream flows during non-monsoon for a number of hours in a day. The stored water would then be able to generate peaking power for a few hours when required. It is a diversion structure with a low crest at the river bed fitted with high-head gates above it. It is described in detail in chapter of Diversion Structures.

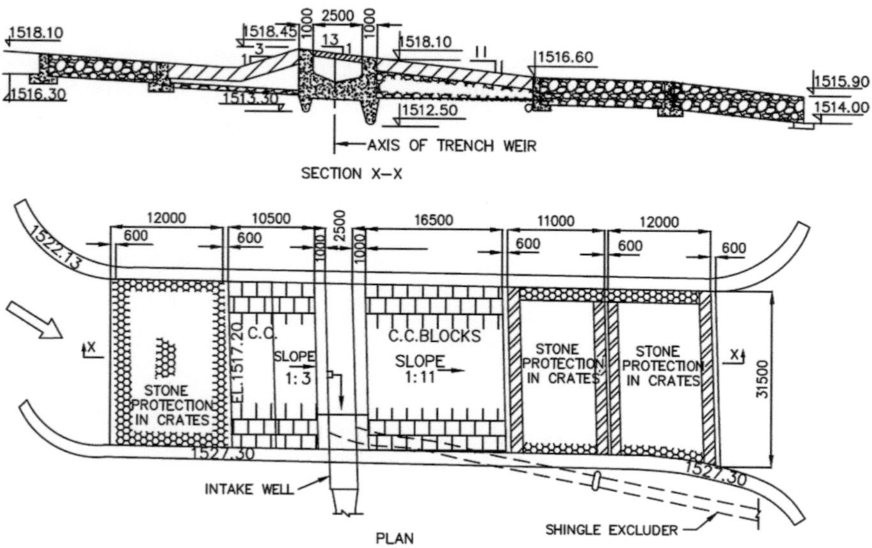

Fig. 2.2 KARNAH H.E. project-plan and section of trench weir. (CBI&P Pub 175)

(iv) Other types of weir: Some other types of weirs that have not been extensively used in India are as below:

(a) Weir with Conda-Effect Screen: It is an ogee-type crest weir fitted with a self-cleaning screen. It has been used in SHPs in the USA and other countries. It is shown in Fig. 2.3. The screen opening of 0.1 mm can exclude 95% of 0.5 mm size sediment particles. The disadvantages are high-head loss (1–1.5 m) and the inability to divert all the non-monsoon flow of the stream through the screens for power generation.

(b) Rubber Dams: These are made of rubber or canvas bags that are anchored

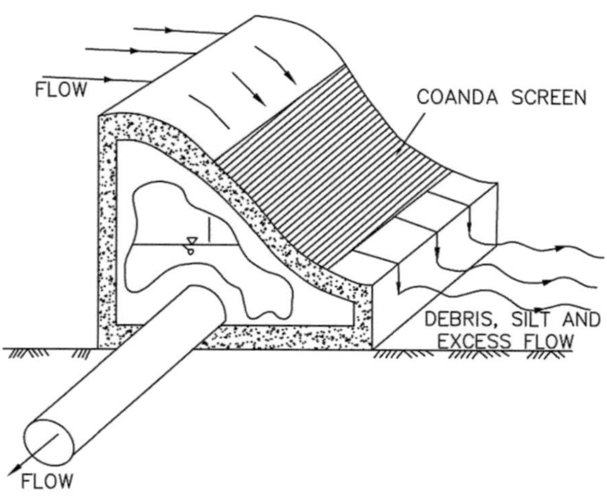

Fig. 2.3 The Conda-effect screen installed on a weir crest

to the base slab or crest of a weir. These have been successfully used in several countries for heights up to 5 m. The anchorage is important. Single-line or double-line anchorage is used as shown in Fig. 2.4a. The bag can be inflated or deflated by water or air filling and emptying, respectively. The bag position during inflating and deflating the bag corresponding to changes in the pond level in the upstream are shown in Fig. 2.4b. For further details refer AHEC–2.2/2.3.

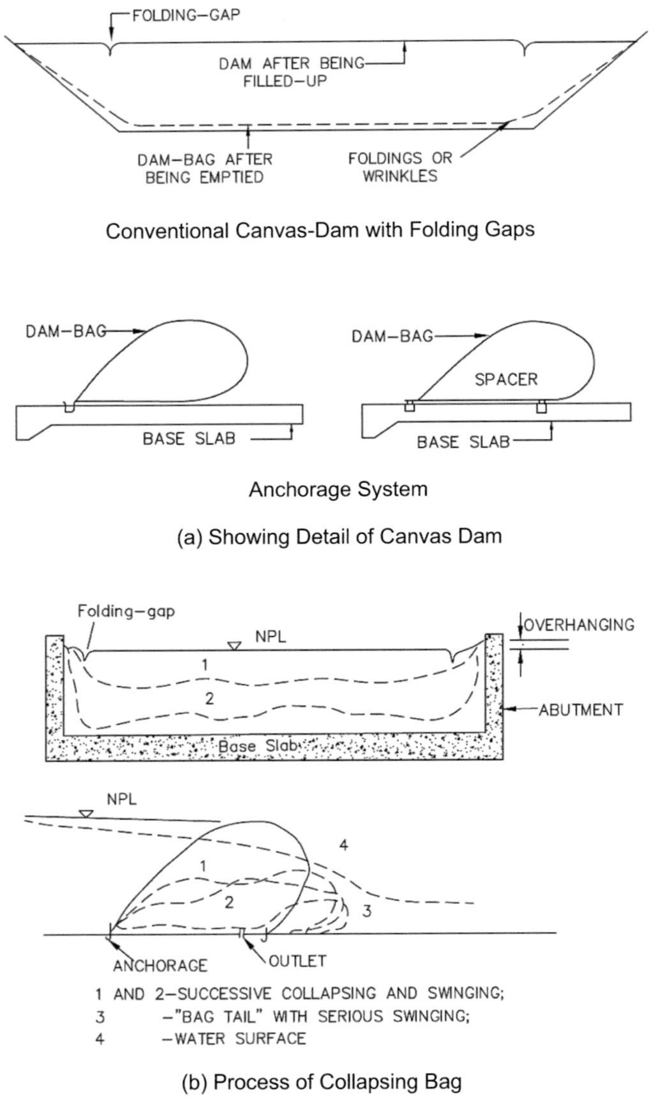

Fig. 2.4 Showing rubber dam

2.2.3 *Intake Structure*

It is a structure to divert regulated flow from the stream to the water conductor. It should be so aligned that it draws minimum sediment. It is located by the side of the weir or barrage in the upstream quite close to the axis of the diversion structure. The typical canal intake is shown in Fig. 2.5. It shows the canal intake location with a weir in plan, and a sediment excluder in one or more of the under-sluice bays. In case, where bed load is not heavy, a settling basin (Fig. 2.6) is provided just downstream of the intake crest with a bypass to pass sediment-laden water to the river.

When the water conductor from the intake is a closed conduit or a tunnel depending on the topography, the typical intake adopted in such conditions is shown in Fig. 2.7.

In the case of a trench weir, the intake is a masonry or concrete well at the end of the trench weir. Water flows from the trench to the well. From the well, the discharge

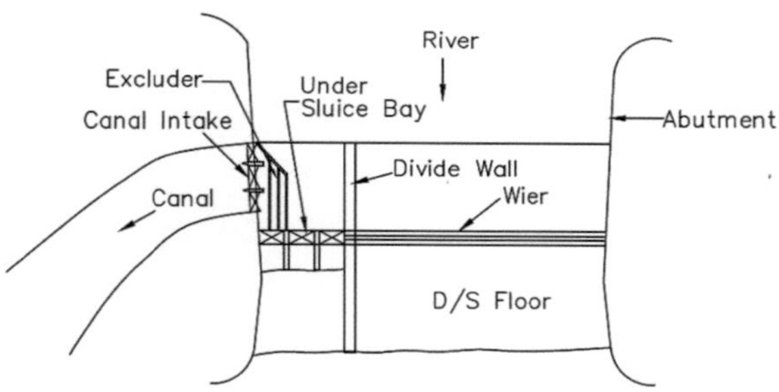

Fig. 2.5 Typical layout of canal head works

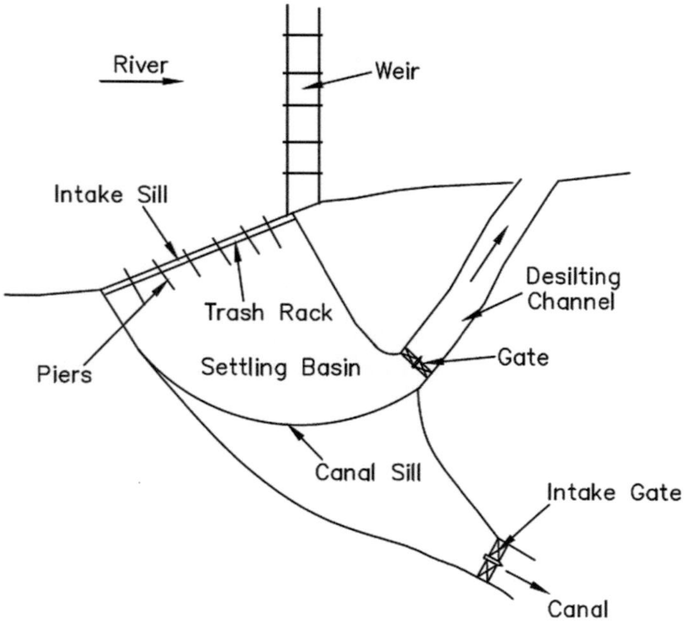

Fig. 2.6 Showing canal intake with settling basin

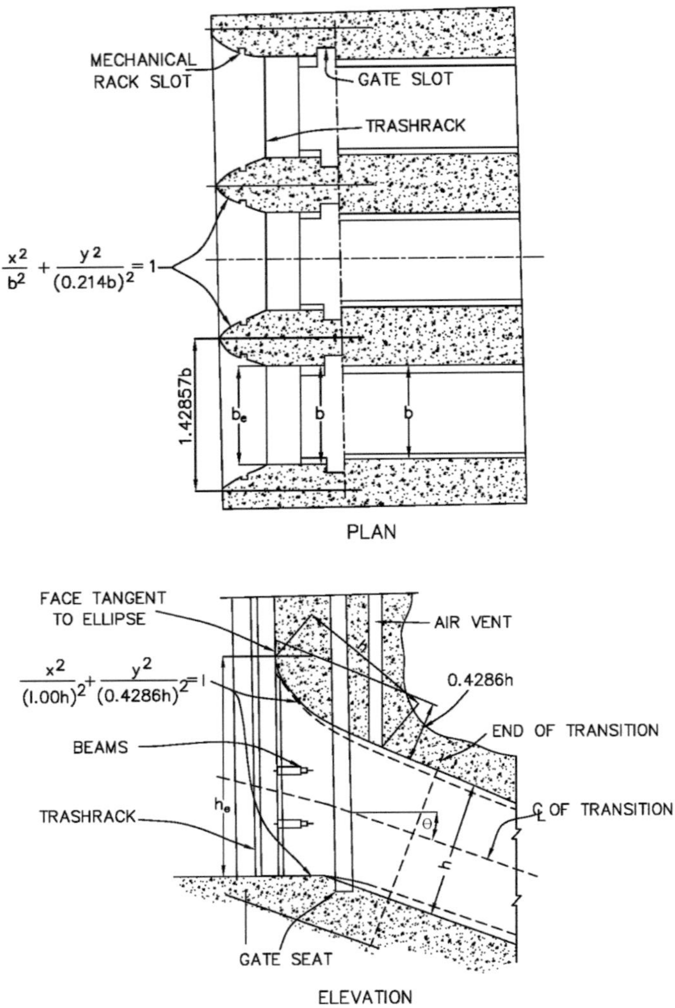

Fig. 2.7 Low-head intake with straight approach (USBR)

is carried to the water channel through a pipe or RCC. Conduit or over a masonry crest. The sediment entering the intake well is flushed through a pipe which is placed in the intake chamber at a level lower than the intake pipe.

2.2.4 Desilting Arrangement

If the water diverted through the intake into the water conductor carries a substantial amount of sediment load then a desilting arrangement should be provided to protect turbines from damage. The desilting arrangement should be located in the water

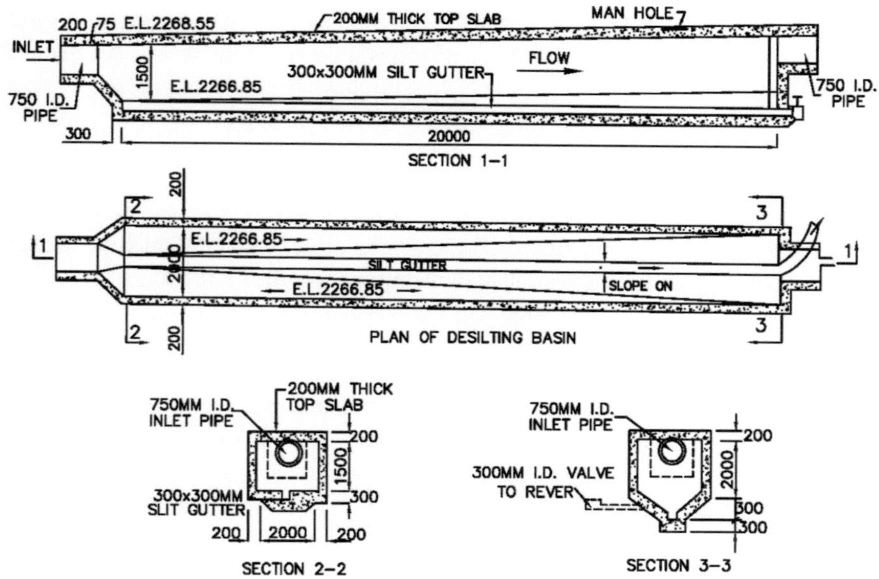

Fig. 2.8 Typical layout of desilting basin. (CBI&P. Pub 305)

conductor as close to the intake as possible but should be sufficiently away so that the sediment size to be extracted settles to the bottom layers of flow. The desilting devices generally used are desilting basins. A typical layout of a desilting basin is shown in Fig. 2.8. The silt flushing outlets from desilting devices should flush silt-laden water in a drain/river above the high flood level. The discharge in the water conductor downstream of the desilting device should be equal to the discharge required for turbine operation. So, the discharge from the intake should be equal to turbine discharge plus the flushing discharge which is generally about 20% of turbine discharge. The other types of desilting measures which are used in SHPs are described in Chapter of Sediment Management.

2.2.5 *Water Conductor System*

Water conductor carries water from the intake to the power house. It can be an open-channel or closed conduit or a tunnel or a combination of these. The selection of the type of water conductor system depends on topography, geology and economy.

The open channel may be lined or unlined. The section may be trapezoidal or rectangular. Generally, the channels are lined to save losses. The higher velocities, which are permitted in lined sections, result in smaller sections as compared to unlined channels for the same discharge thus requiring less land and reducing the cutting of hill side slopes. The open channels and free flow tunnels of long length need a forebay close to the power house. It is an expanded channel section with some storage to feed the turbine when opened after a shutdown till the flow is accelerated in the channel. Some layouts of channels with forebay are shown in Fig. 2.9. Typical unlined channel sections are shown in Fig. 2.10. Lined channel sections for different discharges are shown in Fig. 2.11. These can be used for the preliminary design of the scheme.

RCC, hume pipes, HDPE, GRP or steel pipes are best suited as water conductors in the reaches with steep hill slopes. These can be placed in a cut-and-cover trench without disturbing the hill slopes and natural drainages or the berms. Typical sections of such placement of pipes are shown in Fig. 2.12. These pipes can be designed to run partially or fully under pressure.

Sometimes when geology permits small size tunnels (minimum 3 m diameter) are preferred if the length is long. Tunnels smaller than 3 m in diameter can be made manually if the total length is not large. These can be made D-shaped in section. The tunnels are generally lined to reduce head loss. These may be designed to run either full under pressure or partially as an open channel. Long tunnels need intermediate

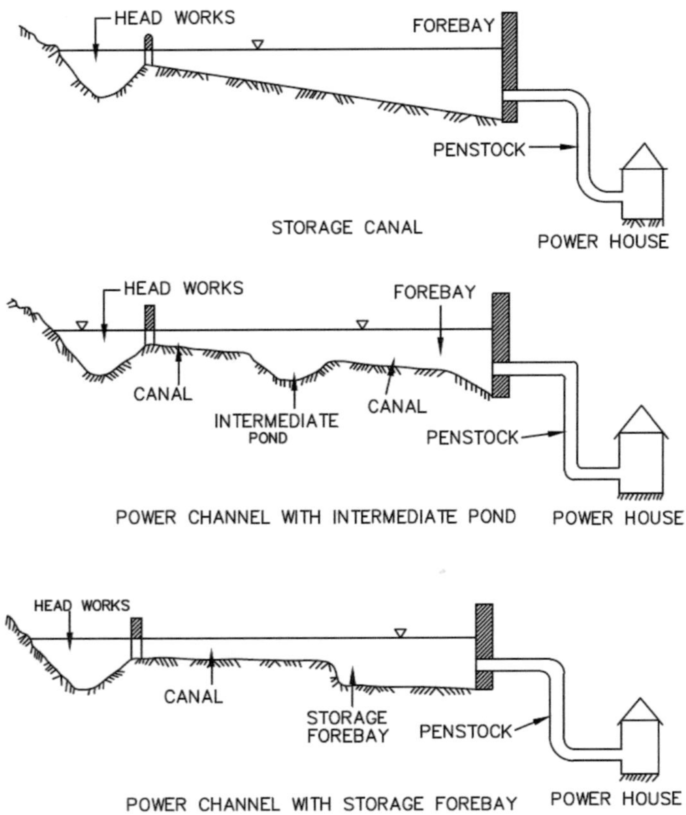

Fig. 2.9 Layouts of channel with forebay

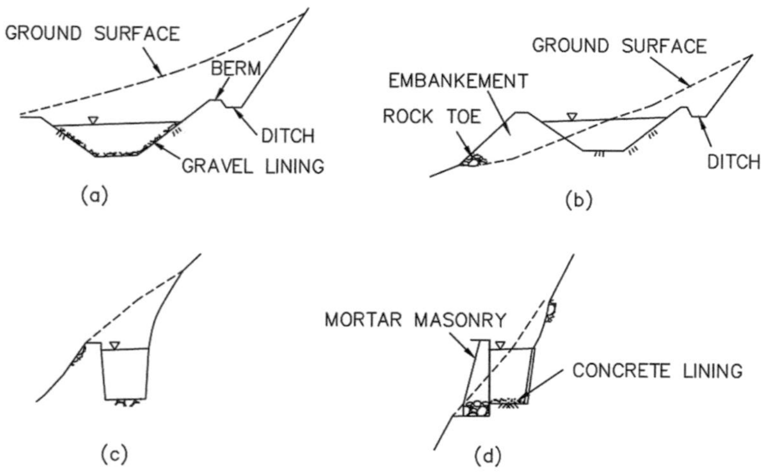

Fig. 2.10 Channel section **a** and **b** in alluvial strata, **c** and **d** in rocky strata

adits for ease of construction. Long tunnels running full under pressure need a surge tank at the end, close to the power house location. A few tunnel alignments with and without a surge tank are shown in Fig. 2.13. A typical tunnel section is also shown in this figure.

The length of an open-channel water conductor is normally more than that of a tunnel. It also requires the construction of several cross-drainage works and crossings (culverts and bridges). The channel is also prone to damage due to landslides. These issues are eliminated in the case of a tunnel. But the tunnel requires geologically strong strata through which it could be excavated.

2.2.6 Forebay and Balancing Reservoir

The forebay is provided at the end of the power channel to absorb surges during load changes of the turbines. It should have the storage capacity to supply water to the turbines for at least 2 min. It should also have an automatic spilling arrangement to pass the flow during the sudden closing of the turbines. An intake for penstocks is provided in the forebay. The intake structure should be fitted with a trash rack and gate or valve. The intake portion of the forebay should be suitably deepened to provide sufficient water cover over the penstock openings so as to eliminate any possibility of air being sucked into the penstocks even at the sudden opening of the turbines. At the upstream end, the power channel should be straight in sufficient length to provide uniform flow to the forebay. A suitable expansion/contraction joint should be provided between the channel and the forebay. A typical sketch of a forebay is shown in Fig. 2.14.

The location of the forebay is governed by the topography and geology of the site and the feasibility of providing the spilling arrangement. The site should be so selected that the length of penstocks is minimum.

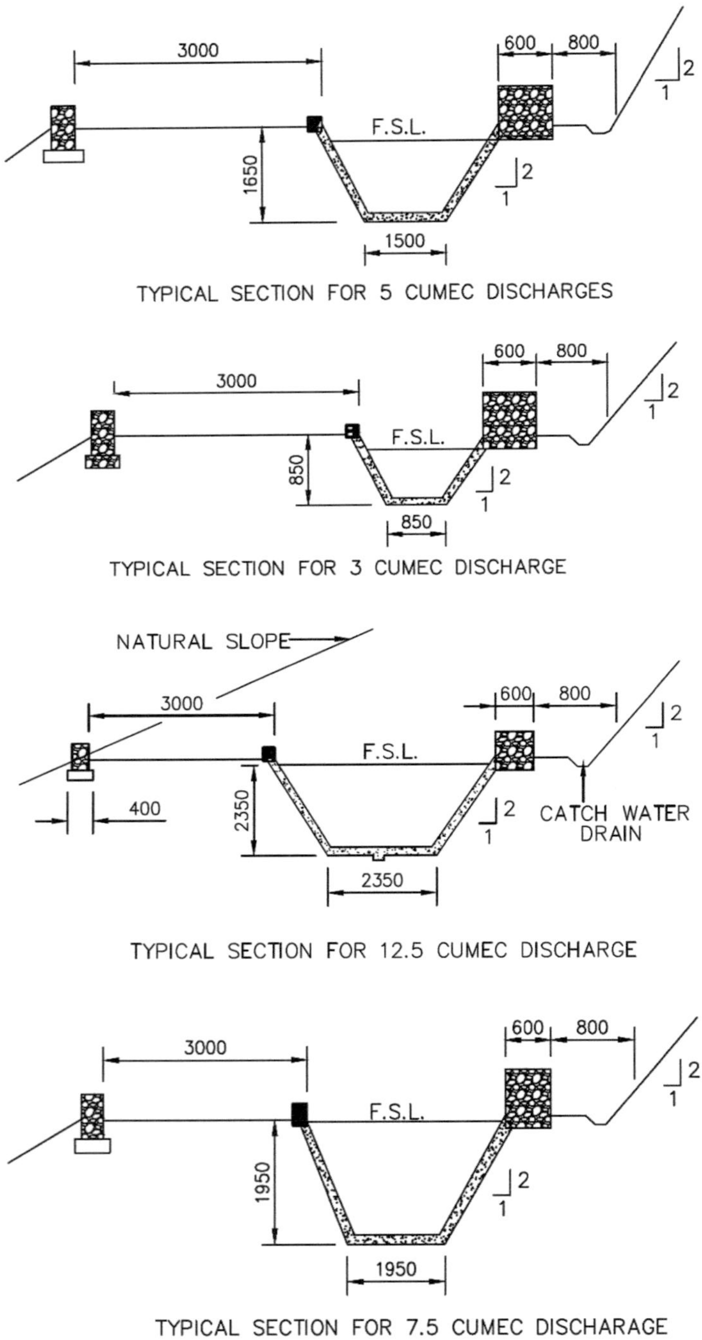

Fig. 2.11 Typical cross section of power channel for different discharges. (CBI&P. Pub 175)

Fig. 2.12 Typical details of placing closed conduits

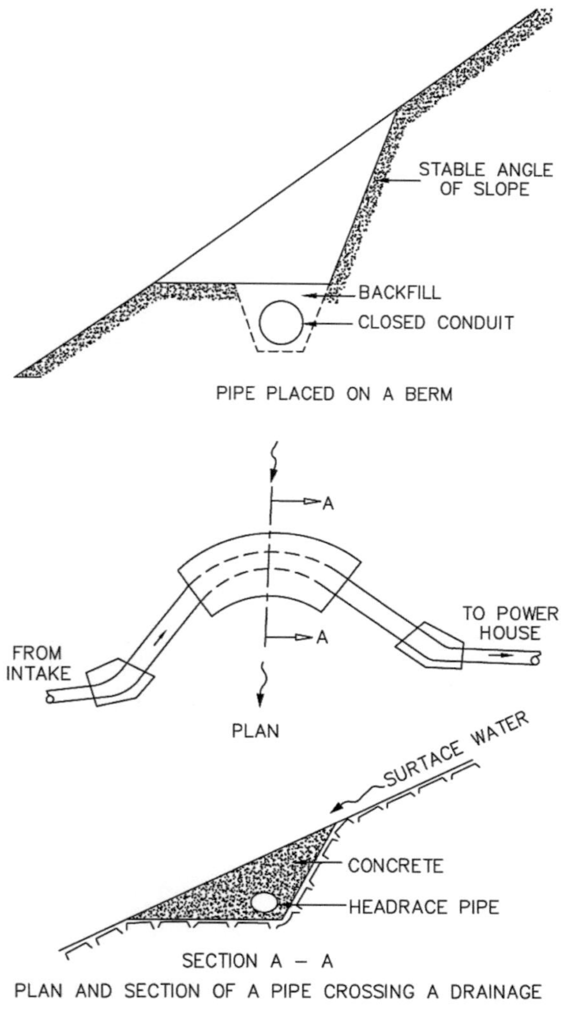

Sometimes a balancing reservoir (preferably a natural pond or depression) is provided in place of or in addition to the forebay to provide storage for a few hours on the diurnal basis to run the power house for peak power supply. Sufficient capacity is created by constructing an embankment around a depressed but practically flat ground. To ensure that the embankment is not overtopped, a spilling arrangement is provided. The inner face of the embankment and the floor of the pond should be lined to prevent seepage. If the balancing reservoir is near the power house, then the forebay may be omitted. A typical layout of a balancing reservoir is shown in Fig. 2.15.

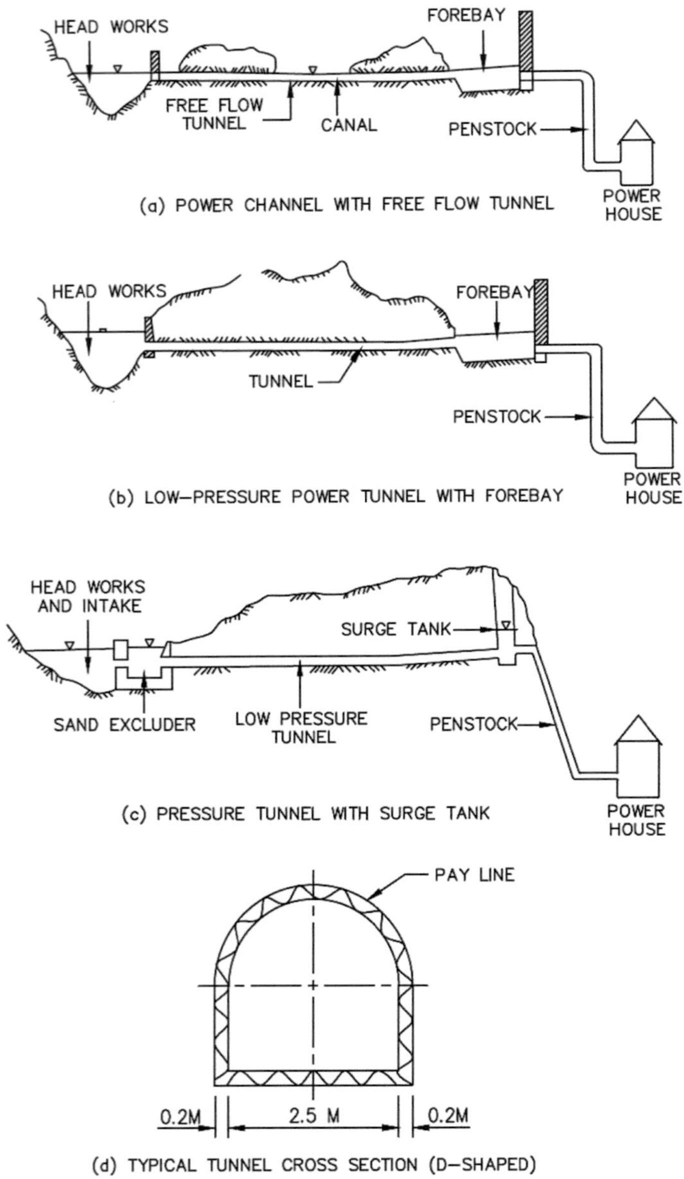

Fig. 2.13 Layout of tunnel as water conductor

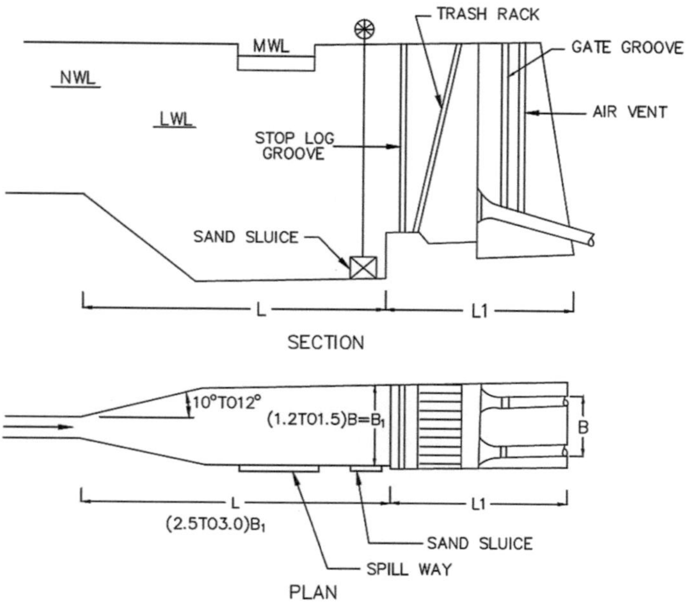

Fig. 2.14 Typical layout of a forebay

2.2.7 Surge Tank

In RoR schemes where the water conductor is a tunnel or a pipe running under pressure and its length is more than five times the head, a surge tank is provided at the end of the tunnel or pressure pipe. A surge tank is required in a project in which a reaction turbine is installed. In the case of an impulse turbine, a surge tank is not generally provided because the pressure surges can be avoided by deflection of the jet at the time of load rejection and by picking up the load at a slow pace at the time of load acceptance. A surge tank when provided should be located near the power house and at a higher elevation.

Surge tank acts like a forebay during load rejection and acceptance. It thus regulates the flow in the turbine. It also reduces water hammer effect on water conductor and helps in regulating the speed of turbine.

The surge tank is a vertical circular shaft/pipe with a height that should be sufficient to contain the highest and lowest surge levels. It is preferable to excavate the shaft in rock strata. The surge tanks are of several types but generally in SHP a simple surge tank is provided. Sometimes it is erected as a vertical pipe or as an inclined pipe laid on the hill slope. A typical vertical shaft surge tank is shown in Fig. 2.16.

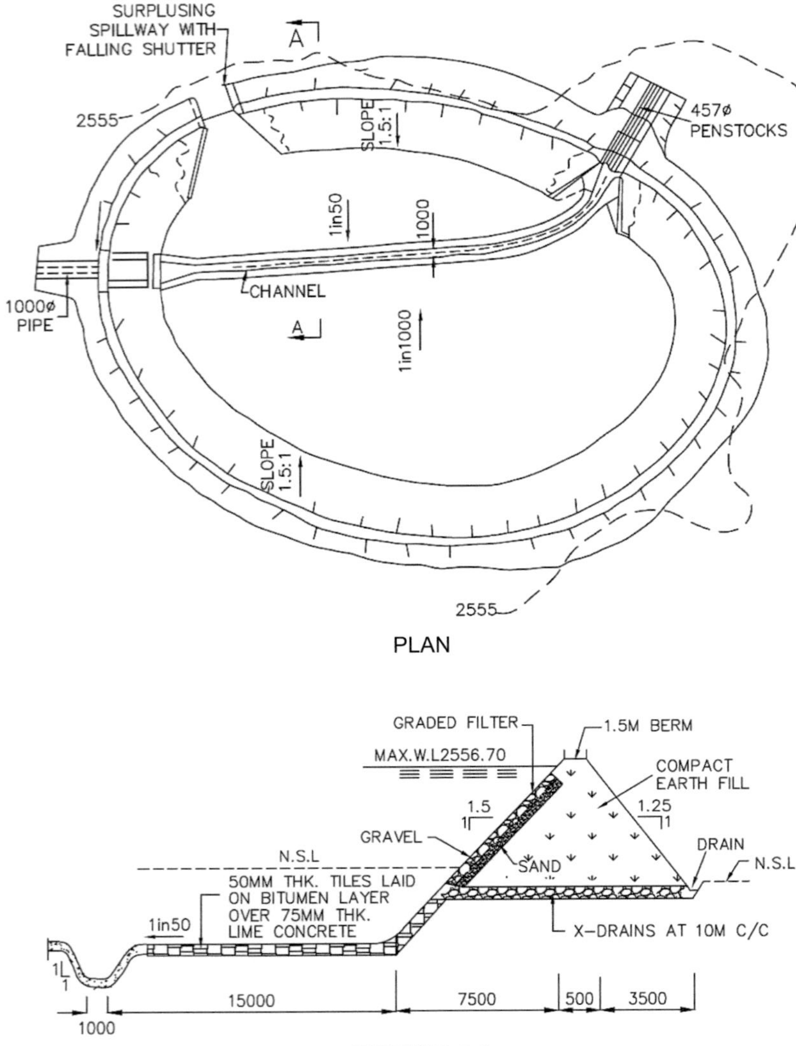

Fig. 2.15 Typical layout of combined forebay and balancing reservoir. (CBI&P, Pub 175)

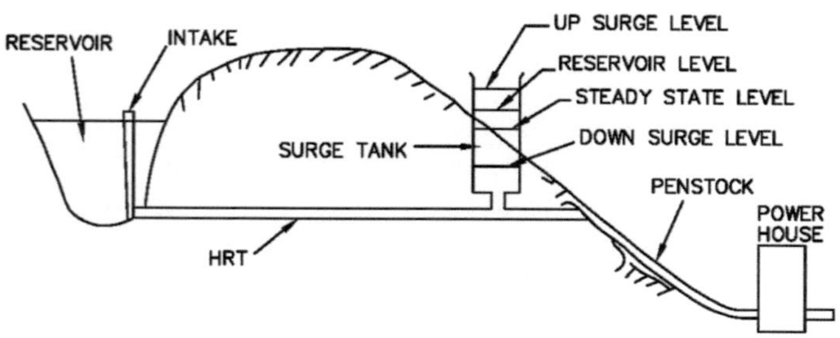

Fig. 2.16 Typical layout of vertical surge tank

2.2.8 *Penstock*

Penstock is a water carrier from the forebay/surge tank to the power house. These are pipes generally made of steel. Hume pipes, HDPE or GRP pipes are also used for small and medium heads (say up to 30 m). Steel pipes are used for high heads. For small heads and discharge, wooden pipes (if wood is available) are also used. These are made of wood staves, butt jointed and held together with steel bands. The pressure in wooden pipes should preferably be less than 0.4 kg/cm^2 to avoid leakage.

The penstock pipe runs under pressure. Its diameter depends on discharge and the velocity permitted in the pipe. The velocity to be allowed depends on head loss and erosion due to sediment. The velocity permitted is generally around 3–5 m/sec. The decision on diameter also depends on transportation considerations. In case of the long penstock, the penstock segments can be made of varying diameters. In such a case, smaller diameter ferrules may be slid into bigger diameter ferrules for case of transportation. During construction, bigger diameter segments are installed at higher reaches and smaller diameter ones at lower reaches of the penstock.

There are two alternatives for the layout of penstocks from the forebay or surge tank. One is to provide independent penstock for each generating unit and the other is to provide one penstock which is bifurcated or trifurcated near the power house for the units. The choice between the two alternatives depends on topography, operational convenience, economy and transport facilities.

The alignment of penstocks should be, as far as possible, straight with minimum bends. The minimum radius at bends should be 5 times the diameter. Before entering the power house the penstock should be straight in a reach of 10 times the diameter for uniform flow to the turbine.

The penstock can be laid either buried below the ground in a trench or exposed along the natural ground profile. The exposed penstocks need supports at regular intervals (6–8 m) and anchor blocks at the bends and also need expansion joints to take care of temperature variation. These structures are not required in the case of buried penstocks. The buried penstocks are laid on a continuous concrete pad. However, due to ease of inspection, maintenance and repairs, the exposed penstocks are generally preferred for a SHP.

2.2.9 *Power House*

It is a structure to house the generating equipment (turbine and generator) and other auxiliary equipment required for operating the generating units. The size and length of the power house depend on the type of turbine and the type of coupling of the turbine and generator which can be through either a vertical or horizontal shaft. It is dealt with in detail in a separate chapter.

The powerhouse complex besides the area for generating equipment has areas for the erection of equipment and control panels. These two areas are positioned around the generating area. The control panels can either be installed at one end or around each generating unit. The erection area may be located at either end of generating area and invariably towards the main entrance of the power house. Transformers may be placed either near the erection area or in the switchyard. A typical plan of machine unit of a power house of a SHP is shown in Fig. 2.17. It shows the generating equipment, associated equipment, control panel, etc., in the machine hall.

The power house complex is generally located on a relatively flat terrace along the river. The power house building foundation should be on firm and strong strata and the excavation should be minimal. The power house building is generally an RCC structure. Generally, a surface power house is preferred for a SHP. However, if the topography and geology of the site do not permit a surface power house then an underground power house may be considered.

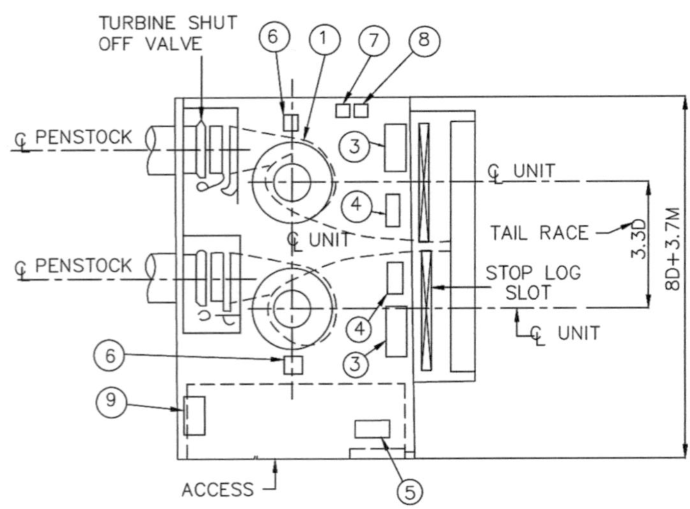

Fig. 2.17 Typical plan of a power house for a SHP. (I.S. 12, 800 Part 3)

2.2.10 *Tail Race Channel*

The tail race channel is provided to carry the discharge from power house to the nearest drainage which may be either the same stream or some other natural water course. The tail race channel is generally an open channel that may be lined or unlined depending on the strata. It may be a free flow cut-and-cover conduit depending on discharge and the topography. Typical sections of a tail race channel are shown in Fig. 2.18 for two discharges.

2.2.11 *Examples of RoR SHP*

(i) Karnah SHP in J & K is a medium-head RoR scheme with an installed capacity of 2 MW (2 × 1 MW). It diverts 7 cumec of water from a trench weir into a 425 m long power channel having a forebay at its end. The layout plan is shown in Fig. 2.19.

(ii) Lower Lagyap Scheme is a typical high-head inter-basin SHP located in Sikkim. A total of 4.7 cumec of water is diverted from three streams by constructing trench weirs. The water conductors end in an inclined 2.3 m diameter steel pipe simple surge tank laid on the hill slope. From the surge tank, a 2 km long surface penstock of 1 m diameter carries water to the power house to generate 12 MW utilising a head of 540 m. The layout plan of the scheme is shown in Fig. 2.20.

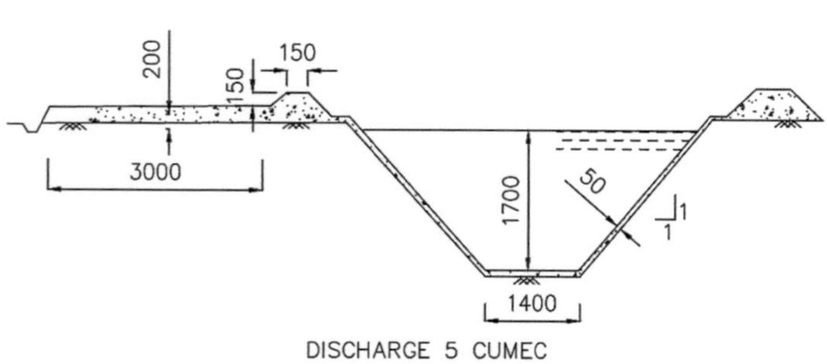

Fig. 2.18 Typical details of Trail Race Channel. (CBI&P Pub 175)

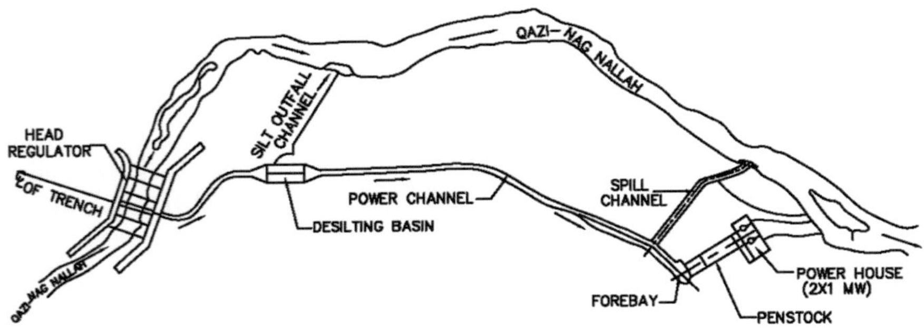

Fig. 2.19 Karnah H.E. project-layout plan

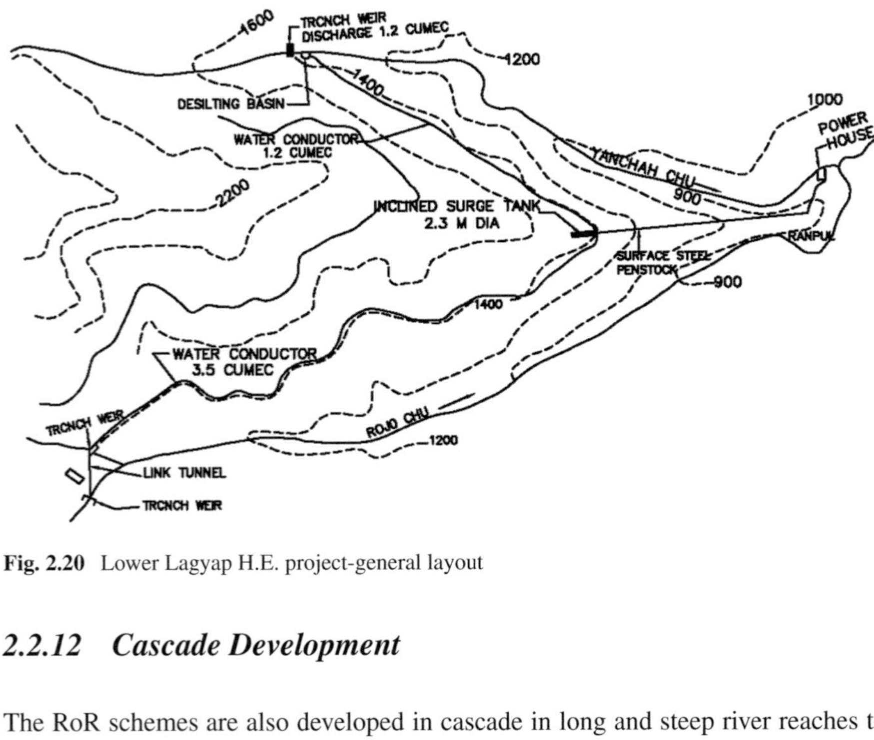

Fig. 2.20 Lower Lagyap H.E. project-general layout

2.2.12 *Cascade Development*

The RoR schemes are also developed in cascade in long and steep river reaches to utilise the river slope and the catchment runoff in the best possible manner. All the cascades are hydraulically and mutually connected (the tail water level of the upper cascade may or may not be the head water level of the lower cascade). This type of development gives flexibility and ease in construction as well as the operation of power plants. A conceptual sketch showing this type of RoR scheme development is given in Fig. 2.21. There may be a gap between the two cascades. The RoR schemes in this type of development may be combined with some storage dam schemes. The last cascade may be planned as a regulating pond for the downstream water requirements. The cascade development has been adopted on several Himalayan streams.

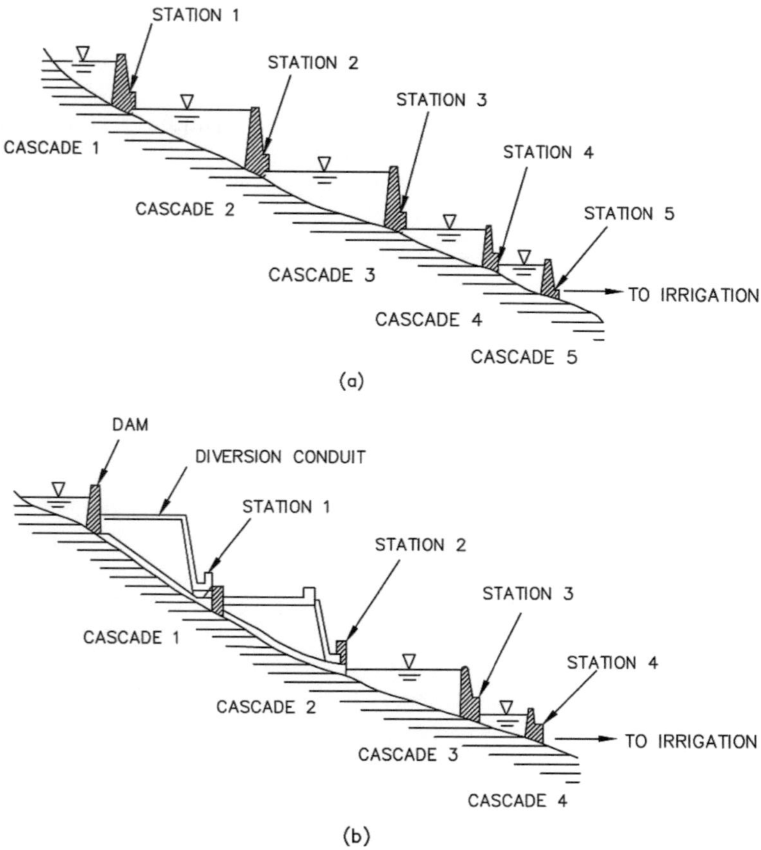

Fig. 2.21 Showing cascade development. (AHEC – 2.1)

2.2.13 In Stream RoR Schemes

In this type of RoR scheme, the power plant is located in the stream/river by the side of a weir/barrage constructed across the stream to raise the water level during non-monsoon. This type of scheme is categorised as large discharge and low head type of power development. The power generation in such schemes may be associated with navigation through locks and log chutes or other uses such as irrigation and other economic utilizations. A layout showing such a scheme is shown in Fig. 1.3. For some examples of such developments, AHEC Report 2.1 may be referred.

2.3 Dam-Based SHP Scheme

In a dam-based RoR scheme, generally, the head is created by the pond level in the storage dam. The dam stores excess water during monsoon and releases it during non-monsoon, when river flows are low or negligible, for power generation. The flow after power generation can be further used for irrigation, drinking or industrial purposes. The power house is normally located at the toe of the dam. A typical dam-based powerhouse is shown in Fig. 2.22.

In the dam toe RoR scheme, the intake is generally a part of the main dam. The water is carried to the turbine through a penstock which is laid inside the dam body. The flow after power generation goes to the river again through a short length tail race channel. The dam may be of concrete or earth or a composite dam consisting of a concrete spillway and earthen embankments. The head available for power generation in a dam scheme is the difference between the reservoir level and downstream river water level. The variation in reservoir level with time shall be accounted for in power studies. Generally, a weighted average reservoir level is adopted for planning purposes. The dam-based RoR scheme, thus, normally comprises a dam, an intake, a penstock, a power house building and a tail race.

Depending on topography and economy, the power house may be located away from the dam also. In that case, water from the dam is carried to the power house either through a power channel or a tunnel. If the length of the water conductor is long, then either a forebay or surge tank shall be provided at the end of the water carrier and then the flow will be taken to power house through a penstock from the forebay/surge tank. In this case, the head available for power generation will be the difference between the reservoir level and the water level in the river where the tail race joins it. This type of scheme normally requires all the structures needed in a RoR scheme described in para 2.2 above. This type of scheme can be planned on existing dams also. In that case, the assessment of the availability of discharge and head should be estimated from available records/drawings, etc. The structural stability of old structures should also be assessed before taking up such a scheme.

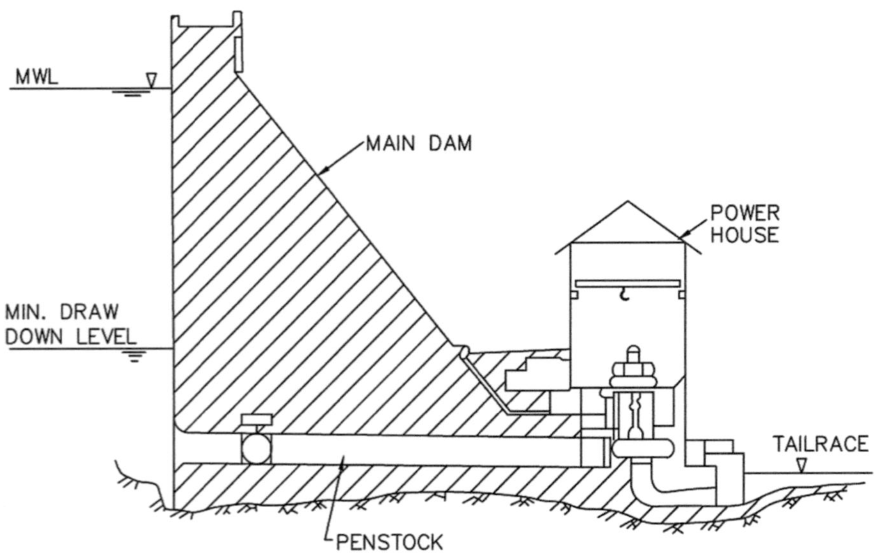

Fig. 2.22 Typical layout of dam Toe SHP scheme

The dam-based schemes have been successfully constructed on a number of existing dams in south India where the rivers/streams carry practically no flow during non-monsoon season. In cases where an outlet exists in the dam to feed the power house it may be used to lay the penstock. But the power house should be located sufficiently away from the dam so that power house foundation excavation would not affect the dam. In many cases, new intake and outlet are required. It is generally difficult to provide the intake in a concrete or masonry dam. In that case, the possibility of a syphon intake or the construction of a new intake through the abutment should be examined. In the case of the new intake, the location of the intake in the abutment, head race tunnel, penstock and the power house would have to be planned depending on topography and geology. The earthen embankment of moderate height (6–8 m) may be cut to provide the new intake and the penstock/RCC barrel/pipe. The typical layout of the above alternatives is shown in Fig. 2.23.

2.4 Canal Fall-Type Scheme

2.4.1 Layout

Generally, irrigation canals are provided with falls to follow the normal ground level to irrigate maximum area in the canal command. The drops on the falls vary from 0.5 m to several metres (say 4–5 m). Rarely the falls are of larger heights. Small drops are also available at the heads of the branches and the distributaries which may also be harnessed. The pond level at the barrages and weirs which are constructed to feed the canal system also provide sufficient drop for power generation utilising the environmental releases during non-monsoon season. Such a scheme is of a small head and large discharge and has significant energy potential. India has a very large network of irrigation canals. Some canal systems are centuries old. Many falls on old canal systems have been harnessed for hydro power generation in the past. Eight falls of the upper Ganga canal were harnessed for power generation from 1912 to 1936 and many of these are still functioning after renovation. Many more such projects are still in the planning and construction stage. Such schemes are easy to plan due to assured discharge and head. These are easy to construct due to their simple layout and are easily accessible because they are located in plains. Other advantages are a small gestation period, no problem of submergence, rehabilitation and environmental impact. The disadvantage is that in many cases private agricultural land by the side of the canal has to be acquired. Such a canal fall scheme normally comprises of:

- A bypass channel with or without regulating structure.
- Intake
- Power house—generally intake and power house are one composite structure. A typical composite canal intake and power house structure is shown in Fig. 2.24.

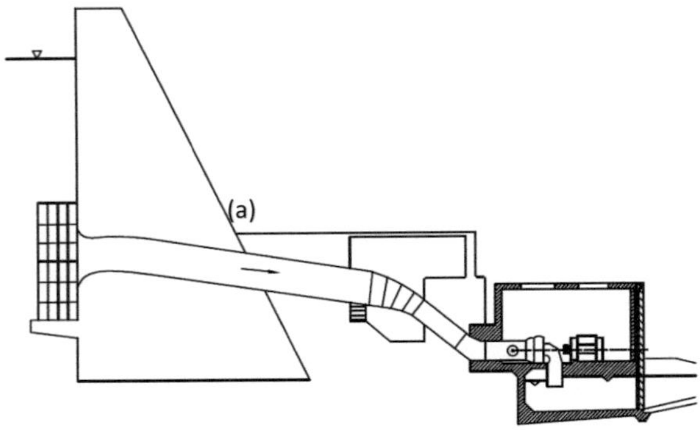

(a) Concrete Dam With Existing Conduit

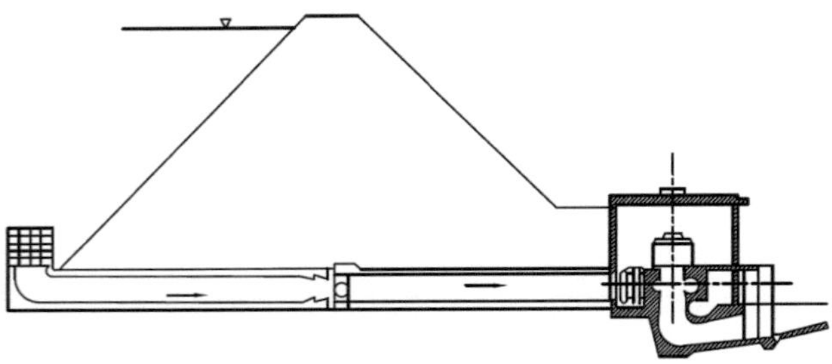

(b) Earth Dam With Existing Outlet Work

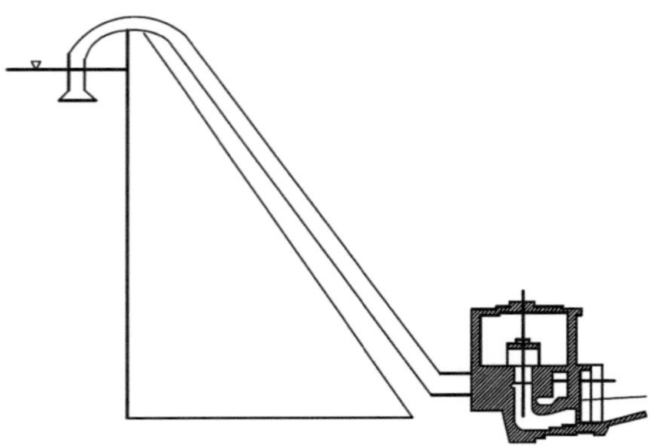

(c) Syphon intake over a dam without having any existing outlet

Fig. 2.23 Typical layouts of SHP on existing dam scheme. (AHEC—2.1)

- Tail race channel joining main canal.
- A spilling channel—it shall off-take from upstream of the power house and join the main canal so that the canal flows are maintained when the power house is closed.

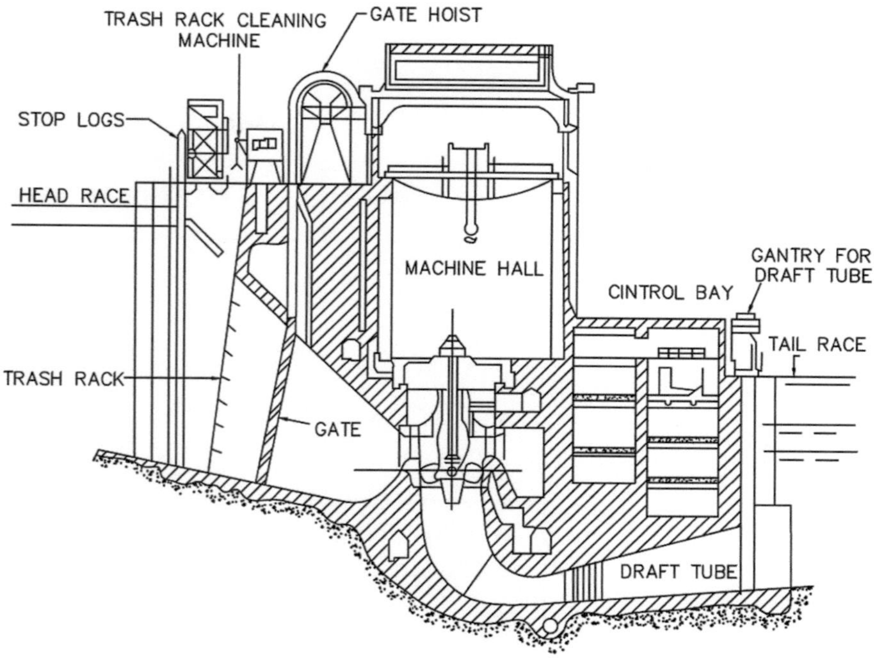

Fig. 2.24 Typical canal intake with power house building. (Adopted from E Mosonyi: Water Power Development-Low Head Power Plants)

These structures should be located sufficiently away from the main canal so that they may be easily constructed without interfering with the operation of the main canal and affecting the safety of existing structures. If the power house is located very close to the existing canal fall due to land constraints then in many cases the problems encountered are of dewatering the power house foundation pit, foundation design due to the low bearing capacity of soil and the effect of large uplift on power house structural stability. The foundation of a power house in weak strata is generally a raft with other measures such as sheet piles, drainage arrangement etc.

Instead of using a single canal fall for power generation, sometimes it is feasible to combine a number of falls, if these exist within a short distance, to increase the head at the power house. If the power house is planned on a new canal under construction, it can be accommodated in the canal bed itself by increasing the bed width at the power house site. In case the drop at the head of the branch canal is to be used for power generation, the power house is located on the link channel to be constructed to join the main canal with the branch. The conceptual layout of these alternative situations is shown in Fig. 2.25.

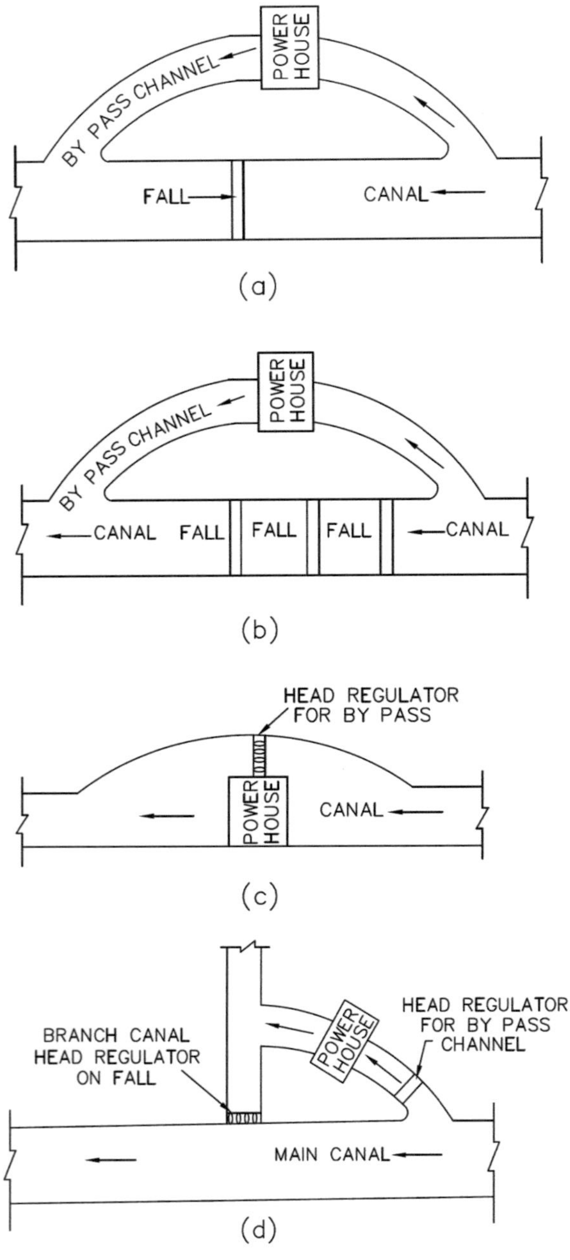

Fig. 2.25 Typical alternate layouts of canal fall-based power houses. (AHEC—2.1)

References

CBI & P.: Small hydro stations standardization", Pub. 175, New Delhi, (1985)
CBI&P.: Manual on planning and design of small hydro electric schemes, Pub. No. 305, New Delhi, (2009)
Guidelines—Small Hydro Development, Standard No. 2.2, 'Civil Works—Hydraulic and Structural Design", AHEC, IIT, Roorkee, (2013)
Guidelines for Small Hydro Development, Standard No. 1.2 (Past I, II, III) 'Planning of the projects on dams, canal falls and Run of river, Projects', AHEC, IIT, Roorkee (2013)
Guidelines for Small Hydro Development, Standard No. 2.1, 'Layouts of SHP Projects', AHEC, IIT, Roorkee, (2013)
I.S. 12800, (Part-3), "Guidelines for Selection of Hydraulic Turbine, Preliminary Dimensioning and Layout of Surface Hydroelectric Power Stations", BIS, New Delhi (1991)

Chapter 3
Electro-mechanical Equipment

3.1 General

In a SHP, civil structures are provided to carry water in required quantity and quality from the source of water to the power house and electro-mechanical equipment are required to be installed to generate energy. These equipment essentially comprise a turbine coupled with a generator. Other auxiliary equipment are also required to regulate and operate the turbine and generator. To control all the operations required for power generation, control panels are also installed. All these equipment are placed in the power house building commonly known as the machine hall.

The flow from the penstock impinges on the vanes/buckets of the runner (turbine) which rotates under the water pressure and converts the potential energy into mechanical energy. The turbine through the connecting shaft rotates the rotor of the generator into the stator (the magnetic field) to generate electrical energy. Thus, the turbine is a runner containing the vanes/buckets and the generator is made of the rotor and stator and the two, i.e. turbine and generator are the basic machines of the power plant to generate electrical energy. The layout and dimensions of power house building depend on the type and size of the turbine and generator and the configuration of their coupling. These are described in the following paras of this chapter and the layout of power house is dealt with in the next chapter.

3.2 Turbines

A turbine is a machine that converts the potential/kinetic energy of water into mechanical energy. This has been evolved from the water wheel which was developed long back before the Christian era. Bourdin and Fourneyron (1824–27) developed the first radial water wheel called a water turbine in France. This was further developed as axial flow and inward flow turbines. These were slow moving turbines that were

B. N. Asthana, *Planning and Design of Small Hydro Power Projects*,
https://doi.org/10.1007/978-3-031-96271-4_3

used for driving mills. The use of these turbines coupled with a generator was later developed for electrical power generation and the first power generating plant was constructed in Wisconsin in 1882 to generate a few kws only. Since then development has taken place for higher and higher speed runners of larger capacity. Now turbines for hydro power generation working under a large range of heads (from a few metres to 2000 m) and giving an output from a few to half a million horse power or more are being manufactured.

In general, the common practice is to design the turbine for each case to suit the conditions of varying head and discharge. The site-specific design of a turbine needs a lot of time and money and its efficiency is normally tested on a model. This increases its cost. In order to reduce cost and time in small hydro projects standard turbines and generators, produced by the manufacturers, are being widely used. So, the turbines available are of standard runner diameters. The use of such standard turbines in SHP often does not ensure maximum efficiency.

3.2.1 Types of Turbines

Basically, turbines are classified into two main classes depending on hydraulic action (i) impulse and (ii) reaction. Since turbines influence the dimensions and layout of the power house, the characteristics and the selection criterion of different types of turbines are discussed below:

(a) **Impulse Turbine**

In an impulse turbine, the available head of water is first converted into kinetic energy or high velocity jets through one or more nozzles. The free jets emerging from the nozzles strike the runner, which consists of buckets, and the runner rotates. It is a tangential flow type turbine. The discharge coming from the nozzle, while flowing through the bucket and after passing the bucket remains at the atmospheric pressure. Hence, this type of turbine is set above the highest tail water level.

The impulse turbines depending on shape and the arrangement of water flow are further classified as below:

(i) Pelton wheel.
(ii) Turgo.
(iii) Cross-flow.

(i) **Pelton Wheel**

In this type of impulse turbine, water from the distributor comes out either through one or more nozzles at high velocity (the entire water head is converted into velocity). The runner on which the high velocity jet strikes, consists of a disc on whose periphery a number of buckets are mounted. The striking jet produces a force on the buckets and rotates the runner. The shape of the buckets and their orientation on the periphery determine the efficiency of the turbine. The buckets are so designed and rotations of the runner are so governed that the exit velocity of the jet is almost zero and the maximum kinetic energy is used in rotating the runner. The buckets are double hemispherical cup shaped and diverge the flow on both sides. The edge between two hemispherical parts of the bucket is sharp and appropriately shaped at the outer edge. The jet first strikes the dividing edge. It is shown in Fig. 3.1.

Pelton turbines are both single jet and multijet types and can be mounted horizontally or vertically. For small hydro projects horizontal shaft Pelton turbines with one or two jets are more suitable where small discharges are utilised. The cross section of a power house with a horizontally mounted Pelton turbine of two jets is shown in Fig. 3.2. Pelton turbines are suitable for use at high heads above 300 m.

Control of the turbine is accomplished by a deflector mounted on the nozzle assembly which deflects the jet at the time of sudden load rejection and power output

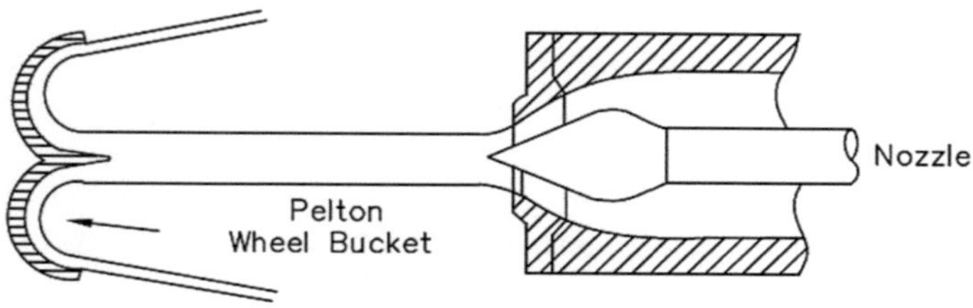

Fig. 3.1 Pelton wheel bucket and nozzle

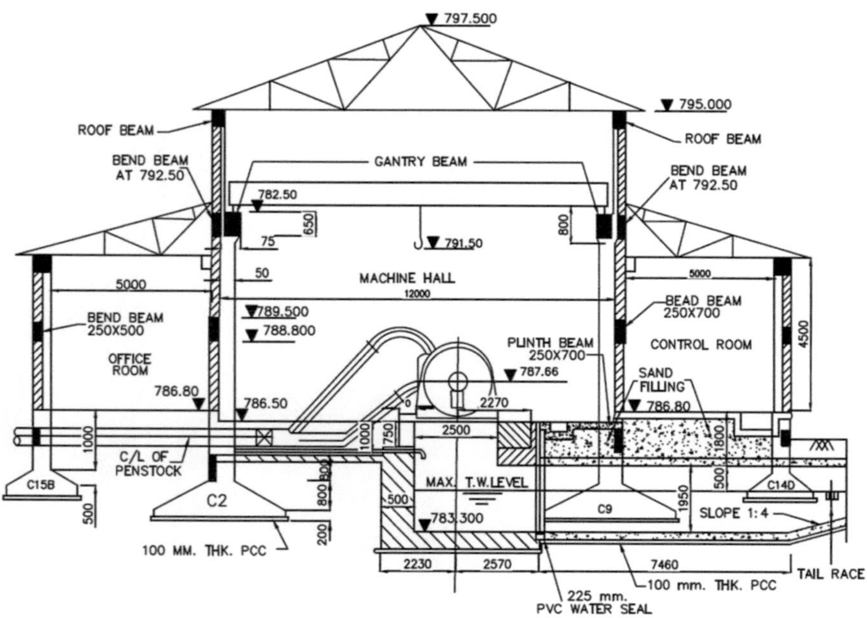

Fig. 3.2 Section of a power house with horizontal impulse turbine (1500 kw). (AHEC–3.1)

and speed of the turbine are controlled by adjusting the opening of the nozzle by a spear shaped needle.

(ii) **Turgo Impulse Wheel**

Another type of impulse turbine is the Turgo turbine. The difference between Pelton wheel and Turgo is that, in Turgo, the jet strikes on one side of runner and exits on the other side. The jet in Turgo turbine strikes on several buckets simultaneously, whereas in a Pleton wheel jet from the nozzle strikes on one bucket at a time. The Turgo operates at a higher speed than Pelton. A typical Turgo runner is shown in Fig. 3.3. The head range of Turgo unit is 15 m–300 m. These are used in SHPs for unit output up to 7500 kW. The efficiency varies from 82 to 83%.

(iii) **Cross-Flow Turbine**

The cross-flow turbine is also an impulse turbine. The water in this type of runner is directed through a rectangular jet into a cylindrical runner and passes from the periphery towards the centre and then after crossing the open centre moves outwards. Hence, it is a turbine with partial air admission. The typical view of the cross flow turbine is shown in Fig. 3.4. The entry angle is about 70° and the peripheral speed is about 0.48 $\sqrt{2g.H}$. The peak efficiency of this turbine is less than other turbines. It is about 60–65%. However, the efficiency remains almost the same at the operating range down up to about 15% of the rated output. These turbines are particularly suitable for low head with fluctuating discharge. The blades are curved radially.

Fig. 3.3 Typical Turgo runner and spear. (CBI&P. Pub. 305)

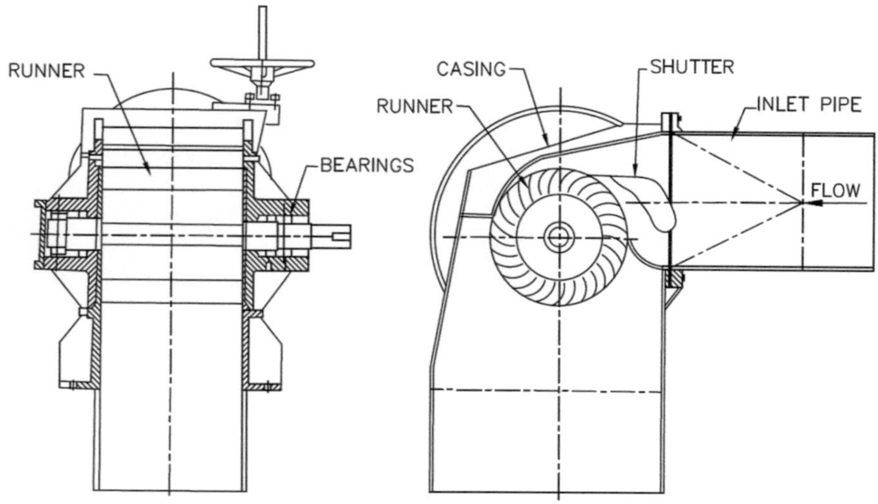

Fig. 3.4 Typical view of cross-flow turbine. (CBI&P. Pub 305)

Hence, there is no axial thrust and so the thrust bearing is eliminated in this type of turbine.

(b) Reaction Turbines

Reaction turbines operate under the head of water. The runner contains the blades which receive water under pressure uniformly through the wicket gates which are inside the guide vanes. The guide vanes are placed in the speed ring and are cast integrally with it. The opening of the wicket gates is adjusted as per load requirements by a mechanism called a servomotor which controls the quantity of water and the speed of turbine. The torque, which develops because of water pressure against the runner blades, rotates the runner. The water leaves the runner under pressure through a passage up to the tail race. Since the runner is under pressure, it can be placed both above and below the tail water level depending upon turbine speed.

Reaction turbines are classified as Francis, propeller and Kaplan. Francis is a mixed flow turbine and propeller and Kaplan are axial flow type. These reaction turbines can be mounted either vertically or horizontally. The propeller turbine may be mounted in an inclined position also. These are briefly described below.

(i) Francis Turbine (Radial–Axial Flow Type)

Francis turbine runner has fixed blades. The water enters the runner through guide vanes under pressure in the radial direction with respect to the shaft and discharges in an axial direction. The discharge from the runner is taken to the tail race through a draft tube. The guide vanes placed in the speed ring receive water from the penstock through a water supply casing called a spiral casing. The entire water passage from penstock to the tail race as shown in Fig. 3.5a is under pressure. The arrangement to regulate the flow through the turbine is shown in Fig. 3.5b.

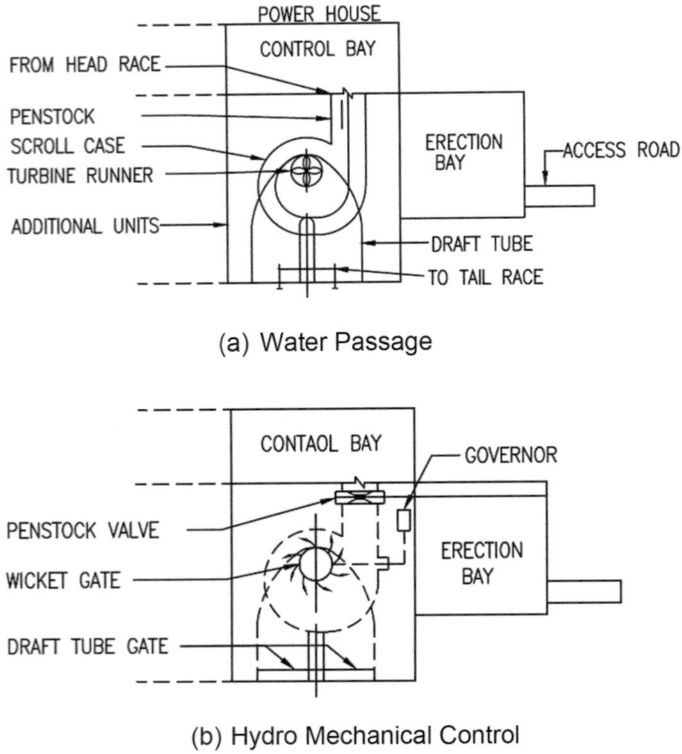

Fig. 3.5 Showing Francis Turbine in Plan (Nigam)

Francis turbines are normally used up to a head of 300 m. This can operate in a range of 40–110% of rated discharge and 65–125% of design head. Operation beyond these limits may cause vibration and power surges, etc. The efficiency of the Francis turbine is about 94–95%.

Francis turbines may be installed horizontally or vertically. Vertical installations are cheaper and require less space and are preferred for high capacity units. Horizontal installations are more economical for small capacity units up to 5 MW for which standard horizontal generators are available. Therefore, horizontal units are preferred for small hydro projects.

(ii) **Propeller and Kaplan Turbines (Axial Flow Type)**

The propeller and Kaplan turbines are axial flow turbines in which the flow enters the blades axially. These turbines have a runner with four, five or six blades. If the blades are fixed it is called the propeller turbine and if the pitch of the blades is adjustable it is called the Kaplan turbine. The efficiency of Kaplan over a wide range of output is better than propeller but it is costlier. The principal components of the axial flow turbines are the same as those of the Francis turbine, i.e. the water supply

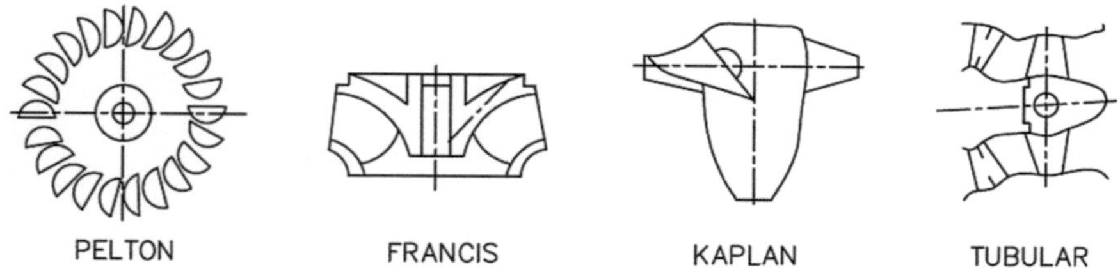

Fig. 3.6 Showing types of runners. (Nigam)

casing, wicket gates, a runner and a draft tube. The axial flow turbines depending on the movement of wicket gates and the runner blades are classified as below:

(1) Adjustable wicket gates and fixed runner blades—Propeller.
(2) Fixed wicket gates and adjustable runner blades—Semi-Kaplan.
(3) Adjustable wicket gates and adjustable runner blades—Kaplan.
(4) A variant of Kaplan turbine with adjustable runner blades with axis at an angle to the shaft axis is called Deriaz turbine.

These turbines are generally used up to a design head of 60 m and power output up to 25 MW. These are normally installed vertically. But, for small capacity units, they are set horizontally or in a slant position (S-type). Various types of turbine runners are shown in Fig. 3.6. The figure includes tubular turbine runner which is described in more detail in the following para.

(iii) **Tubular Turbines**

These are axial flow propeller turbines placed in a horizontal or inclined (S-type) position. These are known as tubular turbines because the turbine is encased in a tube which is the water passage. It is also called S-type because a bend in the water passage is provided to permit the installation of a generator outside the water passage. These are suitable for a head range of 15–25 m and where discharge is comparatively more. These generally do not have wicket gates and the shut up and start up functions are performed by an automatically operated valve. The turbine and generator are connected through a gear box (speed increaser).

In the case of a tubular turbine, the flow is straightened from the upstream up to the tail race. It is confined substantially in more or less axial direction in a steel tube. The shaft axis is generally horizontal. In this case, the water passage is shortened and power house structure becomes compact. The excavation for the foundation is less in this case as compared to the vertical propeller turbine. A typical S-type tubular turbine is shown in Fig. 3.7.

(iv) **Bulb Turbine**

Bulb turbines are axial flow propeller turbines placed horizontally. The runner is directly connected to the generator. In this turbine, the generator, and speed increaser

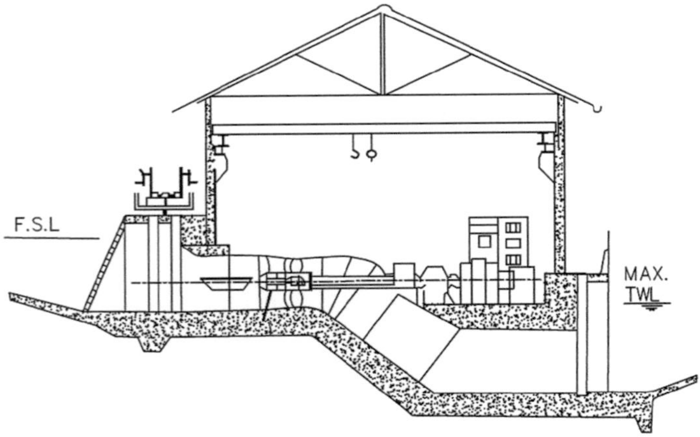

Fig. 3.7 Showing tubular turbine. (AHEC – 2.2/2.3)

(if required) are enclosed in a water tight steel casing. The unit assembly (generator and runner) is placed in the centre of water passage. The runner for bulb turbine has adjustable or fixed blades. The performance characteristics of a bulb turbine are similar to the tubular turbine. These have been generally used for low heads and as small as 50 kW unit output. This has been used for as large as 71,000 hp output with a runner of 7.45 m diameter under a head of 12.4 m in the USA. A section of power house with a bulb turbine is shown in Fig. 3.8. Its floor area and foundation depth are less than the tubular turbine. The disadvantage is that the generator is not accessible for maintenance without dewatering the bulb. An access passage is generally provided in large unit bulb to approach the generator.

3.2.2 Turbine Characteristics

The turbine characteristics that are used in the selection of a turbine for a specific project are head, specific speed and efficiency. These are briefly described below.

(a) **Head**

The effective head is the net head available to the turbine for power generation. Determination of the rated head, design head, maximum and minimum net head is important for a reaction turbine for its efficiency and cavitation characteristics. These are defined below and shown in Fig. 3.9.

(i) Gross Head

It is the difference in elevation between the water level either in the reservoir or in the forebay and the water level at the tail race.

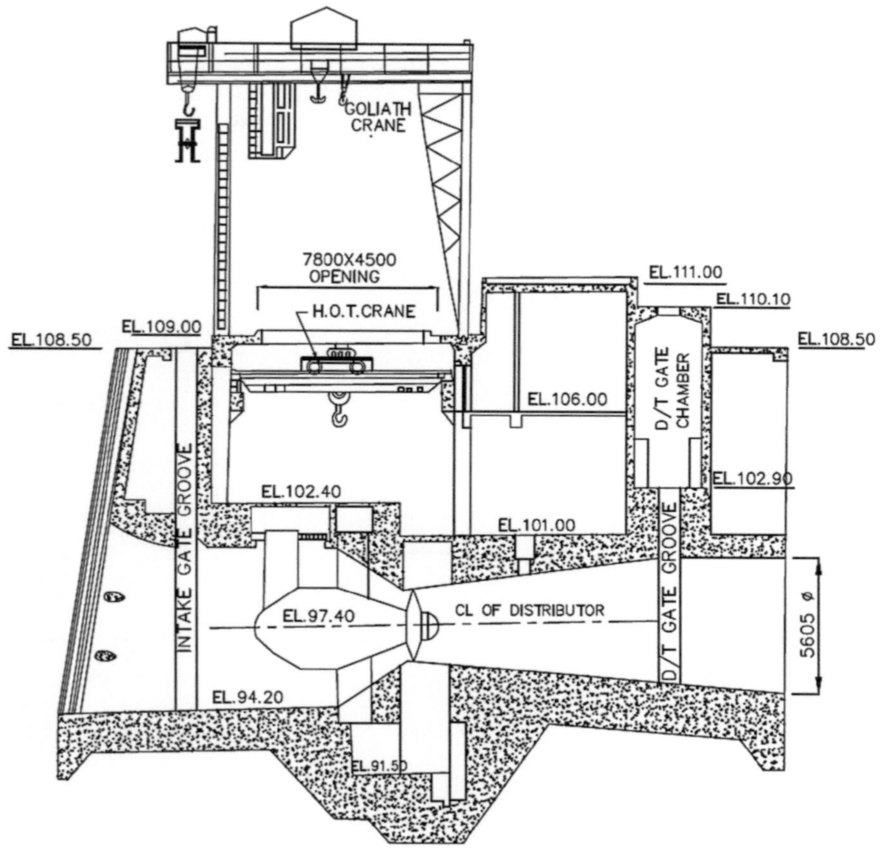

Fig. 3.8 Cross section of Sone Western Link canal H.E. Project Power House (bulb turbine)

(ii) Effective or Net Head

It is the gross head minus all the hydraulic losses in the water conductor system except those chargeable to the turbine. The losses in the spiral casing and the draft tube are chargeable to turbine losses. Hence, all hydraulic losses in the intake and the water carrier (tunnel, channel, penstocks, bends, etc.) should be evaluated as hydraulic losses. The losses in intake, bends, gates, valves, etc., are expressed as $KV^2/2g$, where coefficient K can be adopted from standard texts or IS Codes. The friction losses in penstock and tunnel depend on the length and the permissible velocity. The channel loss is slope multiplied by length.

(iii) Rated Head

It is the net head at which the full gate output of the turbine produces the generator rated output in kW. The turbine name plate rating is usually given at this head.

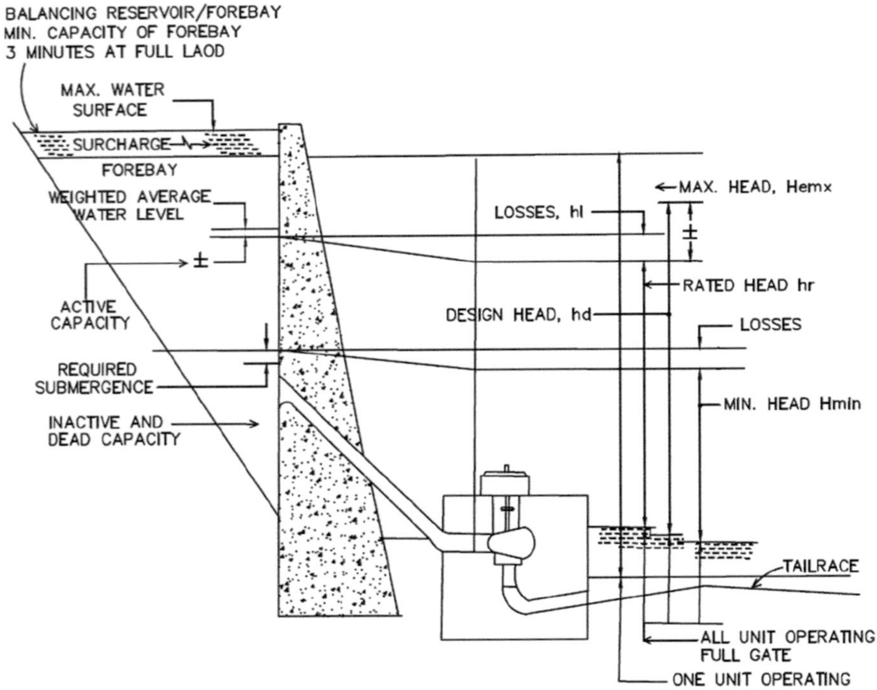

Fig. 3.9 Definition of head in reaction turbines. (AHEC – 3.1)

(iv) Design Head

It is the net head at which peak efficiency is desired. This head preferably approximates the weighted average head but must be within the maximum and minimum permissible operating limits of the turbine. This is the head that determines the basic dimensions of the turbine and the power plant.

(v) Weighted Average Head

It is the net head determined from reservoir operation that will produce the same amount of energy (kWh) between that head and the maximum head as will be developed between that head and the minimum head.

(vi) Maximum Head

It is the gross head resulting from the difference between the maximum reservoir level (not MWL) and the tail race level (spillway not operative) with one-unit operative at no load (turbine discharge is about 5% of the rated flow). In this case, hydraulic losses are negligible and are disregarded.

(vii) Minimum Head

It is the net head resulting from the difference between the minimum reservoir level and the maximum tail water level (all turbines operative at full gate opening) minus the losses.

(b) **Specific Speed**

The specific speed (N_s) of the turbine is the speed in revolutions per minute (RPM) of a turbine homologically reduced in size to develop one horse power (HP) at full gate opening under a head of 1 metre. Mathematically, it is expressed as

$$N_s = NP^{1/2}/H^{5/4},$$

where N = Speed (RPM) of turbine.

P = Power in metric horse power (hp) (1 kW = 0.86 metric hp).

H = Rated head in (m).

The ideal specific speed of different types of turbines may be worked out by using the following empirical relations (P.S. Nigam) in the MKS system:

(a) Francis Turbine $Ns = \frac{2400}{\sqrt{H}}$.
(b) Fixed blade propeller $Ns = \frac{1080}{H^{1/4}}$.
(c) Adjustable blade propeller (Kaplan) $Ns = \frac{1475}{H^{1/3}}$.

USBR (Monograph 20) has given slightly different values for N_s of various turbines.

The above relations show that in reaction turbines the specific speed is inversely proportional to the rated head H. Thus, low-head units have high specific speed and high-head units have low specific speed.

In the case of the Pelton wheel, the specific speed is mainly dependent on the diameter ratio D/d (D is the wheel diameter and 'd' is the diameter of the jet). It is approximately given by

$N_s = \frac{54}{D/d}$(for an average efficiency of 0.85).

The specific speed against the head for various types of turbines is shown in Fig. 3.10 which can be used as a guideline.

(c) **Runaway Speed**

The turbine is coupled with the generator. The electric load on the generator reduces the speed of turbine. When the load is suddenly put off the generator, the turbine with wicket gates fully open rotates at high speed. This speed is called runaway speed.

Thus, the runaway speed of the turbine is defined as the speed at the maximum head with a full wicket gate opening with no load on the generator. The turbine is designed safe for this speed.

The ratio of runaway to normal speed varies from 1.8 to 2.8. The lower value is for impulse turbines and the maximum value is for propeller turbines.

(d) **Efficiency**

The efficiency of the turbine is the ratio of power output and the input. The inputs are the discharge and the net or effective head. The output is measured by the braking load or the output of the generator plus losses. The efficiency of the turbine is mainly dependent on its type, size, head, load and gate opening. The maximum efficiency

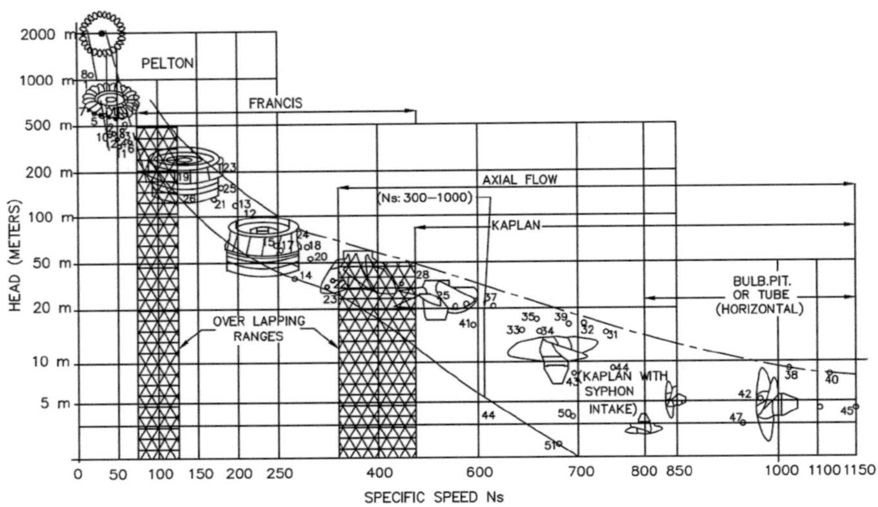

Fig. 3.10 Showing specific speed versus head for various turbines. (AHEC–3.1)

is ensured at the rated output. It varies differently for various types of turbines with the variation in the load. This is shown in Fig. 3.11.

The efficiency curves reveal that Pelton wheel and Kaplan turbines have better efficiency at part load operation than other turbines. Propeller (fixed blade) is not

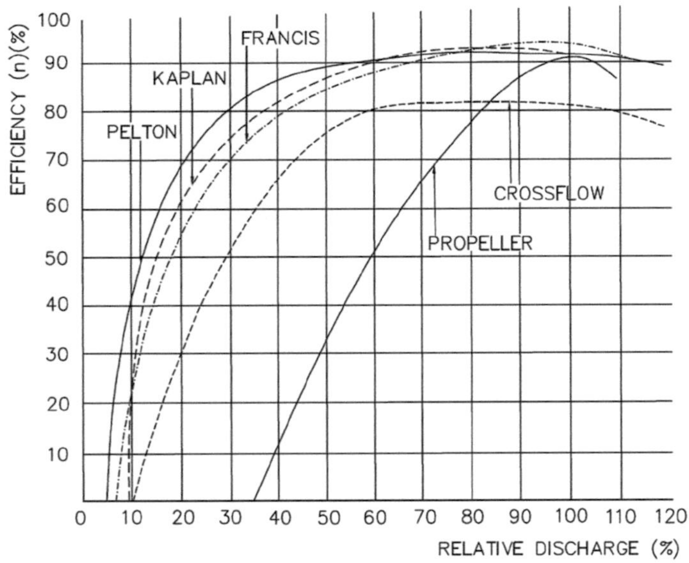

Fig. 3.11 Typical efficiency curves for different types of hydro power turbines. (AHEC–3.1)

Table. 3.1 Criteria for selection of a turbine (IS: 12, 837)

Type of turbine	Head variation % of rated head	Load variation % of rated output	Specific speed	Peak efficiency in %
Pelton	120–80	50–100	15–65	90
Francis	125–65	50–100	60–400	93
Deriaz	125–65	50–100	200–400	92
Kaplan	125–65	40–100	300–400	92
Propeller	110–90	90–100	300–800	92
Bulb	125–65	40–100	600–1200	92

suitable for part load operation. In all types of turbines, the efficiency reduces when these are operated over and above the rated capacity.

3.2.3 Selection of Turbines

The maximum net head is one important parameter for the selection of a hydraulic turbine. The criteria are presented below.

(i)	Pelton wheel	Suitable for high head	Above (300 m)
(ii)	Francis	Suitable for medium head (used up to 600 m)	20–400 m
(iii)	Propeller and Kaplan	Suitable for low head	10–60 m
(iv)	Bulb	Suitable for very low heads	3–20 m

The criteria for selection of turbine based on specific speed (as defined above) and variation in head and load as per IS: 12, 837 is given below in Table. 3.1. It is for medium and large power stations.

The turbine selection criteria for small hydro power projects based on IEC 6116 are shown in Fig. 3.12.

Detailed analysis should be carried out in overlapping range and cost should be optimised. The civil cost varies with type of turbine but it is generally around 50% of total cost of power house. Besides cost, the ease in operation and maintenance of the generating equipment should also be considered while selecting the turbine.

3.2.4 Setting of Turbines

The impulse turbines (Pelton wheel) are set above the maximum tail water level. The high specific speed reaction turbines (Francis and Propeller/Kaplan) are generally required to be placed below the tail water level to avoid cavitation of the runner blades and runner chamber. The cavitation is caused due to sub-atmospheric pressures developed in the runner chamber. The depth of setting of the runner is computed using

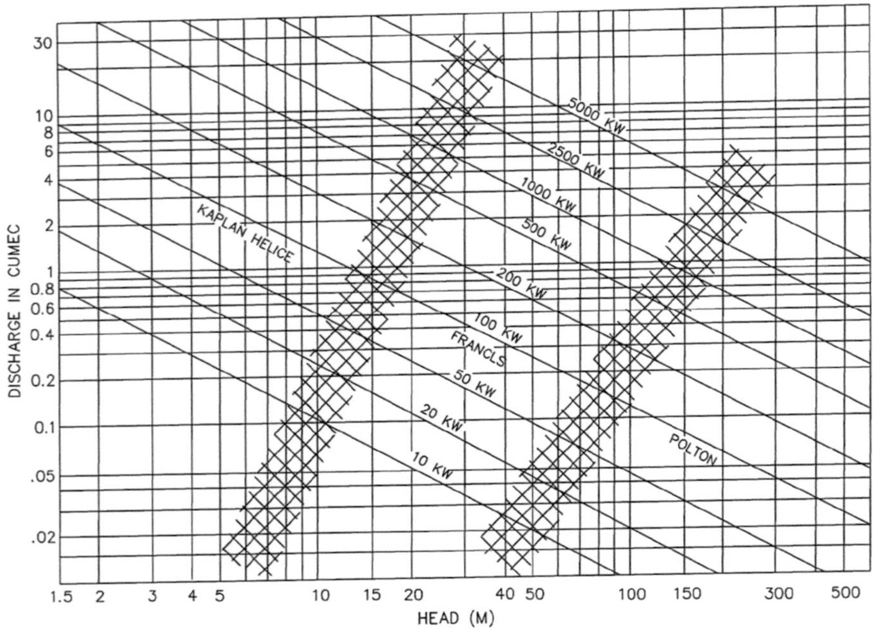

Fig. 3.12 Turbine operating regimes. (Based on IEC:61, 116)

the cavitation coefficient which is related to the specific speed. The relation for depth of setting is as below (IS 12800):

$$H_s = (H_a - H_v) - \sigma.H,$$

where,

H_s = Depth of centre line of runner with respect to minimum tail water level.
H_a = Atmospheric pressure in metre water column at plant elevation.
H_v = Vapour pressure in metre.
σ = Cavitation coefficient.
H = Head on turbine (m).

Value of σ is related to the specific speed. Several expressions suggesting the relationship between σ and N_s are available in the literature. Some of them are listed below:

$$\sigma = 7.54 \times 10^{-5} N_s^{1.41},$$

$$\sigma = (N_s)^{1.64} / 50327, \qquad N_s \text{in metric units (after USBR).}$$

The following recommended values of σ may be used for deciding the setting of a reaction turbine (Table. 3.2).

Table. 3.2 Value of cavitation coefficient

Specific speed at rated capacity	Value of σ
50	0.01
100	0.03
150	0.07
200	0.13
250	0.21
300	0.29
350	0.40
400	0.50
450	0.68
600	0.85

In the case of low-speed turbine, H_s will be positive and the turbine will be above tail water level. In high-speed turbines, H_s will generally be negative and so the turbine will be low setting, i.e. below tail water level. The placement of turbine is shown in Fig. 3.13.

The centre line of the distributor is worked out with respect to the turbine runner level. As shown in Fig. 3.13, the centre line of the distributor is 'b' above the runner level and 'b' is generally worked out by

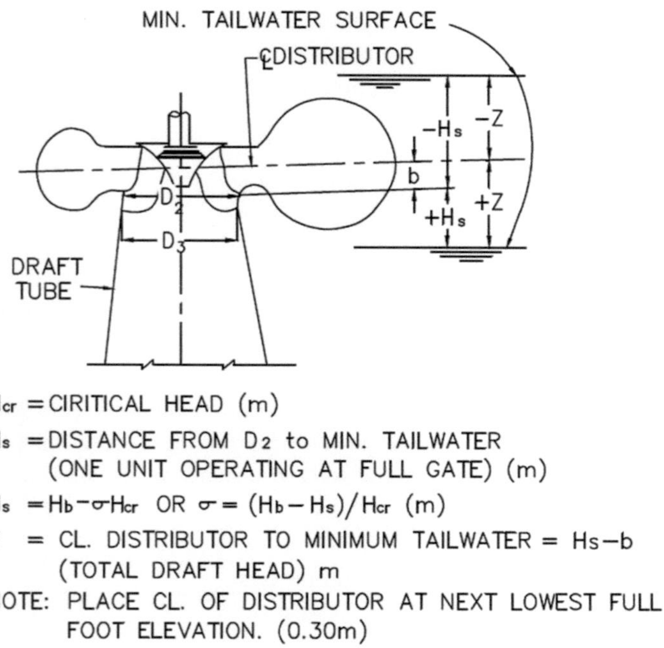

Fig. 3.13 Reaction turbine setting. (*Source* USBR Engineering Monograph No. 20)

$$b = \frac{N_s^{0.34}}{10.7} D_3 \qquad \text{(Francis)}$$
$$= 0.41\, D_3 \qquad \text{(Propeller)}$$

where D_3 is the discharge diameter of runner (m). USBR Engineering Monograph No. 20 may also be referred.

H_s is termed as draft head and can be utilised for power generation through properly designed draft tube.

3.3 Generators

The turbine is coupled with the generator to convert the mechanical energy of the turbine into electrical energy. Generally, three-phase AC generators are used in hydro power plants. The generator consists of a rotor and a stator. The rotor is the rotating part of the generator and is coupled with the shaft of the turbine. The rotor has the exciting field winding and the stator has the armature winding. The rotating exciting field winding induces an electromotive force (emf) in the armature windings.

The speed and rating of the generator depend on the turbine characteristics. The speed of the two is often matched through a speed increaser. In general, the generator is rated for turbine output at rated head. The rating in kW matches with the turbine. The kVA rating depends on power factor which is between 0.8 and 1.0. The generators are AC three-phase with a frequency of 50 cycles.

The efficiency of generator varies from 90% to 96% depending on the speed and capacity of the machine. It remains approximately the same between $1/4^{th}$ and $3/4^{th}$ of rated capacity.

3.3.1 *Speed of Generators*

The speed of generator is generally determined by the turbine speed. The turbine speed for maximum efficiency corresponds to an even number of generator poles. This is called synchronous speed.

The synchronous speed of turbine is the speed at which it is possible to couple the generator to produce the required power. After finding the rotational speed of the turbine the synchronous speed of the generator is found and the specific speed of the turbine is corrected.

$$\text{Trial rotational speed of turbine } N = \frac{120 \times \text{frequency}}{\text{Number of poles}}$$
$$= \frac{6000}{\text{Number of poles}}.$$

Find the number of poles required for the rotational speed of turbine N. Generally, the standard generators are available as multiple of two poles (pair of poles). The number of pair of poles is adjusted subject to the following condition:

If the head is expected to vary less than 10% from the design head the next greater speed is chosen. A head varying more than 10% from the design head the next lower speed should be adopted. With the corrected speed N, the specific speed of the turbine is corrected because it is the basic characteristic of the turbine selection.

Instead of a generator operating on the turbine speed, the generators of higher speed say 500, 600, 750, 1000 or 1500 rpm may be connected to low-head turbines through a speed increaser (gear box). The use of a speed increaser reduces the generator efficiency by 1–2%. It is recommended that the option of use of speed increaser should not be used with unit size of above 3 MW.

3.3.2 *Types of Generators*

There are two types of generators: synchronous and asynchronous (induction) types. The choice depends on the characteristics of the grid and operational requirements. Synchronous generators can be connected to any type of network. Induction generators are used in small isolated units that are connected to strong grids (10 times capacity of unit). These are used up to a capacity of 3 MW. In the case of a SHP in remote areas where the grid is not available, a synchronous generator is preferable. For units of capacity of more than 3 MW, synchronous generators are preferred. Induction generators have lower efficiency than synchronous generators. The merits and demerits of synchronous and induction generators are given in Table. 3.3.

3.3.3 *Generators for SHP*

The SHP units are generally limited to a capacity of 5 MW. Both type of generators are used. Generally, 'off-the-shelf' available generators manufactured by the vendors conforming to IS 4722 are used. These are factory assembled and transported to the site in two separate parts, i.e. rotor and stator. The small capacity (up to 100 KW) generators are assembled and transported as one piece. This reduces assembly work at the site and is economical.

3.3.3.1 Generator Terminal Voltage

The generator terminal voltage should be as high as economically feasible. The SHP generators (up to 5 MW) are generally designed for 6.6 kV, 3.3 kV or 415 V depending on requirement. The economical terminal voltage for SHP generators recommended by CBI&P Manual 305 (2009) is as mentioned in Table. 3.4.

Table. 3.3 Merits and demerits synchronous v/s induction generators

Sl. No	Item	Syn. Generator	Ind. Generator
1	Rotor construction	Salient pole type	Squirrel cage type
2	Excitation	Required	Not required
3	Isolated operation	Possible	Not possible
4	Stability	To be maintained by excitation control	No problem
5	Maintenance	More because of excitation and control equipment	Less because of squirrel cage rotor
6	Efficiency	High	Low
7	Inertia	High	Low
8	Cost	High	Low
9	Power factor	Adjustable by excitation control	Not adjustable determined by load
10	Suitability for highly fluctuating loads	Ideal	Not suitable
11	Loads	Inductive	Highly inductive
12	Reactive power generation	Possible	Not possible
13	Voltage variation	Possible	Not possible

(*Source* AHEC–3.2)

Table. 3.4 Generator voltage

Capacity of generator	Terminal voltage
Up to 750 KVA	415 V
751–2500 KVA	3.3 kV
2501–5000 KVA	6.6 kV
Above 5000 KVA	11 kV

3.3.3.2 Generator Placement

For all type of turbines, the generator can be coupled with the turbine shaft either horizontally or vertically. Generally, a horizontal coupling is required with a tubular turbine. A vertical shaft coupling of the generator is appropriate with vertical turbine installations. For the generators placed horizontally, refer to Fig. 3.7 and, for the vertically placed generator, refer to Fig. 2.27.

The type of coupling of turbine and generator affects the power plant structure and its cost. The vertical unit requires deeper excavation for the foundation and a higher height of the power house. But a horizontal unit requires more width of power house building and lesser height and foundation depth.

The generator for a bulb turbine is located inside the bulb itself.

3.4 Auxiliary Equipment

Besides the main power generating equipment various auxiliary equipment are associated with the turbine and generator to operate and regulate the two. Control systems to control power generation and transmission are also required. These are operated from the control room or through control panels. Various other kinds of equipment are required to operate and control the protection, cooling and drainage systems of the generating equipment. The two main associated equipment required are the inlet valve and the power transformers.

3.4.1 Inlet Valve

Inlet valves are provided upstream of the turbine on long water conductors (penstocks) for the closure of the generating unit. It is operated in fully open or shut condition and is not used to regulate the flow. Butterfly valves are generally used for heads up to 300 m and beyond this head, spherical valves are used. If sediment is erosive and its concentration is high, a spherical valve is preferred. These valves are dealt with in a separate chapter.

3.4.2 Power Transformers

The power transformers step up the generating voltage up to the transmission voltage which is generally 11 kV or 33 kV.

These transformers are generally placed outside the power house. Each generating unit is connected to a transformer through a cable/bus bar/duct. The preferable location of a transformer is between the generator and the switchyard.

References

1. CBI&P.: Manual on planning and design of small hydro electric schemes. Pub. No. 305, New Delhi (2009)
2. Guidelines for Small Hydro Development, Standard No. 3.1, 'Selection of Turbine and Governing System, AHEC, IIT Roorkee (2013)
3. Guidelines for Small Hydro Development, Standard No. 3.2, Selection of Generators and Excitation Systems. AHEC, IIT Roorkee (2013)
4. Guidelines for Small Hydro Development, Standard No. 2.2 'Civil Structures—Hydraulics and Structural Design,' AHEC, IIT, Roorkee, (2013)
5. IEC. 61116.: Electron mechanical equipment guide for small hydroelectric installation (1992)
6. I.S. 12800 (Part-III).: Guidelines for selection of hydraulic turbines preliminary dimensioning and layout of surface hydroelectric powerhouses. B.I.S. Manak Bhawan, New Delhi, (1991)

7. I.S. 12837.: Hydraulic turbines for medium and large power Houses—guidelines for selection. BIS, Manak Bhawan, New Delhi, (1989)
8. I.S. 4722.: Rotating electrical machines. BIS, Manak Bhawan, New Delhi, (1992)
9. Nigam, P.S.: Handbook of hydro electric engineering. Nem Chand and Sons, (1985)
10. USBR.: Selecting hydraulic reaction turbine. Engineering Monograph No. 20

Chapter 4
Layout and Dimensioning of Small Hydro Power Plant

4.1 General

Hydro power plant (or power house) is basically a structure to house generating equipment, i.e. turbine, generator and other equipment associated with the operation, regulation and control of the generating equipment. It is normally a multistorey RCC building with the foundation for the generating equipment. It is also called the machine hall. The handling of the equipment normally needs a crane. Generally, either an EOT or a HOT crane is provided which travels on crane girders along the length of the power house. The crane girders are supported on the columns of the building. The crane hook is placed at a level where it can carry the equipment of one unit over the installed equipment of the other unit. Above the crane, a roof is provided with sufficient clearance for the men to move on the crane. The roof generally consists of the sheet (GC or GI) placed over the trusses which are supported on columns. The building is enclosed with panel walls between the columns. Depending on the height, the columns are suitably braced with cross beams. The structure comprising columns, cross beams, roof and panel walls is called superstructure. The structure below the foundation of columns supports the generating equipment and contains the spiral casing and is called the intermediate structure. Below it is substructure which contains water passage (called draft tube). A power house comprising the above structures is called an indoor power house. It is shown in Fig. 4.1. In case the heaviest load to be handled is small (say less than 5 tonnes), instead of an EOT crane a chain pulley block can be used with the help of the roof truss.

Power stations have been built without a superstructure. These are called out door power stations. Such power stations do not have a roof, over the generating equipment. There are some semi-outdoor type power stations also which have a low roof or a cover over the generators. To access the machines a hatch with a removable cover is provided. Because of extreme climatic conditions, indoor type power stations are preferred in India for maintenance and erection of generating equipment throughout the year without any interruption due to weather.

B. N. Asthana, *Planning and Design of Small Hydro Power Projects*,
https://doi.org/10.1007/978-3-031-96271-4_4

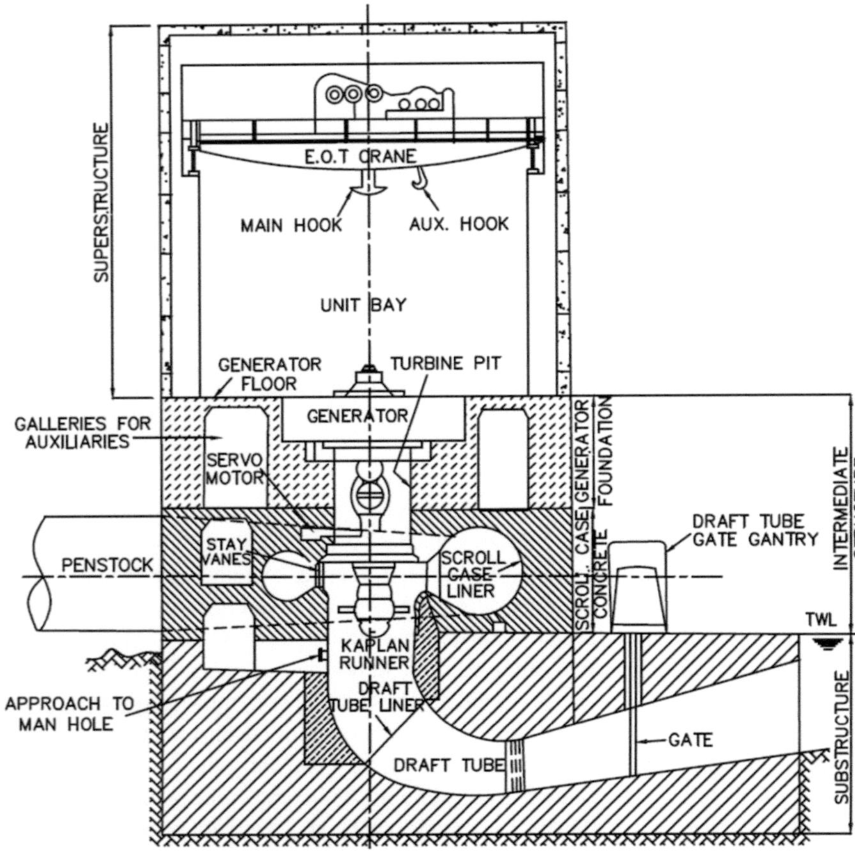

Fig. 4.1 Typical section of an indoor station. (Nigam)

4.2 Layout of Power House

Besides the machine hall or unit bay the other essential constituents of a power house are the erection bay or service bay and the control room. The erection bay floor which is used for erection, repair and maintenance of generating equipment and the transformers are usually placed at the level of the approach road to the power house. The equipment are unloaded on this floor with the help of an EOT crane. Thus, the erection bay is located in continuation with the unit bay/machine hall on the side which is approachable through the approach road. The length of the erection bay is about one to one and a half times the length of a unit bay.

The control room is the space where all the control panels are placed. Through the control room, the operators can supervise the operation of the power house and can control all the activities and operations of the equipment and the power generation of machines. It is generally located either upstream of the unit bay above the penstocks or downstream over the extended portion of the draft tube. Sometimes the control room is housed in a separate building adjoining the machine hall.

The power transformers to step up voltage are generally placed in the open outside the machine hall. But these should be close to the generator. These can be placed in the upstream or the downstream of the machine hall either on the penstock area or the draft tube top slab. If not possible at these locations then they are placed in the switchyard. The location of the transformers should be such that these could be easily pushed to the erection bay for maintenance and repairs.

The switchyard is the area from where the transmission of generated power is carried out. It houses all the equipment necessary for the transmission of power to either a sub-station (11/33 kV) or to a suitable location on a high voltage line passing near the power house.

Thus, the machine hall, erection bay, control room, transformers and switchyard which are necessary major components of a power house complex should be located in relation to each other as a compact and composite entity considering the availability of space, economy and operational convenience. A typical layout is shown in Fig. 4.2. It is a plan with a vertical set generating unit. It shows the machine hall with three units, the erection bay as the extension of the machine hall, the control room in the upstream of machine hall and the transformers on the draft tube deck. The switchyard, in this case, may be located on the bank of the water course downstream of the power house. In case the transformers are placed on the upstream of the machine hall, the switchyard should preferably be in the upstream of power house.

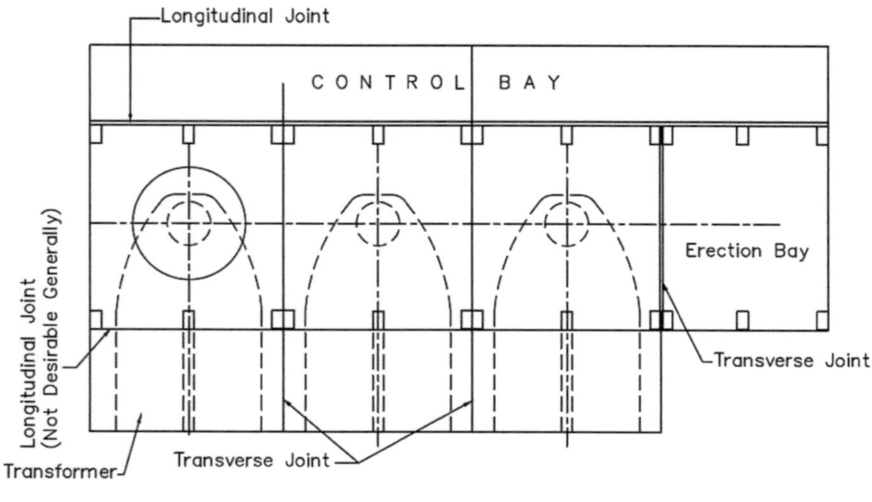

Fig. 4.2 Plan of a power house complex

4.3 Layout of Machine Hall

The machine hall houses the generating equipment, i.e. turbine, generator and the other associated equipment required for their operation. The layout of the machine hall depends on the type of turbine and the orientation of the generator which is coupled with the turbine either horizontally or vertically. Generally, horizontal coupling is adopted with tubular turbines and also with other turbines up to a capacity of 3 MW. For higher capacity units generally, vertical coupling of turbine and generator is preferred. In the case of the vertical setting of the turbine and generator, the height of power house and depth of the foundation are more and the width is less as compared with the horizontal units.

The water passages for the two types of turbine and generator coupling are different and that also affects the layout of the machine hall structure. The water passage for the Francis turbine needs a scroll case or spiral case to be connected with the penstock for uniform distribution of flow to the turbine. The discharge from the turbine goes to the tail race channel through a draft tube. The water passage with a horizontal setting Francis turbine is shown in Fig. 4.3 and with a vertical setting it can be seen in Fig. 4.1. In the case of a Pelton wheel, the penstock is connected with a manifold from which nozzles deliver high velocity flow jets to the buckets of the wheel. The flow after striking the bucket falls freely into a pit which discharges flow into the tail race channel. In the horizontal setting, the Petton wheel, which is generally used in SHPs, one or two nozzles are used as shown in Fig. 3.2. The multijet arrangement is made in vertical setting unit. The water passage with a vertical setting Pelton wheel is shown in Fig. 4.4.

The figures cited above also show the typical locations of the control room and transformers vis-à-vis the machine hall.

The water passage with a tubular turbine with a horizontally coupled generator can be seen in Fig. 3.7. Since a tubular turbine is a reaction turbine it needs a draft tube to discharge water but a spiral casing is not required in this case. The turbine is connected with tail race channel through a conical inclined draft tube.

The bulb turbines have the generator inside a bulb in the water passage of the turbines. A typical layout with a bulb turbine with water passage can be seen in Fig. 3.8.

4.4 Dimensions of a Power House

The power house dimensions depend on the size of the turbine and generator and on the dimensions of the water passage from the penstock to the tail race. The dimensions of the water passage for a vertical reaction turbine, i.e. the spiral case which may be of steel for high heads and of concrete for heads up to 40 to 50 m and the draft tube are related to the discharge diameter of the turbine for planning purposes. Similarly,

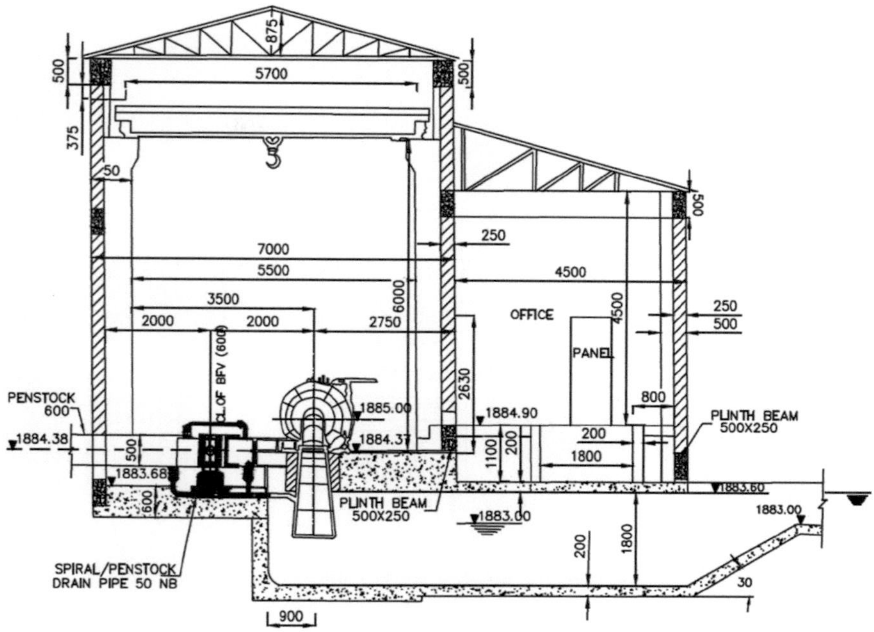

Fig. 4.3 Typical section of a hydroelectric power station with horizontal shaft reaction turbine

the runner diameter of the Pelton wheel decides the size of the pit and the tail race. The final dimensions are provided by the equipment supplier.

4.4.1 Runner Diameter

a) **Reaction Turbine**

In the literature, there are a large number of relations for the discharge diameter of both Francis and propeller turbines. These are all empirical or statistical relations. Some are given below

(i) As per I.S. 12,800 (Part 3) and P.S. Nigam the discharge diameter is as below.

The discharge diameter $D_3 = \frac{84.6f\sqrt{H}}{N}$, (in meter)
where H is head (m), N is rotational speed of turbine and ϕ is peripheral coefficient which is equal to the peripheral velocity at the inlet of turbine divided by sprouting velocity. Several expressions are given by different authors relating ϕ to N_s (specific speed of turbine). One simple expression for Francis turbine is

$$\phi = .0197N_s^{2/3} + .0275.$$

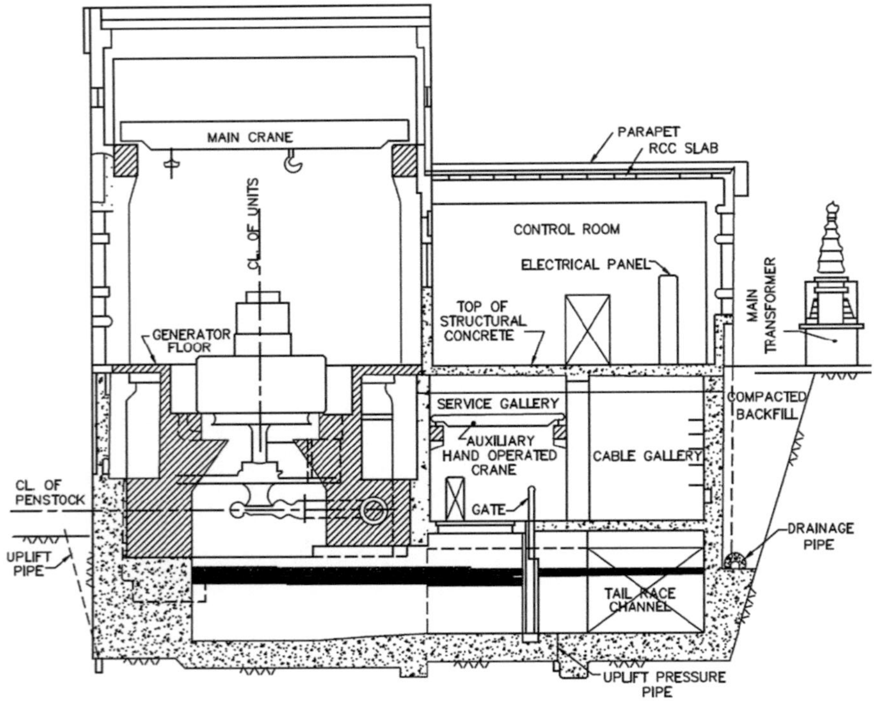

Fig. 4.4 Typical cross section of a surface hydroelectric power station with Pelton wheel. (I.S. 4247 Part-1)

The value of ϕ for propeller turbine can be worked out from the relation

$$\phi = .0242 N_s^{2/3}.$$

I.S. 12,800 (Part 3) recommends the expressions for Francis turbine as $.0211(N_s)^{2/3}$ and ϕ for the Propeller turbine as $.0233(N_s)^{2/3}$.

b) **Pelton Wheel**

(i) Nigam has suggested for the runner diameter of a Pelton wheel as below

$$\frac{\pi DN}{60} = \phi\sqrt{2gH}$$

or

$$D = \frac{60\phi\sqrt{2gH}}{\pi N}.$$

ϕ generally varies between 0.44 and 0.46. Using an average value of ϕ,

Table. 4.1 Details of Pelton turbine for SHP in the head range 150 m–360 m

Head (m)	Turbine output per jet (kW)	Specific speed N_s per jet	Speed	Discharge approx. per jet (m^3/sec) diameter (PCD) (mm)	Runner diameter (Pitch circle)	Set rating	
						1 Jet	2 Jets
150	535	22.03	500	0.408	975	500	1000
190	269	23.2	1000	0.1641	550	250	500
230	789	23.6	750	0.3983	800	250	500
230	1064	21.85	600	0.57	1100	1000	2000
250	315	17.85	1000	0.1437	644	300	600
270	377	17.71	1000	0.1583	630	350	700
360	1570	16.99	750	0.4961	1050	1500	3000

$$D = \frac{38\sqrt{H}}{N}.$$

(ii) CBI&P (Pub. No. 305) has given parameters of the Pelton turbine to select the runner diameter of the wheel. These details are given in Table 4.1. A wheel on this basis can be selected for a specific case.

(iii) I.S. Code 12,800 (Part 3) suggests that

$$D = \frac{84.6\phi H^{1/2}}{N},$$

where $f\frac{d}{D} = .0019N_s$,

d is diameter of jet.

Normally, $\frac{d}{D} = .055$.

Any of the above criteria may be adopted for working out the runner diameter for the type of turbine for the purpose of preliminary dimensioning of power house. The supplier of equipment gives the final dimensions. The CBI&P Pub. No. 175 contains the details of standard runners of Francis turbine and Pelton wheel of different manufacturers. These may also be considered.

4.4.2 Dimensions of Water Passage

(a) **Reaction Turbine**

The water passage of the reaction turbine is made of a spiral casing and a draft tube. The dimensions of these are related to runner discharge diameter D_3. The units with a capacity of about 5 MW and more are vertical setting type, i.e. the generator is placed vertically above the turbine and the shaft connecting the rotor and turbine is vertical. The spiral casing of a Francis turbine is made of steel and the spiral casing of a Kaplan turbine, because of low head, is often made of concrete. The steel spiral casing and its various dimensions can be obtained from Fig. 4.5 (IS 12800 Part I). The practice of BHEL for dimensioning the steel spiral casing is shown in Fig. 4.6.

The dimensions of the draft tube in terms of D_3 for various sizes of discharge diameter of runner (D_3) are given in Fig. 4.7.

The concrete casing is used with a Kaplan turbine up to a maximum head of about 50 m. Usually, the concrete spiral is trapezoidal. The dimensions of the concrete casing required for design are supplied by the turbine supplier. The preliminary design of the scroll case may be made on the basis of the dimensions recommended by Mosonyi (See Mosonyi Vol – 1) which are shown in Fig. 4.8.

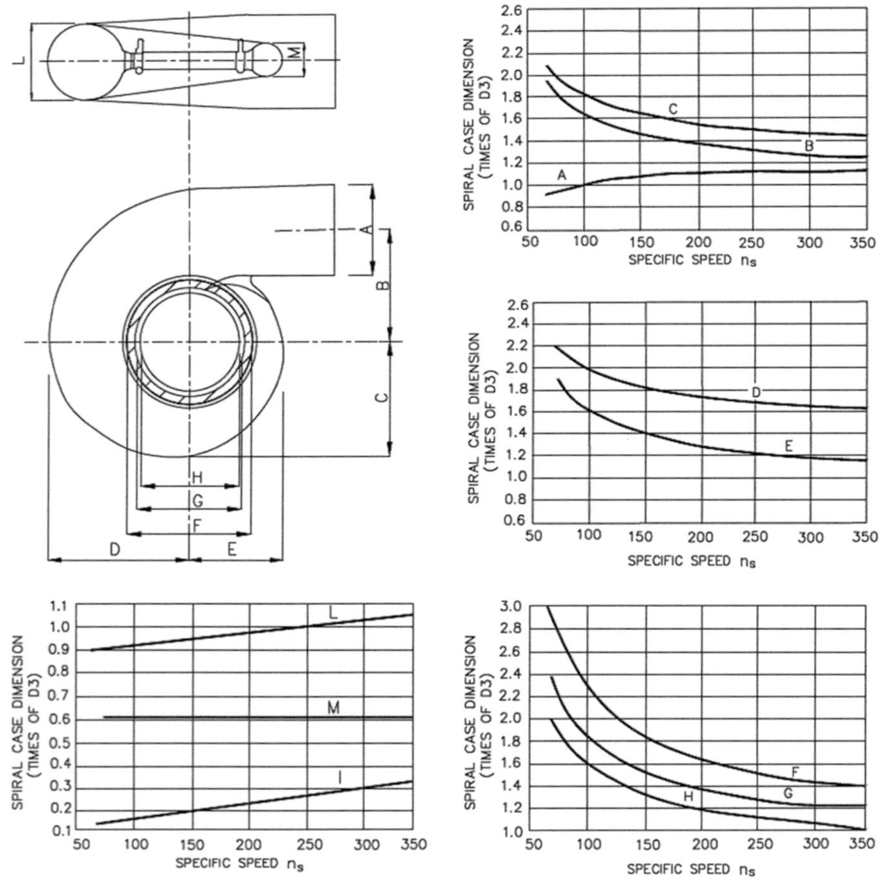

Fig. 4.5 Spiral case and its dimensions

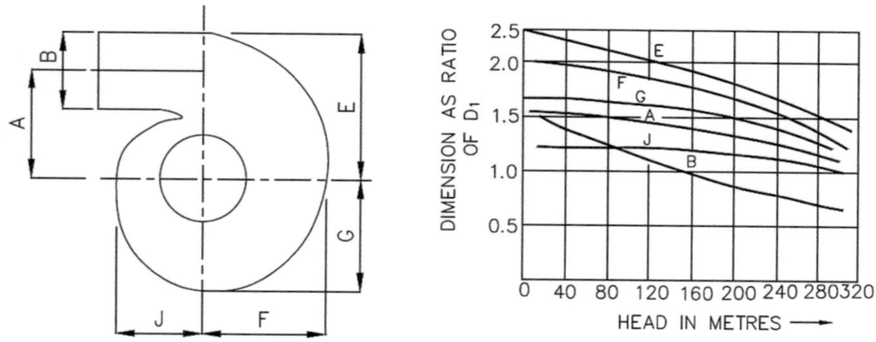

Fig. 4.6 Spiral casing dimension (BHEL)

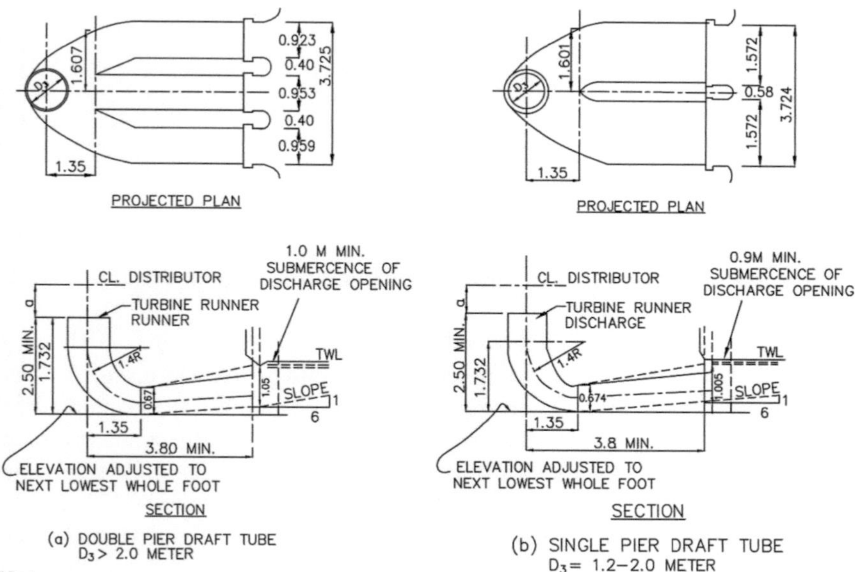

Note:

1. Section beyond pier nose may have a max. floor slope of 10:1 while maintaining similar areas.
2. Structural requirements may alter pier width and therefore overall width.
3. All dimensions proportioned to D_3.
4. TWL is tail water level with one unit full gate.
5. Slope of draft tube is normally kept equal to the maximum slope of stream bed.
6. In case $D_3<1200$, no pier is provided other dimensions are practically same as for D_3 equal to 1.2M.

Fig. 4.7 Draft tube and its dimensions (USBR)

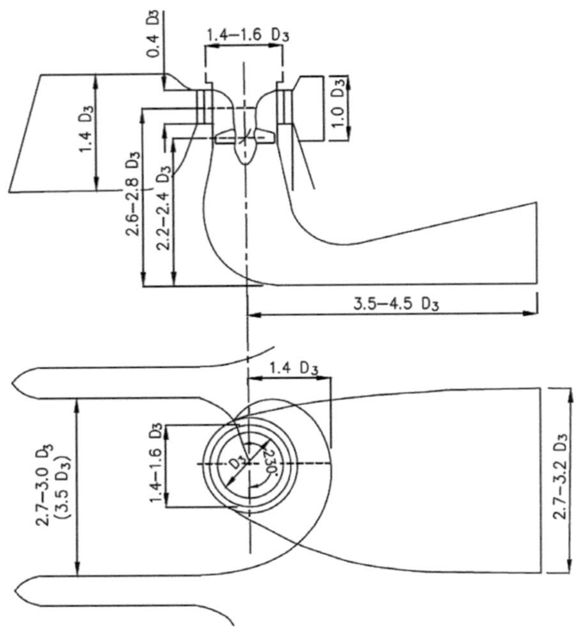

Fig. 4.8 Informative dimensions for preliminary design of power house equipped with Kaplan turbines

The above dimensions of the scroll case and draft tube are normally used with vertical units of reaction turbine.

The arrangement of a horizontal shaft unit with a Francis turbine is shown in Fig. 4.9. It shows the dimensions of scroll case and draft tube in terms of runner discharge diameter. In this case, the turbine with scroll case is a free-standing item. The bottom entrance portion of the scroll case is firmly fixed with concrete through pedestals or can be encased with concrete. The draft tube exists in the side with a removable bend connecting to flaring steel shaft discharging into tail race below tail water level. This part may be straight down as shown in the figure or inclined or elbow type depending on size. On the side of scroll case is the generator with the associated operating mechanism.

(b) **Pelton Wheel**

The water passage for a Pelton wheel comprises a pit connected with tailrace in which a high velocity jet after striking the buckets of the wheel falls above the tail water level. The pit is required in both vertical and horizontal Pelton wheel installations. The dimensions of the pit depend on the pitch diameter (D) of the wheel which has been discussed in para 4.4.1 (b). The outer diameter of the wheel is equal to D plus two times the length of bucket which is 2.75 to 3.5 times d (d is diameter of jet). In a vertical unit (See Fig. 4.4), the pit diameter should be at least

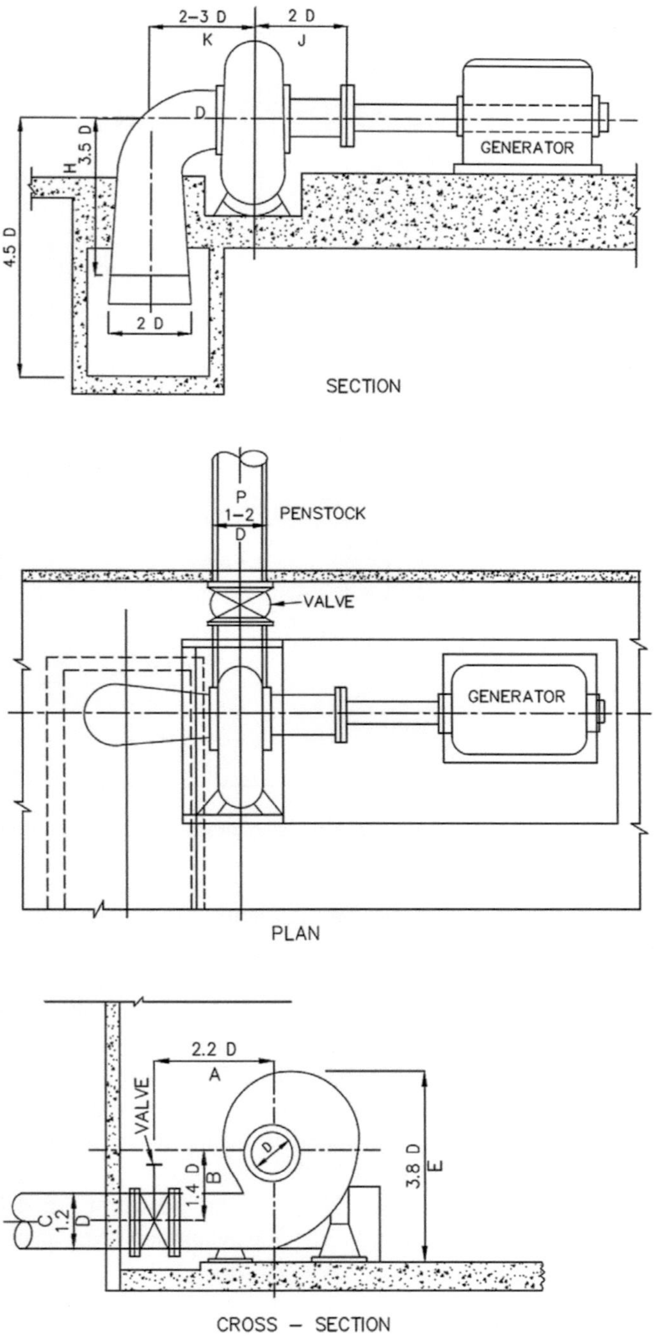

Fig. 4.9 Dimensions and arrangement of horizontal Francis turbine. (CBI&P Pub. 305)

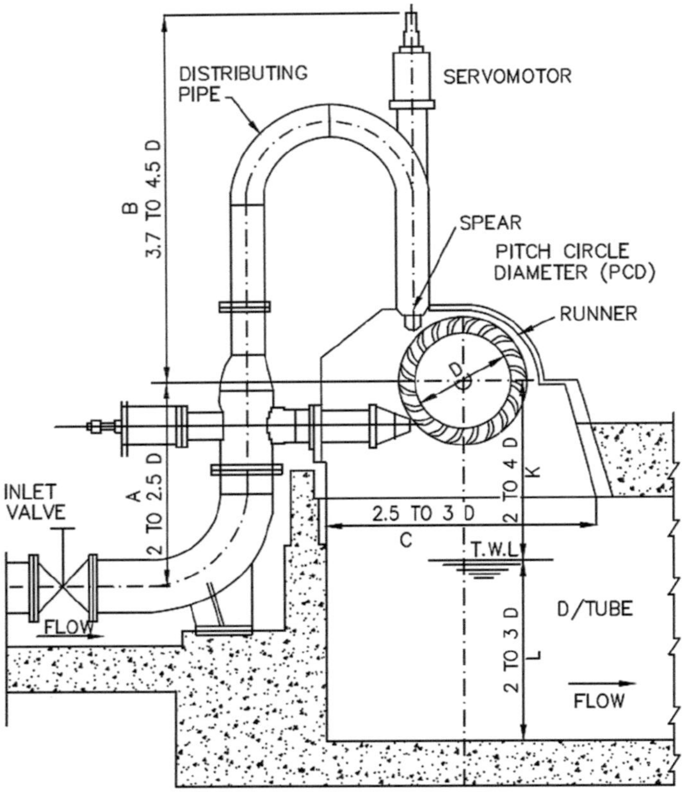

Fig. 4.10 Dimension & arrangement of horizontal Pelton turbine. (CBI&P Pub. 305)

2.5D. The width of the tail race is also kept the same, i.e. 2.5D with an exit velocity of 0.6 to 1.2 m/sec at minimum water level. The minimum height of the wheel above the maximum tail water should be 1.2 m and 2.4 m above the normal tail water level.

In the case of horizontal units of Pelton wheel, the housing above the lower jet is kept to a minimum to reduce windage losses but the wheel pit should be large enough to ensure free discharge from the buckets. The lower horizontal tangent to the pitch diameter of the horizontal unit should be well above maximum tail water level. The minimum height is recommended as 1.2 m, above maximum tail water level and 2.4 m above normal tail water level.

The dimensions suggested in CBI&P Pub. 305 are shown in Fig. 4.10 where a two jet horizontal Pelton wheel is shown. The dimensions are given in terms of pitch diameter of wheel.

4.4.3 Dimensions of Generator

(a) **Vertical Shaft Arrangement**

In this type of arrangement, the shaft connecting the turbine and rotor is vertical. The generator in this arrangement is placed vertically above the turbine. In this type, two types of generators are used. These are the suspended type and umbrella type. In the suspended type, the thrust bearing is at the top above the rotor and in the umbrella type the thrust bearing is below the rotor and alternatively can be placed at the turbine cover. The load bearing bracket makes the difference in the height of the two types of generators. A dimensioned sketch of a vertical generator is shown in Fig. 4.11. The dimensions depend on air gap diameter (D_i) and active length of core (l_t) as shown in Fig. 4.11. These two dimensions are related to the output and synchronous speed of generator and are given in Fig. 4.11.

The other dimensions may be worked out with the following expressions.

$$D_a = D_i\left(1 + \frac{\pi}{P}\right)$$

where P is number of poles and D_a is outer core diameter of stator (m).

D_f (stator frame diameter) $= D_a + 1.2$ m.

D_b (inner diameter of generator barrel) $= (D_f + 1.6$ to $2.0)$ m.

Length of stator frame $l_f = (l_t + 1.5$ to $1.6)$m.

Height of load bearing bracket h_j is given by

$h_j = k\sqrt{D_f}$ For suspended type

$= k\sqrt{D_i}$ For umbrella type

$$\begin{aligned} K &= 0.85 \text{ for 50 t/arm load on bracket} \\ &= 1.0 \text{ for 100 t/arm load on bracket} \\ &= 1.1 \text{ for 150 t/arm load on bracket} \end{aligned}$$

Number of arms depends on the total load on thrust bearing which is the hydraulic thrust on the turbine runner plus the weight of rotor.

(b) **Horizontal Shaft Arrangement**

In this arrangement the shaft connecting the turbine and generator is horizontal. The generator is placed by the side of the turbine and the draft tube may be below or on the other side of turbine. The generators used in this arrangement are of the following types:

(i) Synchronous generators with static excitation.
(ii) Synchronous generators with brushless excitation.
(iii) Induction generators.

The above generators and their dimensions are given for various capacities in CBI.Pub.305. These may be referred to for preliminary design.

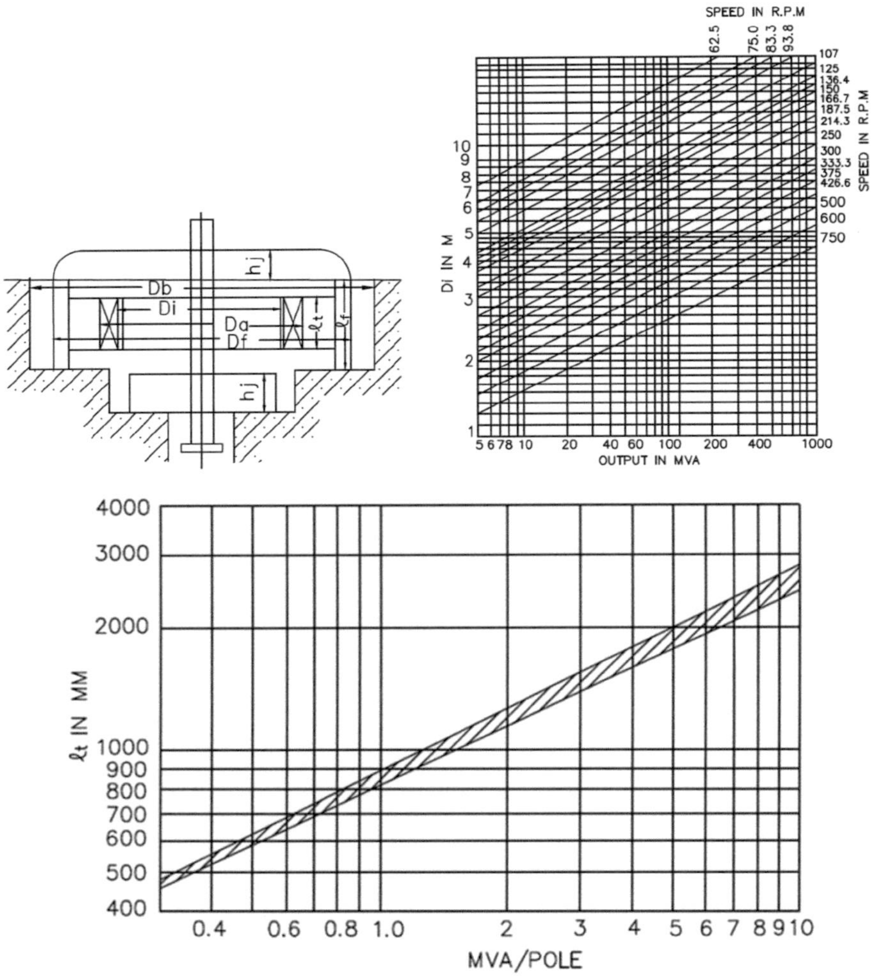

Fig. 4.11 Vertical generator and its parameters. (P.S. Nigam)

4.4.4 Power House Dimensioning with Vertical Unit

(a) **Length of Power House**

It depends on unit size, number of units, length of erection bay and length required for EOT crane handling in the last unit.

Unit size:

For determining the length of one unit, a plan should be drawn showing the overall dimensions of scroll case, draft tube and the outer dimension of generator with the vertical axis of the machine as the centre. The outer diameter of generator barrel is taken as D_b + 1 to 1.5 m (for D_b refer para 4.4.3). A clearance of about 3.0 m on either side of the extremities of the above drawn figure will determine the unit size. This clearance should ensure a minimum concrete thickness of about 1.5 to 2.0 m on either side of the scroll case and the draft tube.

The length of erection bay is generally equal to the size of unit and may be more depending on erection requirements and loading and unloading of equipment and the total number of units. The total length ‘L’ of power house will be as below:

L = Unit size × number of units + length of erection bay + length for EOT crane handling in last bay which is usually about 3.0 m.

(b) **Width of Power House**

The plan drawn in sub-para ‘a’ above to determine unit size should be used for determining the width of power house. The superstructure columns should be clear of upstream and downstream extremities by about 1.5 to 2 m. In the upstream, a concrete cover of 1.5 to 2 m should be ensured for the scroll case. Generally, an inlet valve pit to accommodate the main inlet valve is placed in the upstream. It should be clear from the generator extremities and the upstream superstructure columns so that the EOT crane may handle the valve. The outer dimensions of the width of power house will be obtained by adding to the width required by the above exercise plus the length of the upstream and downstream superstructure columns. Generally, column length (in plan) depends on EOT crane specifications. These are of the order of 0.7 to 1.2 m. A typical plan of the unit is shown in Fig. 4.12.

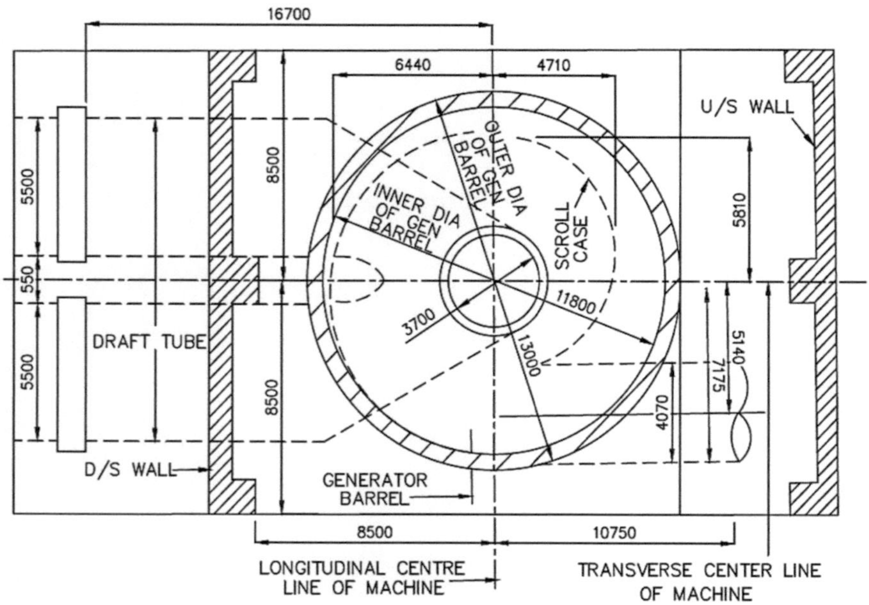

Fig. 4.12 Typical plan showing main dimensions of a unit bay

(c) **Height of Power House**

For determining the height of power house, it is essential to estimate the height of centre of distributor (spiral casing) above the bottom of the draft tube (H_1) and the height above the centre line of the distributor up to the top of the generator (H_2). These are shown in Fig. 4.13. H_1 can be determined by the dimensions of the draft tube and the setting of turbine. Generally, a thickness of 1 to 2 m of concrete below the lowest point of the draft tube is assumed to work out the height above the foundation.

Height H_2 is determined as below.

$H_2 = l_f + h_j + K$. (for l_f and h_j refer para 4.4.3)The valve of K is assumed as 5.5 to 7 m.

The height of power house above the generator depends on the level of the EOT crane hook and the crane girder level and the clearance required above the top ceiling of the crane cabin.

The EOT crane hook level should be such that the equipment of one unit, i.e. turbine runner and rotor assembly could be hauled over the other unit which has been erected or is in operation. It should also be governed by the erection floor level and the clearance required to unload and haul the main transformer. A minimum clearance of 0.3 m between the top of the crane cabin and the bottom of roof should be ensured.

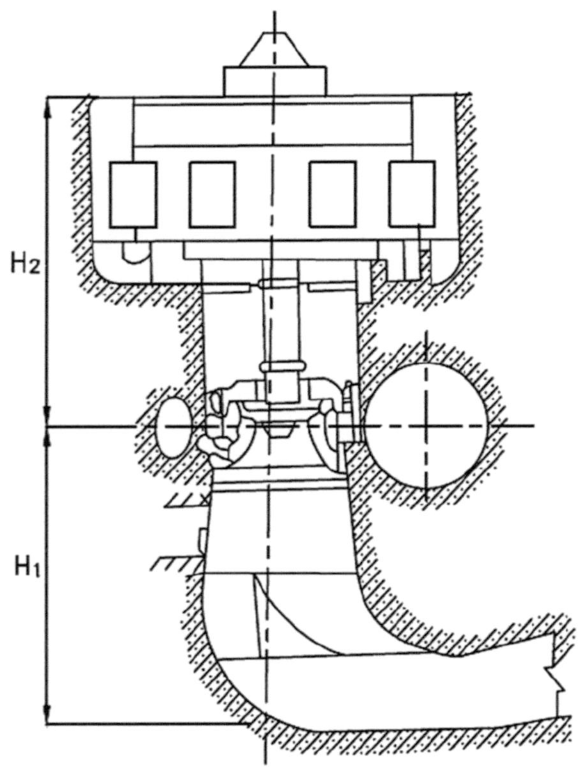

Fig. 4.13 Cross section through generating unit (showing height). (I.S. 12,800 – Part 2)

4.4.5 Dimensioning of Power House with Horizontal Units

The unit size will depend on the type of turbine and its placement, the type of generator and the draft tube (refer Figs. 4.9 and 4.10). The dimensions of the turbine, generator and draft tube for the horizontal arrangement have been described in paras 4.4.2 a and b and para 4.4.3 b. The clear spacing beyond the equipment on either side and the minimum concrete cover to the draft tube may be taken in the same manner as described for a vertical unit. The length of power house shall be worked out in the same manner as for a vertical unit.

The width of power house can be determined by providing a minimum clearance of 3 m up to the wall from the extremities of the turbine and the generator.

The depth of the substructure and the height of equipment placed above it shall be worked out. Thereafter, the height of power house above the equipment should be worked out in the same manner as described for a vertical unit.

4.4.6 Dimensioning of Power House with Tubular Turbine

Tubular turbine, which is a propeller turbine, is commonly used in small hydro projects.

The typical arrangement of a tubular turbine with the generator which is horizontally coupled with the turbine shaft is shown in the plan and section in Fig. 4.14. The dimensions for the length and width of the unit for different sizes of the runner are also given in Fig. 4.14. The height of power house above the equipment can be worked out in the same manner as described for the horizontal unit.

4.4.7 Standard Layout and Dimensioning of a SHP Powerhouse

Guidelines to fix the layout and the dimensions of the power house of a SHP with various types of turbines and their different orientations with respect to the generator are given in figures of IS 12800 (Part 3). The dimensions are given in terms of the discharge diameter of the runner. These figures show the power house dimensions for Francis vertical and horizontal units, Pelton wheel vertical, bulb turbine, etc. The dimensions for a power house given in IS 12800 (Part 3) are very helpful for developing a quick preliminary design and the cost estimate of the power house of a SHP.

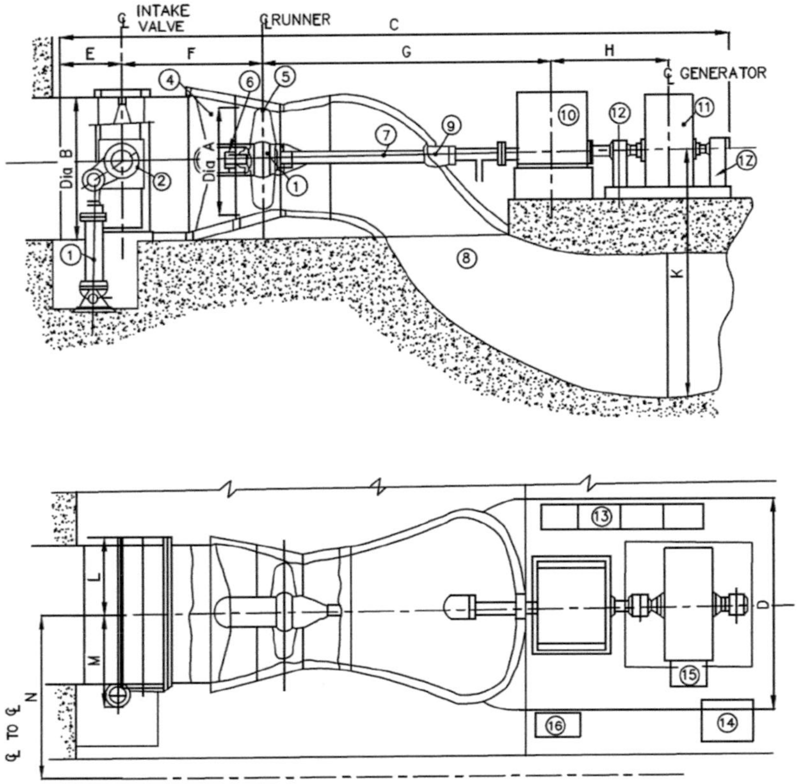

BASIC DIMENSIONS

A=RUNNER DIAMETER IN mm ALL OTHER DIMENSIONS ARE IN PROPORTION FROM RUNNER DIAMETER

A	1200	1500	1800	2000	2200	2500
B	1700	2100	2500	2800	3000	3500
C	10000	12500	14500	16000	17750	19800
D	2500	3150	3800	4200	4650	5300
E	1200	1500	1800	2000	2200	2500
F	3000	3750	4500	5000	5500	6250
G	3200	4000	4800	5400	6000	6600
H	1450	1800	2150	2400	2650	3000
J	1200	1500	1800	2000	2200	2500
K	1950	2450	2950	3200	3550	4000
L	1000	1250	1500	1700	1900	2200
M	1250	1500	1800	2000	2300	2600
N	3700	4500	5400	6000	6500	7000

NOMENCLATURE

1.RUNNER 2.MAIN INLET VALVE
3.VALVE SERVO MOTOR
4.STAY VANE. 5. RUNNER CHAMBER
6.TURBINE BEARING. 7. SHAFT
8.DRAIN TUBE. 9. SHAFT SEAL
10.GEARBOX. 11. GENERATOR
12.GENERATOR BEARING
13.CONTROL AND PROTECTION CUBICLE
14.NG RESISTOR
15.TERMINAL BOX 16. LAVT CUBICLE

Fig. 4.14 Typical dimension of tube turbine. (I.S. 12,800−Part 3)

References

1. CBI & P.: Manual on planning and design of small hydroelectric projects. Pub. No. 305, New Delhi, (2009)
2. CBI & P.: Manual on planning and design of SHP, Pub. No. 280, New Delhi, (2001)
3. Guidelines for small hydro development standard No. 2.2, Civil Works—Hydraulic and structural design. AHEC, IIT, Roorkee, (2013)
4. I.S. 12800 (Part 1, 2 & 3), Guidelines for selection of hydraulic turbines, preliminary dimensioning of SHP. B.I.S New Delhi, (1991)
5. I.S. 4247 (Part 1).: Structural design of surface hydraulic power stations. BIS, New Delhi, (1993)
6. Mosonyi, E.: Water power development, vol. 1. Publ. House Hung. Acad. Sci., Budapest, Hungary (1963)
7. Nigam, P.S.: Hand-Book of hydroelectric engineering. Nem Chand and Bros, Roorkee, (1985)

Chapter 5
Planning a Small Hydro Project

5.1 General

Small hydro project needs careful planning for a techno-economically feasible project formulation. Such planning for a project requires various surveys, investigations, collection of data and information regarding site topography, geology, hydrology, material availability, environmental and ecological aspects, etc. This requires considerable investment and a lot of time and effort from various agencies. Hence, project planning is generally carried out in stages and the site surveys and investigations are also divided into various stages of planning. These are briefly detailed in the following section:

5.2 Stages of Project Planning

(i) **Identification Stage**

At this stage, all available maps, data and documents regarding topography, hydrological data, geology and previous studies, if any, shall be collected. With this data and information, a study shall be carried out for evaluation of the possible availability of head and discharge for power generation. Viable alternative locations of the site and layout of schemes shall be developed and marked on the available topographical maps. A rough assessment of power generation and the cost of each alternative shall be worked out. A list of site data requirements for further studies shall also be prepared.

The report of this stage of planning, i.e. identification report helps the owner/decision-maker in deciding whether the site shall be further explored.

B. N. Asthana, *Planning and Design of Small Hydro Power Projects*,
https://doi.org/10.1007/978-3-031-96271-4_5

(ii) **Pre-feasibility Stage**

The planning study of the project at this stage depends on the identification report and further data and information required from the field visits. The team for the field visits should comprise a hydrologist, a geologist and a hydro power engineer. The team during visits should involve local people of the area because their information could also be valuable.

The team during field visits should collect information regarding topography, geology, river flow conditions, flow data and availability of infrastructure facilities of feasible alternative sites. The team may identify the possible locations of various structures such as diversion structure, desilting arrangement if required, suitable alignment of the water conductor, the power house and power transmission facilities. The elevations and coordinates of various locations can be determined by portable GPS. The field visit report should be supplemented with drawings and site photographs.

On the basis of the identification report and the field visit report, the pre-feasibility stage study should be carried out and a pre-feasibility report should be prepared. It should contain technical analysis regarding hydrology, geology and power potential. It should also describe preliminary details for the availability of construction materials and environmental issues. The report should contain the preliminary design of civil structures and electro-mechanical equipment for different feasible alternatives and their cost estimate. Some preliminary financial analysis should also be given.

The report so compiled should give the merits and demerits of different feasible alternatives and should give adequate indicators to decide the best feasible alternative for further investigations or to drop the project as not feasible. If the study at this stage suggests an alternative for detailed investigations, then a scheme of further investigations should also be given in the report.

(iii) **Feasibility/Detailed Project Planning Stage**

The report of this stage of planning is generally called Detailed Project Report (DPR). It should give the estimated cost of the project which should be at the least variation from the actual cost of execution. It should contain sufficient technical details, cost estimate and financial analysis and should be comprehensive enough for quick evaluation at technical and administrative levels for final investment decisions for execution by the owner/developer. The preparation of a DPR follows the outcome of the detailed surveys and field and laboratory investigations. Generally, the following surveys and investigations are carried out for the preparation of DPR:

(1) Topographical surveys.
(2) Hydrological and meteorological investigations.
(3) Geological mapping and foundation explorations with investigations.
(4) Construction material survey.
(5) Load survey and power evacuation facility.
(6) Survey for infrastructural facilities.
(7) Survey for environmental aspects.
(8) Socio-economic survey.
(9) Muck disposal survey.

The above surveys and investigations required for a DPR are discussed in a separate chapter. These require a lot of investment and time to complete. Depending on the remoteness of the site, these investigations may take a couple of years. So, these shall be carefully planned and capable and technically competent agencies should be engaged.

It has been experienced that generally for a SHP no observed discharge data of the stream at the selected site is available. So, for hydrological studies, the discharge data at the project site is indirectly derived from the data of other sites. Sometimes that too is not found feasible. Hence, it is advisable that as soon as a feasible site is identified in the pre-feasibility report, a discharge observation site shall be established at the site so that discharge observations of a couple of years could be available by the time the DPR is prepared.

Since DPR is a comprehensive report on the project for the final investment decision, it should contain details of the physical features of the site and the results of all surveys and investigations and the technical and financial analysis. It may be divided into the following chapters:

Chapter-1	Need for the project
Chapter-2	Basic data
	It shall contain details of the river basin, catchment area, topography, hydrological and meteorological data, type of scheme, layout of scheme, etc. It should have a brief description of how and why this scheme was selected
Chapter-3	Surveys and Investigations
	It should contain brief details of all surveys and investigations giving the methodology adopted and the field and laboratory test results. The details should be supplemented with necessary maps, drawings, photographs, etc.
Chapter-4	Geological Studies
	It should contain the results of geological, geophysical, if any, and geotechnical investigations. It should contain a description of regional geology, the geology of the project site for each and every component of the project, recommendations regarding the geological suitability of the project site and details of further investigations and the treatments that would be required to be carried out during execution
Chapter-5	Hydrological Studies
	It should contain the compiled discharge data at site (may be observed or derived from other sites), analysis of data to obtain water availability and flood discharge. It should give a stage–discharge curve at diversion and tailrace sites and also give sediment intensity and its characteristics in the flow at the diversion site
Chapter-6	Power Studies
	It should contain studies to give optimal installed capacity, unit capacity and assessment of power potential

(continued)

(continued)

Chapter-7	Design of Civil Works
	It should contain the hydraulic and structural design computations for each component of the scheme. A list of structural components of a run-of-river scheme is given below. This may vary with the type of scheme Diversion structure (weir, barrage, dam) and the intake. Water conductor system—it may be a canal or tunnel or pipe. Desilting arrangement—it may be an ejector or desilting basin. Surge tank or forebay. Penstock. Power house. Tail race channel. The design should be detailed enough to arrive at a realistic cost estimate of the project.
Chapter-8	Electro-Mechanical Equipment
	It should contain details about type and size of turbine and generator and the number of units along with all the associated equipment for operation and control of generation. It should also include the equipment for the switchyard and power transmission
Chapter-9	Hydro-Mechanical Equipment
	It should contain details (type, size, number, etc.) of regulation gates and valves required to regulate flow from the diversion structure up to the tailrace
Chapter-10	Construction Planning
	It should contain details about the availability and procurement of construction materials, construction methodology including river diversion scheme, construction schedule, requirement of construction equipment, muck disposal plan, infrastructure and communication requirements
Chapter-11	Environmental Aspects
	It should contain details of environmental impact assessment and mitigation measures as well as rehabilitation requirements, if any
Chapter-12	Project Organisation:
	It should contain details of the organisation required for the construction as well as the operation of the project. It should contain details of both technical (Civil, Electrical, Mechanical) and administrative personnel which would be required
Chapter-13	Cost Estimate:
	It should have a detailed cost estimate of various components of the project and the total cost of the project
Chapter-14	Financial and Economic Analysis
	It should give a detailed analysis for arriving at unit generation cost and financial returns as per applicable norms and procedures
Chapter-15	Recommendation
	It should contain the details of the benefits of the project, the cost and the financial returns. It should also give justification for the project

At the beginning of the DPR, an executive summary should be given in brief containing the project details, hydrology, geology, power studies, type of civil structures and their special features, electro-mechanical equipment, special construction and environmental aspects and cost and financial returns for the consideration at the highest level of decision-making. The executive summary should be followed by the salient features (a format is enclosed as an appendix).

The above should be the contents of the report volume of the DPR. The second volume should contain detailed cost estimates including the quantities and schedule of rates. The third volume should contain maps and drawings showing the details of civil, electrical and mechanical works. The drawings should give enough details for a realistic cost estimate of the project. Details of surveys and investigations (field and laboratory tests) should form an additional volume for DPR.

The above is a general format for a DPR of a run-of-river project. This may need some changes for other types of SHPs.

Appendix

Salient Features for Power Projects (Run-of-River Diversion Scheme).

S.no	*General*	
I	**Project Data**	
1	**Name of the Project**	
2	**Location of Project**	
	a. State	
	b. District	
	c. Village	
	d. Access to site	
	Nearest Rail Head	
	Nearest Airport	
3	**Geographical Coordinates**	
	a. Longitude	
	b. Latitude	
	c. Altitude	
	d. Reference toposheets	
	(Survey of India)	
4	**Hydrology**	
	a. Name of stream	
	b. Catchments area	sq.km
	c. Main river	

(continued)

(continued)

S.no	*General*	
	d. Maximum flood discharge	Cumecs
	e. Type of stream	
	f. 100% dependability flow	Cumecs
	g. 75% dependability flow	Cumecs
	h. 50% dependability flow	Cumecs
	i. 25% dependability flow	Cumecs
II	**Project Features**	
1	**Weir and Intake**	
	a. Type	
	b. Shape	
	c. Crest Level	EI…..m
	d. MWL/FRL	EI…..m
	e. Number of Bays	
	f. Intake structure (Type and size)	
2	**Feeder Channel**	
	a. Shape	
	b. Material	
	c. Bed slope	
	d. Bed width	….m
	e. Water depth	….m
	f. Length	….m
	g. Design discharge	....m^3/s
	h. Free board	….m
3	**Desilting Tank**	
	a. Size of tank	Length – m, width – m
	b. Transition length (upstream)	….m
	c. Transition length (downstream)	….m
	d. Material	
	e. Particle size to be removed	….mm
	f. Silt disposal outlet	….mm dia pipes
	g. Design discharge	….cumecs
	h. Free board	….m
4	**Power Channel**	
	a. Shape	
	b. Material	
	c. Bed slope	
	d. Bed width	m

(continued)

(continued)

S.no	*General*	
	e. Water depth	m
	f. Length	m
	g. Design discharge	m^3/s
	h. Free board	m
5	**Head Race Tunnel (If Applicable)**	
	a. Length	m
	b. Diameter/shape	
	c. Type & thickness of lining	
6	**Surge Tank (If Applicable)**	
	a. Type	
	b. Diameter	m
	c. Height above orifice	m
	d. Orifice diameter	m
7	**Penstock**	
	a. Type/Material	Steel grade
	b. Main Penstock	
	(i) No. of pipes	
	(ii) Length	m
	(iii) Diameter	m
S.no	*General*	
	(iv) Thickness	mm
	(v) Design discharge	cumec
	c. Branching near power house	
	(i) No. of Pipe	nos
	(ii) Length	m
	(iii) Diameter	m
	(iv) Thickness	mm
8	**Power House**	
	a. Type	Surface power house
	b. Gross Head	m
	c. Net Head	m
	d. Design Head	m
	e. Installed Capacity	kW
	f. Size	---m---m---m
	g. Machine Hall Floor Level	---m
	h. Erection Bay Floor Level	---m
9	**Turbine**	

(continued)

(continued)

S.no	*General*	
	a. Type	
	b. Number	
	c. Capacity	kW (at generator terminal)
10	Type of Generator	
	a. Type	Synchronous
	b. Nos	
	c. Capacity	kW (each generator)
11	**Tailrace Channel**	
	a. Length	---m
	b. Size	---m---m rectangular
12	**Annual Energy Generation (in MU)**	--- MU at --% load factor
13	**Estimates of Cost**	
	a. Total Cost	Rs. -----million
	(i) Cost of Civil works	Rs. -----million
	(ii) Cost of E & M Equipment	Rs. -----million
	(iii) Other expenses	Rs. -----million
	(iv) Transmission system	Rs. -----million
	b. Cost of installation per kW	Rs.....
14	**Generation Cost Rs./kWh**	
	(i) Without Subsidy/incentive (ii) With Subsidy/incentive	Rs..... Rs......

Reference

AHEC, IIT Roorkee.: Guidelines for Small Hydro Development, Standard No. 1.2 (Part III), 'Planning of the Run-of-River Projects (2013)

Chapter 6
Field Investigations

6.1 General

The field investigations in detail are required to prepare a realistic DPR for a SHP. These investigations are of the following type:

- Hydrological.
- Topographical.
- Geological.
- Construction material.
- Infrastructural facilities.
- Muck disposal.
- Environmental aspects.
- Power evacuation.

These are briefly described in the following paras except the hydrological investigations which are covered in the Chapter on Hydrological Studies. There are three types of SHPs: run-of-river, canal-based and dam-based schemes. The topographical and geological investigations for each type of scheme are somewhat different because the structures of each of these schemes are different.

6.2 Topographical Surveys

Detailed topographical surveys of various components of the scheme are generally carried out for the final alternative which is fixed after inspection, reconnaissance and the study of available data. However, the detailed topographical surveys should have the scope for minor adjustments and modifications in the location of the structures during the implementation of the project.

B. N. Asthana, *Planning and Design of Small Hydro Power Projects*,
https://doi.org/10.1007/978-3-031-96271-4_6

The first requirement of the topographical survey is to establish a benchmark at the project site linking it with the nearest survey of India reference bench mark. For SHPs on old canals or at the toe of an existing dam/barrage, the existing benchmarks may be used.

Surveys should be carried out to prepare the general layout of the whole scheme as well as to identify and locate various structures of the scheme. The surveys for the general layout for a run-of-river scheme should extend from about 2 km upstream of the diversion structure to about 1 km downstream of the confluence of the tail race with the stream. The river L-section should be prepared at a scale of 1: 10,000 and X-sections at an interval of 200 m extending up to at least 10 m above HFL at a scale of 1 in 500. Contour plans for general layout should be prepared at a scale of 1 in 10,000, and for individual structures at a scale of 1: 200 or 1: 500. The contour interval may vary depending on the terrain. In hilly terrain, the contour interval may vary from 2 to 5 m. In plain terrain, the contour interval may be as small as 0.5 m. The contour plan for each component of the scheme should cover an area sufficient enough to accommodate the possible changes in the alignment and location if required to optimise the scheme.

The SHP schemes planned for peaking purposes would need a certain storage capacity in the upstream of diversion structure which may be either a dam or high gated barrage. For the estimation of the capacity of the reservoir, surveys are carried out, to prepare a contour plan of the reservoir and its sections are developed from the contour plan. For further details, AHEC Manual No. 1.12 and IS 5497 may be referred.

6.3 Geological Investigations

Geological investigations include the geological mapping of the entire area of the scheme, identification of the geological structures such as shear zones, faults, weak zones, etc. in the project area and in the vicinity, geology of the area of specific project components, foundation investigations for the extent of overburden and the type of strata such as rock type/soil and testing of its properties such as bearing capacity, permeability, 'c' and 'phi' values. All these investigations are very time-consuming and costly involving an experienced engineering geologist. Hence for an SHP where generally, the civil structures are small, geological investigations should be carefully planned to economise on time and cost.

Initially, a geological map making use of the contour plan of the scheme should be developed on the basis of available geological information about the area and the field inspection by an experienced engineering geologist. He may identify the problem areas and suggest a plan for further investigations and field and laboratory tests.

IS codes are available for conducting geological investigations of the project area and the foundations of the structures. For more information, AHEC Manual 1.12 and CBI&P Pub. No. 305 may also be referred.

6.4 Construction Material Survey

The civil structures of a SHP are made of reinforced cement concrete. The construction materials for making concrete are coarse aggregate, sand, cement & steel. Hence the survey for the availability of the required quantity and quality of aggregate, boulder and sand in the project area or in the nearby area shall be carried out. If natural aggregates are not available, a suitable stone quarry should be located and suitable aggregates coarse and fine should be obtained after crushing the stone from the quarry. Generally, run-of-river projects on hilly streams get the required construction material for concrete from the bed of the stream. The availability of construction materials should be as near as possible to the project site because the distance affects the cost of the project.

The available construction material should be tested for quality. The testing procedures are given in I.S. Code 2386. The important parameters for testing aggregates are gradation and alkali-aggregate reaction. The stone or boulder to be used for protection works should be strong enough to resist weathering due to alternate wetting and drying.

The nearest location for the procurement of cement and steel should also be identified.

6.5 Infrastructural Facilities

A detailed survey for the availability of necessary infrastructural facilities is of great importance for the construction of the project because this may cut down the time of project construction and its cost. These facilities include an approach to the project site, availability of construction power and nearness to a township.

6.6 Muck Disposal

A survey should be carried out to identify the areas for dumping unusable excavations from the construction of various structures of the project. These should be barren areas or depressions and should be adequate enough in size to accommodate the estimated quantity of muck. These locations should be marked on the general layout plan of the project and should be approved by the competent authorities.

6.7 Environmental Aspects

Normally environmental impacts of an SHP are negligible. If it is necessary to evaluate the environmental impact, surveys of the project area including areas for project facilities such as colonies, stores and workshops should be carried out to find the loss of flora and fauna and the displacement of people. The environmental impact report should also contain measures for mitigation of adverse impact.

6.8 Power Evacuation

Power evacuation depends on whether the proposed scheme is for isolated power requirements or whether the power from the plant will be connected to the grid.

If the plant is isolated and is expected to supply power to the adjoining area, then a survey should be carried out for the present supply position in the region and the expected demand for both firm power and secondary power. This will need a market survey. It requires the collection of data for the number of users and type of load which may be one or more domestic/industrial/commercial/irrigation types. The demand pattern should be worked out for each type and total energy consumption should be estimated. The demand should be compared with the power generation pattern from the proposed scheme. This market survey would help in finalising the size and number of units in the proposed power plant.

If the power plant of the proposed scheme is to be connected to the grid, the survey should be carried out to find the nearest substation where the power produced can be evacuated. A necessary survey for the transmission line from the power plant to the substation should be carried out and a plan showing the proposal for evacuation should be prepared.

References

1. CBI&P Manual 305.: Manual on Planning and Design of Small Hydroelectric Scheme, CBI&P, New Delhi (2009)
2. Guidelines for Small Hydro Development, Standard No. 1.12, Site Investigations, AHEC, IIT Roorkee (2013)
3. IS 5497.: Guide to Topographical Surveys for River Valley Projects, BIS, New Delhi, (1985)

Chapter 7
Hydrological Studies

7.1 General

The power potential at any site depends directly on discharge 'Q'. Therefore, a reliable estimation of water availability in the stream at the project site and its variability with time in a year is necessary for planning the project. This requires discharge data at project site for a reasonable length of time to assess the annual pattern of stream flows. Generally long term observed discharges at the proposed SHP site are rarely available. The hydrological studies are therefore carried out for water availability based on the short period observed data at the site and on the data obtained from other sources using appropriate indirect methods suitable for the project site.

In addition to water availability, other hydrological inputs required for the design of project components are design flood and sediment characteristics.

The above hydrological studies and sediment data are required for the type-I scheme, i.e. run-of-river project. For other types of schemes, i.e. schemes on canal fall and at the toe of a dam, the flows available for power generation are worked out from the discharge data of the canal flows and the releases from the dam. These are directly used to work out water availability of different dependability (50, 75, 90%). Hence this chapter deals with the data, its collection methods and hydrological studies generally required for run-of-river-type SHP schemes.

7.2 Data Requirement

In a run-of-river type SHP scheme, where observed discharges at the proposed diversion site are not available, the water availability is generally assessed by using indirect methods for which the data of rainfall and discharges of other sites either on the same stream or of the sites in adjoining areas of a similar catchment are collected.

B. N. Asthana, *Planning and Design of Small Hydro Power Projects*,
https://doi.org/10.1007/978-3-031-96271-4_7

The catchment characteristics of the stream are also collected. The data required is listed below

- Catchment area map showing (i) altitude, (ii) rain gauge locations, (iii) river system (iv) land use pattern (v) snow cover (if any), (vi) existing projects and (vii) stream gauging sites (if any).
- Rainfall data of the gauges in the catchment for as many years as possible.
- Discharge and catchment characteristic data of other sites for as many years as available for the discharge measuring sites of adjoining areas or catchments.
- Sediment load data of these discharge sites, if available.
- Water utilisation data of existing projects.
- Climatological information such as temperature, wind velocity, evaporation and cloud cover at the project site.

7.2.1 *Sources of Data*

The sources of the above data are the following:

- Indian Meteorological Department can provide information about rainfall, snowfall and other climatological data through their regional offices.
- Central Water Commission regularly gauges rivers and streams in the country for discharge as well as sediment and has long term data. It can be obtained for project planning after authorisation of the competent authority. The procedure is laid down to obtain the data from CWC. The state water resources departments also observe discharges of their rivers.
- Other departments or agencies which can supply useful data are the following:

  (i) Survey of India.
  (ii) State Irrigation/Water Resources Department.
  (iii) State Agriculture Department.
  (iv) District Revenue Department.
  (v) Forest Department.
  (vi) Snow and Avalanche Study Establishment (SASE).

7.2.2 *Collection of Data*

The data from the above sources should be collected and compiled for hydrological analysis. In addition, the discharge observations should be started at the project site and some rain gauges should also be installed in the catchment, if these do not exist. One rain gauge should be installed at the project site.

For the planning process of a SHP without any direct or indirect discharge data, it has been considered necessary to establish a stream gauging station at or near the proposed project site. For preliminary planning, CEA, CBI&P and IREDA have from time to time recommended that observed discharge data of at least two years covering two lean seasons and one flood season should be collected. These observed data can also be used to establish the accuracy of the indirectly developed discharge data for the proposed project site. The observations for discharge at the established stream gauging site should be continued beyond two years and additional data up to the time of the start of construction of the project should be used to update the project report. Thus, establishing a stream gauging station at the project site and observing its discharges at the earliest is the first requirement from the owner of the project. Details of installing rain gauges and stream gauging sites are briefly described below.

(a) Rain Gauges

Rain gauges are required to measure precipitation at a place. These are of two types (i) non recording or manual type and (ii) self-recording. In manual type, the rainfall occurring during a certain period say in a day can be measured. In the case of a self-recording rain gauge, a graph showing rainfall versus time is obtained. This is helpful in getting the information about rainfall intensity (cm/hr). The rain gauges should be installed and operated conforming to IMD standards.

(b) Stream Gauging

It first requires the selection of a suitable site. The river at the gauging site should be reasonably straight in sufficient length both upstream and downstream of the site. The river should flow between the banks up to the HFL and the banks should not be erodible. Then a gauge to record the water level in the river should be established. There are different types of gauges but broadly these are of two types (i) manually observed and (ii) self-recording. Generally, a simple staff gauge is adequate for a SHP. The self-recording gauge records water level against time on graph paper. It does not require personal attention to record river water level and the recorded graph paper may be taken out after a few days or when required. Hence establishing a self-recording river gauge is economical and reliable for remote and inaccessible sites of rivers in the upper Himalayas. For useful information on water levels, it is essential to connect the zero of the gauge with some permanent GTS benchmark. The zero of the gauge should be well below the minimum water level in the stream.

There are various methods of stream gauging. There is a CWPRS Manual on stream gauging and several I.S. Codes dealing with the details of various methods of stream gauging. These are also described in AHEC Standard No. 1.3. It is a specialised job and some expert should be engaged. A commonly used method of stream gauging is by Area-Velocity method. The method is based on the relation

$$Q = AV \tag{7.1}$$

where Q is discharge, A is the cross-sectional area of the stream and V is the velocity. Area and velocity are observed at the site. The cross section of the stream is divided into a number of segments depending on width. The width of each segment and the average depth give the area of the segment. The mean velocity in the segment, which is measured either by the current metre or by using floats, when multiplied

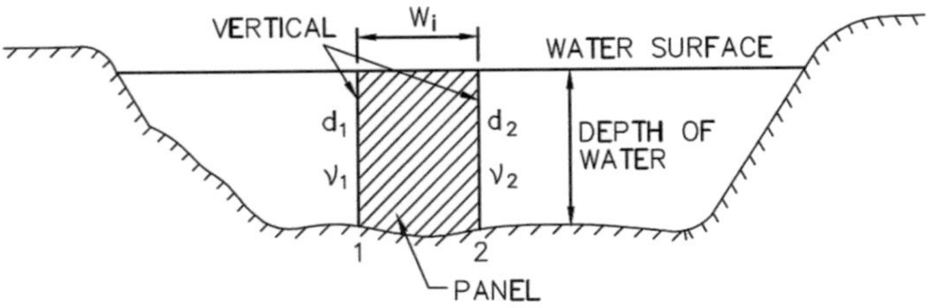

Fig. 7.1 Showing stream gauging (CBI and P 305)

by its depth and width gives the discharge passing through the segment. The sum of discharges passing through all the segments gives the total discharge of the stream. Figure 7.1 explains how the observation of depth and mean velocities at the two adjacent verticals are used to work out the discharge between two verticals of a segment. The observations of depth and mean velocity at the middle of the segment can also be used for working out discharge through a segment.

$$\text{Discharge in the segment}\, q_1 = \frac{v_1 + v_2}{2} \cdot \frac{d_1 + d_2}{2} \cdot w_1. \tag{7.2}$$

The depth is generally measured by a graduated sounding rod in streams with small depths and low velocities either by wading or through a boat moving on a cable. In big streams, eco-sounders fitted in a boat are used. If a bridge is available, it can be used for measuring depth.

The velocity measurement by the current metre for accuracy is made at several points along the depth because the velocity distribution profile is parabolic with a minimum at the bottom and a maximum near the water surface. If depth is less (say about 1 m) one point velocity observation at 0.6 depth from the water surface is adequate for mean velocity at any vertical. If depth is more two-point velocity observations at 0.2 and 0.8 depth from the water surface are considered adequate and the mean of two values gives the mean velocity at the vertical. Generally, the one-point method of velocity observation is preferred in small streams of an SHP project.

If velocity observation by the current metre is not possible, the float method to observe velocity may be adopted. In this case, a float which is generally made of a wooden disc or other material of sufficient weight so that its 3/4th part gets submerged below the water surface, is placed in the stream to float on the water surface, in sufficiently long and straight reach. The time taken by the float to travel a known distance is noted. The total time of observation should be more than 20 s. The distance divided by time gives the velocity of flow. To compute the mean velocity of flow a correction factor is used. It is generally taken as 0.85. The float method is not as accurate as the current metre method. Some of the other methods of stream gauging are as below:

(i) Area—Slope Method

In this computational method, Manning's equation is used for working out the mean velocity of flow. The velocity is given by

$$V = \frac{1}{n} R^{2/3} S^{1/2} \tag{7.3}$$

where V is velocity, 'n' is the regosity of channel, R is hydraulic mean radius and S is hydraulic gradient (slope of flow).

$R = \frac{A}{P}$ where A is cross-sectional area and P is wetted perimeter of the channel section. At least three channel sections should be considered to work out the average value of R. S is taken as the water surface slope in sufficient length of the river reach at the site. The value of 'n', the regosity coefficient depends on the river channel conditions and the nature of bed material. It also varies with the stage of flow in the channel. The observed velocity at some stages of flow (at low stages) may be used to work out the value of 'n' using the above Manning's equation for known values of R and S. If it is not possible then the value of 'n' can be taken from those given in standard texts for different channel conditions. The values recommended in the Manual of Stream Gauging of CWPRS are given below in Table 7.1.

This method can be used with reasonable accuracy in channels with stable boundaries flowing in straight reaches and having relatively coarse material in the bed. It is not suitable for very large and meandering streams. It should be adopted when area–velocity method is not possible.

(ii) Stage–Discharge Method

The discharge measurements by area–velocity method corresponding to certain stages of the river for a considerable period of time are used to develop a relation between stage and discharge. This relation is a straight line when data of stage and discharge are plotted on a log-log paper. This relation can be extrapolated and used for estimating discharge corresponding to the observed river stage in the future.

(iii) Using discharge measuring structures

In small streams flowing between well-defined and stable banks, a weir may be constructed for discharge measurement instead of direct methods. The discharge is

Table 7.1 Values of manning's 'n'

Sl. No	Channel type	Value of 'n'
1	Natural stream flowing smoothly in a clean channel	0.030
2	Natural stream in stable condition	0.035
3	River which meanders and has noticeable aquatic growth	0.045
4	River or stream with rock and boulders, shallow and weedy	0.060
5	Slow moving meandering river with rapids and pools, very weedy and over grown	0.10

worked out by the relation given below:

$$Q = CL\,H^{3/2}, \tag{7.4}$$

where Q is discharge (cumec), L is the length of the weir crest (m) and H is head over the crest in metre. C is the coefficient of discharge which depends on the type of weir (broad crested or sharp crested) and the condition of flow, i.e. whether the flow is free or submerged. The flow is submerged when the downstream water level is above the crest. The value of C can be obtained from the standard texts on Hydrology. Sometimes for measuring very small discharges (say up to 1 cumec), notches may be installed. These are generally triangular or rectangular in shape. These are used generally for free-flow conditions. The head over the crest is measured to work out the discharge. For further details, readers may refer to AHEC Manual 1.12 and CBI&P Manual 305.

(c) Stage-Discharge Curve and HFL

Besides the observations of rainfall and discharge, the stage-discharge curve is also very essential for planning and designing an SHP. Along with discharge observations the stage should also be observed and a stage–discharge curve be developed.

The high flood level (HFL) should be determined from the local inquiry. The high flood discharge corresponding to HFL may be obtained from the extrapolated stage discharge curve. If the design flood is determined from other methods, the corresponding HFL may be obtained from stage–discharge curve.

(d) Sediment

The data collection regarding sediment content in the stream flow is important, especially for a run-of-river SHP on Himalayan streams where the flow is associated with a lot of sediment of an erosive nature. This sediment when passing through the runner blades, guide vanes, etc. causes a lot of damage. The sediment besides erosion causes operation and maintenance problems. To manage the sediment problem, the sediment content (ppm) in flow and its characteristics such as size, shape and hardness are required. Therefore, the samples of sediment in the flow during floods should be collected. Standard sampling bottles are used for this purpose. The suspended sediment samples can be collected at the time of discharge measurement. The samples are collected and analysed for content (ppm), size and other characteristics. It is difficult to collect bed load samples so generally the suspended load content (ppm) is increased by a factor of 1.1 to 1.2 depending on the stage of the river. A curve between discharge and the corresponding amount of sediment in ppm should be prepared for further studies.

7.2.3 *Processing of Data*

(a) Consistency Checks

The discharge data collected should be examined for its reliability and adequacy. The consistency checks on data of discharge collected from other reliable sources should be applied. Standard texts on Hydrology should be referred to for consistency checks. The most common method to check the consistency of discharge data is a double mass curve. In this method accumulated monthly (or any other time unit) run-off values at a station in question are plotted against accumulated values of other reliable gauging stations. The trends and changes in the slope of the double mass curve may be attributed to the change in gauge location or method of measurement or changes in river regime. These should be examined to establish the consistency of data. This method can be used for rainfall data also where accumulated data of the station in question is plotted against the accumulated average values of a group of reliable adjoining gauging stations.

(b) Filling Missing Data

Sometimes, it is seen that rainfall and discharge data are found missing for some days or months in the total period of data. The following techniques are normally used to fill in the missing data.

(i) Using average values observed before and after the missing period.
(ii) By correlating with the data of adjoining stations for the missing period.
(iii) Using rainfall-runoff correlation if rainfall data is available.
(iv) Using normal ratio method.

In this method, the rainfall R_A at station A is estimated as a function of the normal monthly or annual rainfall of the station and those of the neighbouring stations for the missing period. At least three neighbouring stations should be used.

$$R_A = \frac{\sum_{i=1}^{n} \frac{NR_A}{NR_i} \times R_i}{n}, \tag{7.5}$$

where R_A is the estimated rainfall at station A, R_i is the rainfall at adjoining stations, NR_A is normal monthly or annual rainfall at station A, NR_i is normal monthly or annual rainfall at station i and n is the number of adjoining stations involved in analysis.

Missing rainfall data can be estimated easily by using the HEC-4 program of US Army Corps of Engineers.

(c) Discharge Adjustment

The observed discharges at a downstream site must be adjusted for the upstream withdrawals to arrive at virgin flows of the catchment for a rainfall-run-off correlation analysis.

(d) Presentation of Data

The data shall be compiled in suitable format/tables, graphs or plots for either weekly, 10 daily, or monthly time periods as required for hydrological analysis. Commonly data is processed for 10 daily averages.

(e) Processing of Rainfall Data

There may be several rainfall gauging stations in the catchment for which long-term rainfall data may be available. For rainfall–runoff analysis, the mean monthly or annual rainfall from the catchment is required. Generally following three methods are used to work out mean rainfall from the catchment.

(i) Arithmetic mean: In this method, the unit period mean rainfall of all the gauging stations is added and divided by the number of rainfall stations to get the mean catchment rainfall. It is expressed as

$$P_m = \frac{P_1 + P_2 + - - - - P_n}{n}, \tag{7.6}$$

where P_m is unit period mean rainfall of the catchment and P_1, P_2–P_n are the mean rainfall of all the gauge stations in the catchment for the corresponding time period and 'n' is the number of gauging stations.

The accuracy of this method depends on number of stations and their distribution in the catchment. If the number of stations is large and each covers approximately equal area, then the P_m calculated by arithmetic mean is reasonably reliable.

(ii) Thiessen Polygon Method: This is a better method for estimation of mean catchment rainfall when the rain gauge stations are non-uniformly distributed. To draw the polygon for each station the rain gauge stations are joined with adjoining stations in the catchment and then the perpendicular bisectors of these lines are drawn. It is shown in Fig. 7.2. The area of each polygon within the catchment is measured and this area is multiplied with the rainfall of the station falling inside the polygon. The mean rainfall of catchment Pm is worked out by

$$P_m = \frac{P_1A_1 + P_2A_2 - - - -P_nA_n}{A_1 + A_2 - - - - - A_n} \tag{7.7}$$

where P_1, P_2–P_n are the rainfall of a raingauge and A_1–A_n are influence areas of polygons.

(iii) **Isohyetal Method**: In this method, the rainfall of each station is recorded on the map of the catchment. The lines of equal rainfall, i.e. isohyetals are drawn as shown in Fig. 7.3. The area between two isohyetals is measured and then it is multiplied by the average rainfall value of the two isohyetals. The catchment mean rainfall P_m is worked out by

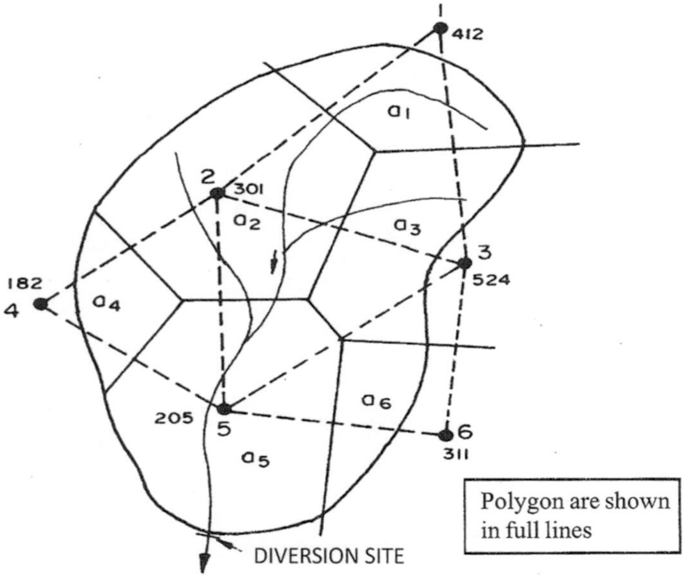

Fig. 7.2 Thiessen polygons

$$P_m = \frac{P_1A_1 + P_2A_2 - - - -P_nA_n}{A_1 + A_2 - - - - - - A_n}, \tag{7.8}$$

where P1, P2–Pn are average rainfall in the area of two successive isohyetals and A1, A2–An are the areas between two successive isohyetals. This method gives more reliable results. HEC-4 computer program can be used to process rainfall data. CBI&P Pub. 305 contains several illustrative examples for processing rainfall data.

7.3 Developing Stream Flow Series at SHP Site

For the estimation of water availability for an SHP, a stream flow series at the diversion site for a long period say 15–20 years is required. If an observed stream flow series at the proposed site for a sufficiently long period of time is available, it can be used for power potential studies without any hydrological analysis. Generally, such observed data is not available, so indirect methods to generate the required series are adopted. These depend on the availability of discharge and rainfall data. These are briefly given below for some data availability cases.

(i) Short term measured discharge data (say 5-years) at the project site and long-term rainfall data of the catchment.
(ii) Short term measured discharge data and no rainfall data.
(iii) Measured discharge for two lean seasons and one flood season.
(iv) No data is available.

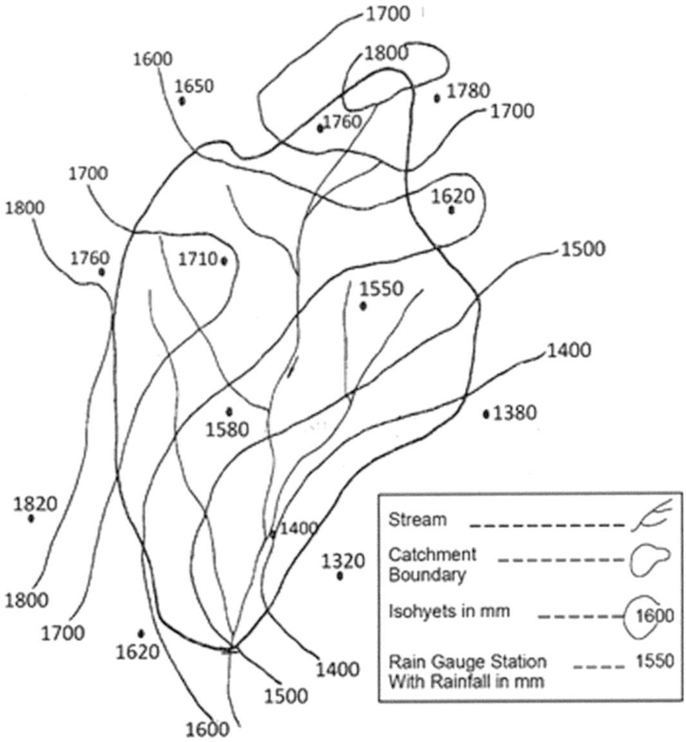

Fig. 7.3 Normal annual isohyetal map

7.3.1 Short Term Discharge Data and Long Term Rainfall Data

The short term observed discharge data at the SHP site is converted into long term stream flow data with the help of rainfall-run-off correlation. The rainfall and run-off data of the concurrent period are correlated and a suitable regression equation is established. Generally, a bivariate linear equation is developed by using the method of least squares.

$$y = a + bx \tag{7.9}$$

where 'y' is runoff (mm) and 'x' is rainfall and 'a' and 'b' are constants. If the coefficient of correlation is obtained more than 0.8 then the relationship is considered reasonable. If the coefficient of correlation is less then multivariate relation as given below may be attempted.

$$y = a + bx_1 + cx_2, \tag{7.10}$$

where y is runoff (mm), 'x_1' is the corresponding mean rainfall and 'x_2' is rainfall of one time period earlier. For SHPs of small catchments, the bivariate linear equation is generally found adequate as x_2 does not affect much for a time unit of one month.

In the least square method for bivariate linear relation, the following two equations are used

$$\Sigma Y = aN + b\Sigma X \tag{7.11}$$

$$\Sigma XY = a\Sigma X + b\Sigma X^2. \tag{7.12}$$

These equations on solving give a and b as below

$$b = \frac{\sum XY - N\overline{X} \cdot \overline{Y}}{\sum XY - N \cdot \overline{X}^2} \text{ and} \tag{7.13}$$

$$a = \overline{Y} - b\overline{X} \tag{7.14}$$

where N is number of data (number of concurrent periods), $\overline{X}$ is mean value of rainfall and $\overline{Y}$ is mean value of run off. The coefficient of correlation is given by

$$R = \frac{\sum XY - N\overline{X}.\overline{Y}}{\sigma_x \sigma_y (N-1)}, \tag{7.15}$$

where σ_x and σ_y are standard deviations for X and Y.

The correlation is generally developed for monthly time periods and used to develop the run-off series for the period for which only rainfall is available. For regression analysis computer programs are available. To carry out this analysis, refer illustrative example given in CBI&P Pub. No. 305.

7.3.2 Short Term Discharge Data and no Data for Rainfall

In such situations, long term discharge data of the adjoining site either on the same stream or any nearby stream is used to elongate the short term discharge data of SHP site. In this case a correlation between the discharges of the two sites for the concurrent period is developed in the form of a bivariate linear equation using the least square method described above

$$Q_1 = a + bQ_2 \tag{7.16}$$

where Q_1 is mean discharge at the short term data site and Q_2 is mean discharge at the other site of long term data. For mean discharges, the unit period may be 10 daily or monthly.

7.3.3 *Measured Discharges of Two Lean Seasons and One Flood Season and Rainfall Data*

The measured discharges of two lean seasons can be used to compute 50, 75 and 90% dependability discharge at the proposed site by using the rainfall data of the catchment. An annual rainfall duration curve (ARD) for the catchment is developed. The annual mean rainfall of the catchment is worked out for all the years for which the data is available. The ARD is the plot between the mean annual rainfall and the percentage of time, it is equalled or exceeded. A typical ARD is shown in Fig. 7.4. The mean annual rainfall of the catchment is worked out for the period for which discharges are measured (i.e. 2 years). The annual rainfall of 50, 75 and 90% dependability are read from the ARD curve and their ratios with mean annual rainfall for the period of discharge measurement are work out as below:

$$r_{50} = \frac{P_{50}}{P_m}, \quad r_{75} = \frac{P_{75}}{P_m} \quad \text{and } r_{90} = \frac{P_{90}}{P_m} \tag{7.17}$$

where P_{50}, P_{75} and P_{90} are 50, 75 and 90% dependability rainfall read from ARD and P_m is the mean annual rainfall of the catchment for the period of discharge measurement (i.e. for 2 years). These ratios are assumed to be applicable for discharges also. The mean monthly discharges of 50, 75 and 90% dependability are computed by multiplying the mean monthly discharges of observed period by r_{50}, r_{75} and r_{90} respectively.

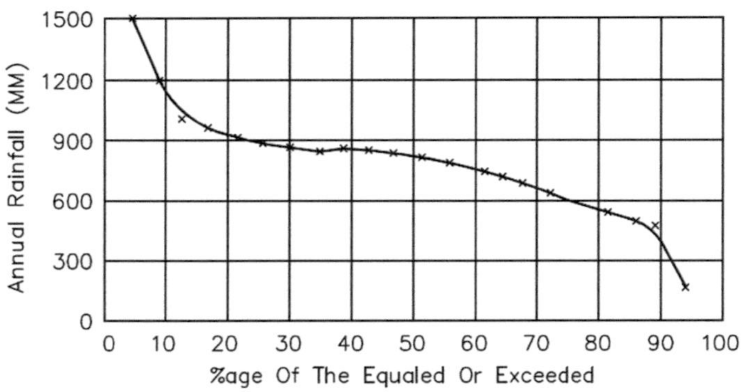

Fig. 7.4 Annual rainfall duration curve. (CBI&P 305)

7.3.4 *Assessment of Stream Flows for an Ungauged Catchment*

In a large number of cases, the situation arises when neither discharge data nor rainfall data of the catchment are available and stream flow assessment has to be made for planning the SHP and its project report. In such a situation, the stream flow at the site may be derived from the measured discharge data of the other sites either on the same stream or on the stream in the adjoining basin having similar characteristics of area and rainfall. It is done in proportion to the catchment area of the two sites.

$$\frac{Q_1}{Q_2} = \frac{A_1}{A_2} \quad \text{or} \quad Q_1 = Q_2 \frac{A_1}{A_2} \tag{7.18}$$

where 1 denotes the ungauged site and 2 denotes the site for which long term measured discharges are available.

7.3.4.1 Regional Approach

A more reliable assessment of stream flows can be made by using data of measured discharges of several similar basins by using the regional approach, in the following two cases.

(i) The measured discharges of two lean seasons and a flood season are available and no rainfall data of the catchment:

In this situation, the regional approach is applied by collecting measured discharge of several similar streams. Monthly specific discharges for each site are computed. The specific discharge is discharge per unit catchment area. The hydrographs of monthly specific discharges of a particular dependability say 50% are plotted for all the sites on a graph sheet as shown typically in Fig. 7.5. A mean hydrograph of all sites is also determined and plotted. It is termed as regional hydrograph. The monthly flows of particular dependability say (50%) at the ungauged site are computed by multiplying the monthly specific discharge ordinate of the regional hydrograph with the catchment area of the ungauged site.

(ii) No data of discharge and rainfall is available:

In such cases, a regional model using observed discharge data from other basins is used. A regional model has been developed for generating flows of different dependabilities at the ungauged site jointly by AHEC, IIT Roorkee and NIH Roorkee for hilly hydro projects under the UNDP programme. This is described below.

In this model, FDC for an ungauged catchment is derived using regionalisation procedure. The regions in India are identified and given in Table 7.2. For each region, based on available data of gauged catchments a mean FDC of that region in terms of Q/Q_{mean} and percentage of time is developed. The regional flow estimated values

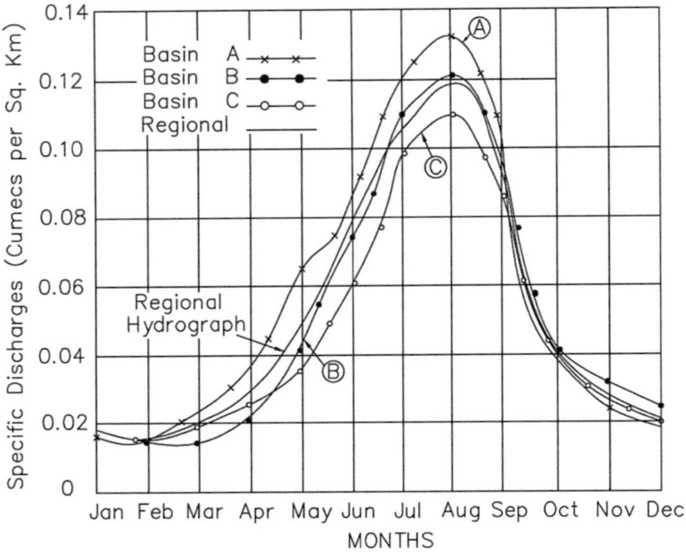

Fig. 7.5 Regional hydrographs with 50% dependability. (CBI&P 305)

for (Q/Q_{mean}) for various dependability levels (D) are given in Table 7.3, Q_{mean} for each gauged catchment is related to the catchment area (A).

$$Q_{mean} = CA^m \tag{7.19}$$

where C is coefficient and m is exponent. The value of C and m are given in Table 7.2.

Table 7.2 Values of the parameters of the regional models for mean flow

S. No	Region	State Covered	Exponent, m	Coefficient, C	Coefficient of correlation (R)
1	A	Jammu and Kashmir (Except Leh and Kargil)	0.06046	3.8189	0.808
2	B	Jammu and Kashmir Leh and Kargil)	$Q/A = (1/2)\ (Q/A)_{Leh} + (Q/A)_{Kargil} = 0.05804$		
3	C	Himachal Pradesh	0.86811	0.1200	0.8759
4	D	Uttar Pradesh/Uttarakhand	0.89075	0.0463	0.8174
5	E	Bihar/Jharkhand	0.74795	0.0652	0.7742
6	F	West Bengal and Sikkim	0.98920	0.0577	0.8467
7	G	North Assam and Arunachal Pradesh	0.26817	2.2807	0.3706
8	H	South Assam and Meghalaya	0.48589	1.4136	0.6820
9	I	Manipur, Nagaland, Mizoram and Tripura	1.22343	0.0151	0.9435

Knowing the area of ungauged catchment (A), Q_{mean} can be worked out using the values of C and m of the region. This value of Q_{mean} multiplied by the factor $(Q/Q_{mean})_D$ for that region from Table 7.3 will give the required dependable flow (Q_D) for that ungauged catchment. After obtaining Q_D for different values of D, the FDC of the ungauged catchment can be plotted for further planning purposes.

Table 7.3 Regional flow estimate for various dependability

Region	Regional values for $(Q/Q_{mean})_D$ for various dependability levels (D)					
	D = 25%	D = 50%	D = 60%	D = 75%	D = 80%	D = 90%
A	1.1562	0.6584	0.5428	0.4011	0.3577	0.2686
B	1.2240	0.8434	0.7360	0.5888	0.5396	0.4304
C	1.1797	0.6609	0.5399	0.3917	0.3466	0.2544
D	1.2828	0.8364	0.7078	0.5315	0.4729	0.3447
E	0.7374	0.2711	0.1974	0.1226	0.1031	0.0675
F	1.0942	0.5089	0.3896	0.2551	0.2171	0.1444
G	1.3075	0.8500	0.7148	0.5270	0.4640	0.3257
H	1.1436	0.4909	0.3551	0.2053	0.1646	0.0913
I	1.2451	0.5511	0.3957	0.2198	0.1716	0.0856

7.4 Water Availability Assessment

After developing 10 daily stream flow series for a number of years at the project site, it is processed for the assessment of water availability for power development. The water availability assessment for power studies is required for particular dependability such as 75 or 90%. The SHPs are generally designed for 75% dependability flows. The stream flow data is processed in two steps.

(i) To Determine the Year of Stream Flow Data of Certain Dependability

In this analysis, the 90 and 75% dependable year is generally calculated by arranging the annual run-off series in descending order and using Weilbul's formula

$$P = \frac{m}{N+1} \times 100 \quad (7.20)$$

where P is dependability percentage, 'm' is the rank of desired dependability and N is the number of data. If P is 90%, N = 19 then 'm' works out as $\frac{90}{100} \times 20 = 18$. Thus 90% dependable flow year is ranked 18 from top. Similarly for 75% dependable flow year 'm' = $\frac{75}{100} \times 20 = 15$, i.e. the year at rank 15 from top. The stream flows annual series of 75% and 90% dependable years are thus processed further for power studies.

In case the stream flow data at the site is for a short period, the average of ten daily flows of all the years are worked out and the stream flow annual series so obtained is used for further processing for power studies.

This average of ten daily discharges for all the years is also used to develop the stream flow series for the projects which are planned on the canals and at the toe of a dam where the recorded daily releases in the canal or from the dam are generally available.

(ii) To Develop the Flow Duration Curve (FDC)

The flow duration curve shows the proportion of time during which the discharge equals or exceeds a certain value. The basic time unit which can be daily, 10 daily or monthly greatly affects the shape of FDC. As compared to daily discharges the averaged values of 10 daily or monthly time units give a flatter FDC. Generally, for a SHP the 10 daily average discharge series is used to develop the FDC. A typical FDC is shown in Fig. 7.6.

If a FDC is to be drawn with 10-daily discharge series then there will be 36 discharge data in a year. These are arranged in descending order with the largest value at the top and lowest at the bottom. Then each discharge value is assigned a rank (M). The largest value will get rank 1 and the lowest value will get rank 36. The exceedance probability (P) of each rank is calculated by

$$P = \frac{M}{N+1} \times 100. \tag{7.21}$$

For M = 1,

$$P = \frac{1}{(36+1)} \times 100 = 2.7$$

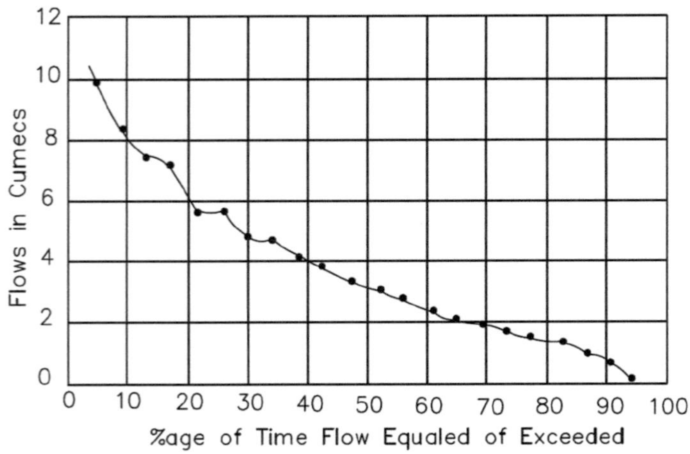

Fig. 7.6 Typical flow duration curve

M = 36

$$P = \frac{36}{37} \times 100 = 97.3.$$

So each discharge value gets a value of exceedance probability and when discharges are plotted on the Y-axis and exceedance probability on X-axis, we get the FDC as shown in Fig. 7.6. For illustrative example readers may refer to AHEC Report 1.3.

The FDC so developed is used to calculate flows of 20, 50, 75 and 90% dependability or of any other dependability at the site of the SHP to carry out the power studies.

7.5 Estimation of Flood Discharge

The estimation of flood discharge is essential for a safe and economical design of the diversion structure of a run-of-river SHP. These structures are generally weirs or barrages or diversion dams with no or small storage. With such structures moderation of river flood is not possible, so the design flood is taken equal to the peak flood discharge. For barrages and weirs generally a flood of 50 to 100-year return period is taken as a design flood from economic and safety considerations and for dams design flood may be adopted as per IS Code 11223–1985. As per this code, for small dams up to a height of 12 m and gross storage up to 10 Mm^3, the spillway design flood may be taken as a flood of 100-year return period.

7.5.1 Flood Frequency Analysis

Past record of annual peak floods is used in flood frequency analysis to predict probabilities of occurrence of floods of different magnitudes. This analysis generally gives satisfactory estimates of 50 and 100 year return period floods if the past record of floods for frequency analysis is available for a period of 20 to 25 years. A large number of methods for flood frequency analysis are available in the literature such as Gumbles, Normal, Log-Normal and Log Pearson Type-III distribution. The method of best fit distribution should be used for flood frequency analysis. The Gumbels method is generally used for SHP. It is as below. (For illustrative example reader may refer to CBI&P Pub. No. 305).

When peak flood flows are plotted against the return period on the Gumbel-Powell probability paper, the relationship will be a straight line. The equation of this straight line can be obtained either by the method of moments or the method of least square or the data plotting method. These are described below

Table 7.4 Values of reduced variate

Return period T	Reduced variate Y
2	0.37
5	1.5
10	2.25
25	3.2
50	3.9
100	4.6

(a) Method of Moments

The flood frequency equation is given by

$$X_T = X + S(0.78Y - 0.45), \tag{7.22}$$

where X_T is flood peak of return period T
X is average value of all annual flood peaks
S is standard deviation of flood peak series
Y is called reduced variate and is a function of T and its value is given in following Table 7.4:

(b) Method of Least Square

The flood frequency equation is given as

$$X_T = A + BY, \tag{7.23}$$

where

$$Y = -\log\log\frac{T}{T-1}$$

and

$$T = \frac{N+1}{m}.$$

'm' is the rank of a peak discharge in the peak flood series arranged in descending order, N is the number of flood peaks considered. A and B are worked out by the method of least square. After determining A and B, X_T for any return period T can be worked out from the above equation.

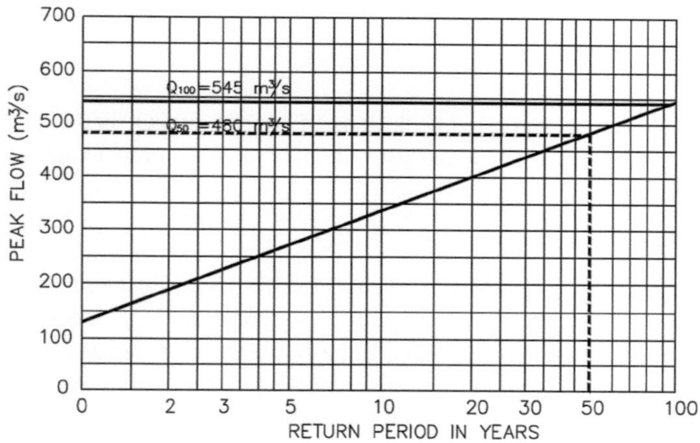

Fig. 7.7 Flood frequency peak flow vs. return period. (AHEC—1.3)

(c) Data Plotting Method

When the annual flood peaks arranged in descending order are plotted on a semi-log paper with flood peaks on the ordinate and the return periods on the abscissa, the plot is a straight line. This can be extrapolated to get the peak flood value of a higher return period say 50 and 100. To determine the best fit line the method of least square may be used to determine the constants of the line of the equation

$$X_T = A + B \log_{10} T. \tag{7.24}$$

For plotting the data, the peak floods are arranged in descending order. If the data available is for 20 years then the highest value will have a return period of 20 years and second highest value will have a return period of 10 years and third value will have a return period of 6.67 years and so on. When these are plotted, the plot will typically be as shown in Fig. 7.7.

7.5.2 *Methods When Peak Flood Data Not Available*

Generally long period record of peak floods at a SHP site is not available. In that case one of the following methods may be used.

(a) Long Term Data of Other Sites Used

When long term record of peak floods of some other site either on the same stream or on a stream of adjoining similar basin is available, the flood frequency analysis shall be carried out by any of the above-described methods to work out peak floods of 50 and 100 year return period or any other desired return period. This shall be transposed to the ungauged catchment in proportion to the catchment areas by the following formula

$$\frac{Q_g}{Q_u} = \left(\frac{Ag}{Au}\right)^{3/4}, \tag{7.25}$$

where Q_g is T year return period peak flood of gauged site, Q_u is the peak flood of same return period of the ungauged site. Ag and Au are catchment areas of gauged and ungauged sites.

(b) Using Unit Hydrograph

The unit hydrograph approach is to convert rainfall of a drainage basin into run off from that basin. It assumes that a storm of say 1-hour duration produces rainfall at a constant rate uniformly over the drainage basin upstream of the gauging station. If the recorded runoff at the gauging station represents 1 cm rainfall over a period of 1 hour, then the hydrograph of recorded run-off is 1-hr unit hydrograph for 1 cm excess rainfall. This also assumes that no infiltration has taken place in the basin and all the rainfall has been converted into run off. The unit hydrograph theory also assumes that 1-hr hydrograph ordinates are in proportion to rainfall excess. If the rainfall excess is 2 cm then the ordinates of 1 cm excess hydrograph can be multiplied by 2.

Thus, the unit hydrograph is defined as a flood hydrograph resulting from a unit (1 mm or 1 cm) rainfall excess distributed uniformly over the basin area at a uniform rate during a unit period (say, 1-h, 3-h or 6-h). Thus a unit hydrograph represents both the hydrologic and physical characteristics of a basin. For the detailed study of the unit hydrograph approach for working design flood hydrograph of certain dependability or PMF, etc., a reference may be made to standard text on Hydrology Varshney 1986.

The unit hydrograph of a basin can be derived if measured flood hydrographs for a few years with corresponding storm data are available. But generally, this data is not available for SHP drainage basins. Therefore, methods have been developed to derive synthetic unit hydrographs. The method adopted by CWC for small ungauged basins to develop unit hydrographs is described below. The parameters of the unit hydrograph are as shown in Fig. 7.8. These are defined and related to basin characteristics in Table 7.5.

The constants a, b, c, d, e, f and g as well as m_1–to m_7, are worked out from an actual hydrograph developed at the site where storm and run-off data is available along with the required basin characteristics L, L_c, S and A, which are defined below

L is length of the longest stream up to the project site, in km.

L_c is length of the longest stream from a point opposite to CG of the catchment up to project site, in km.

S is equivalent stream slope (ratio).

A is catchment area in sq.km.

The Central Water Commission (CWC) jointly with RDSO and IMD has developed the above parameters of a synthetic unit hydrograph (SUH) for the ungauged catchment in India. The country has been divided into 26 hydro-meteorologically homogeneous sub-zones which are shown in Fig. 7.9. For each sub-zone based on the available data of rainfall, runoff and other characteristics of some catchments in

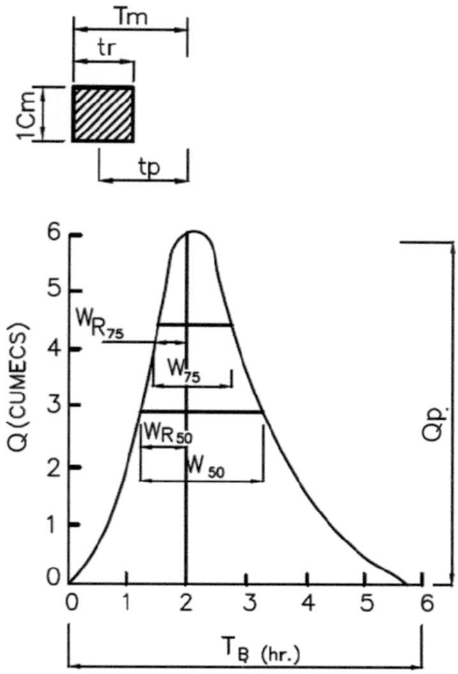

Fig. 7.8 Unit hydrographs parameters defined. (CBI&P, Pub 305)

Table 7.5 Synthetic hydrograph parameters

Parameter	Defined as	Relationship with basin characteristics
t_p	Time in hours from the centre of unit rainfall duration to peak of unit hydrograph	$a\left(\frac{LLc}{\sqrt{S}}\right)^{m_1}$
q_p	Peak discharge of unit hydrograph per unit area of catchment $m^3/sec/km^2$	$\frac{b}{(t_p)^{m_2}}$
W_{50}	Width of unit hydrograph in (hours) at 50% of peak discharge	$\frac{c}{(q_p)^{m_3}}$
W_{75}	Width of unit hydrograph in (hours) at 75% of peak discharge	$\frac{d}{(q_p)^{m_4}}$
W_{R50}	Width of rising limb of unit hydrograph in hours at 50% of peak discharge	$\frac{e}{(q_p)^{m_5}}$
W_{R75}	Width of rising limb of unit hydrograph in hours at 75% of peak discharge	$\frac{f}{(q_p)^{m_6}}$
T_B	Base width of unit hydrograph in hours	$g(t_p)^{m_7}$
Q_p	Peak discharge of unit hydrograph m/sec	$q_p \times A$

the sub-zone, the parameters of SUH have been developed to draw a SUH characteristic of that sub-zone which may be used to work out design flood for the ungauged catchment in that sub-zone. For example, the SUH parameters for Western-Himalayas subzone 7 are worked out as below

$$t_p = 2.498\left(\frac{LLc}{S}\right)^{-0.156}$$

$$q_p = 1.048(t_p)^{-0.178}$$

$$W_{50} = 1.954\left(\frac{LLc}{S}\right)^{0.099}$$

$$W_{75} = 0.972\left(\frac{L.Lc}{S}\right)^{0.124}$$

$$W_{R50} = 0.189(W_{50})^{1.769}$$

$$W_{R75} = 0.419(W_{75})^{1.246}$$

$$T_B = 7.845(t_p)^{0.453}$$

$$Q_p = q_p \cdot A.$$

In order to work out the flood of 50 or100 year return period for an ungauged catchment in this subzone, the basin data, i.e. A, L, Lc and S will have to be collected along with 1 day rainfall of 50 and 100 year return period which can be obtained from IMD. For the computation of flood hydrograph of certain dependability (say 50 or 100 year) by using the SUH, reference may be made to the CWC report of the subzone. One illustrative example is given in AHEC Report 1.3.

(c) Regional Flood Frequency Method

In this approach, the regional relationships are developed between mean annual flood peak and catchment area using the least square approach coupled with L moment approach in the form of

$$Q_T = C_T \cdot A^b \tag{7.26}$$

where Q_T is flood (in m^3/sec) for T year return period and C_T and b are regional (subzone) coefficients and A is the catchment area of ungauged catchment (For details refer to AHEC Report 1.3). The values of C_T and b for different subzones (7 nos.) have been worked out, and for example, the values of subzone 7 are given below.

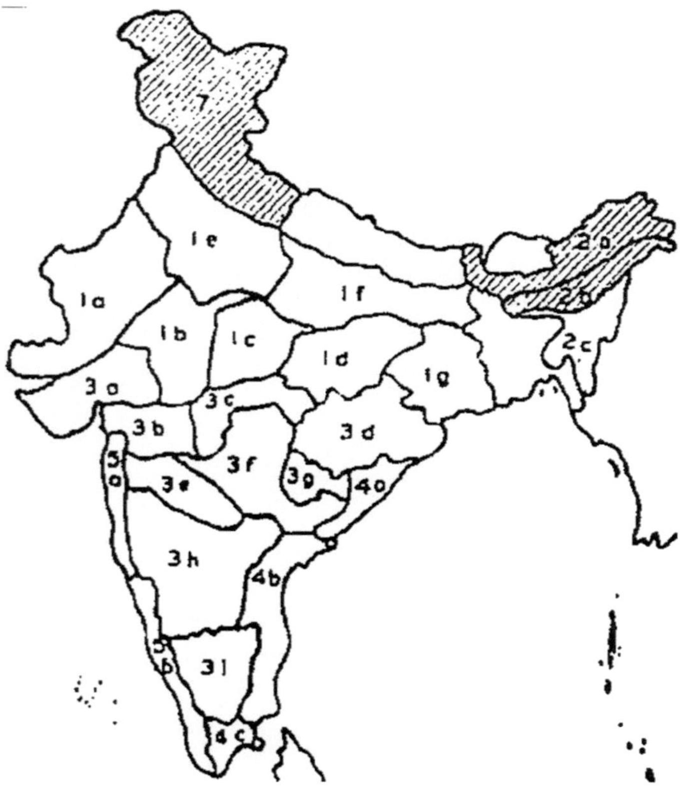

Fig. 7.9 Locations of North Brahmaputra Subzone 2 **a**, South Brahmaputra Subzone 2 **b** and Western Himalayas Zone-7. (AHEC- 1.3)

Subzone	'b'	C_T for various values of T					
		10	25	50	100	200	Years
7	0.772	9.745	12.452	14.709	17.216	20.007	

(d) Empirical methods

(i) Based on field information

The high flood mark at the site may be ascertained by a site visit and local enquiry. The discharge corresponding to this flood level may be read from the stage discharge curve which may be developed by using Manning's equation as discussed in para 7.2.2.

(ii) Using Dicken's Formula

Dicken's empirical formula is commonly used in India. It is given below

$$Q = CA^{3/4} \tag{7.27}$$

where C is a constant that varies from region to region due to variations in rainfall and physical characteristics. Based on long term discharge observations at various locations in the country the value of C is assigned to different regions as shown in the map given in Fig. 7.10.

These empirical methods can be used to expeditiously decide on the location of a hydro power project along a stream.

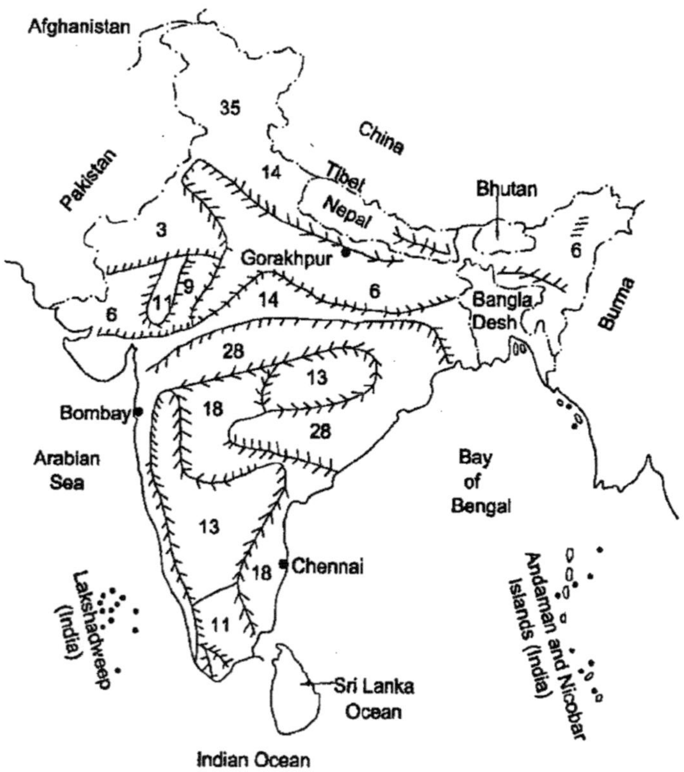

Fig. 7.10 Dicken's constant C for different parts of India. (AHEC—1.3)

7.6 Rating Curve

A rating curve at the diversion site is required for planning the works at site and the proper design of the energy dissipators downstream of the diversion works. This can be developed by plotting the available stage and discharge data and mathematically determining the best fitting curve. If data is not available then it can be developed by the slope-area method described in para 7.2.2 (I.S. Code 2912 may also be referred).

7.7 Sediment Data

Sediment moving with the incoming flow damages the runner of the turbine and other under water parts such as guide vanes. This erosive action of sediment is maximum when a run-of-river scheme is operated during flood season because the flood flows have a maximum concentration of sediment. The studies have shown that the following sediment characteristics are responsible for damage:

- Concentration, generally expressed as ppm (parts per million).
- Hardness: Particles of hardness above 5 (on the Mohs scale) are considered harmful. Quartz with hardness 7 is very damaging.
- Shape and size: Abrasion is moderate for particles less than 0.05 mm, but angular or sharp edged particles of very small sizes are also damaging.

The damage 'W' is directly proportional to the above factors and can be expressed as

$$W \ \alpha \ S_1 \cdot S_2 \cdot S_3 \cdot S_4, \tag{7.28}$$

where S_1, S_2, S_3 and S_4 are coefficients for sediment concentration, hardness, shape and size.

For evaluation of the above sediment characteristics, the collection of sediment samples and their laboratory analysis is very necessary. The suspended sediment samples may be easily collected while observing discharges. Alternatively, this data shall be collected from other neighbouring gauging sites in similar catchments.

References

1. CBI&P.: Manual on Planning and Design of Small Hydroelectric Schemes, Pub. No. 305, New Delhi (2009)
2. Guidelines for Small Hydro Development, Standard No. 1.3, Project Hydrology and Installed Capacity, AHEC, IIT Roorkee

3. Guidelines for Small Hydro Development.: Standard No. 1.12, Site Investigations, AHEC, IIT Roorkee (2013)
4. I.S. 2912.: Liquid Flow Measurement in Open Channels-Slope-Area Method, BIS, New Delhi (1998)
5. I.S. 11223.: Guidelines for Fixing Spillway Capacity, BIS, New Delhi (1985)
6. Varshney R.S.: Engineering hydrology. Nem Chand and Bros, Roorkee (1986)

Chapter 8
Power Studies

8.1 General

Power studies are carried out to fix the installed capacity (MW) of the project, the size and number of units, the benefits from the scheme in terms of annual power generation (MU) and to work out the economical viability. The procedure to carry out the power studies is the same for both major projects as well as SHP. It is dealt with briefly in the following paras:

The power potential at any site is worked out by the following equation:

$$P = 9.81\,Q.H.\eta \tag{8.1}$$

where P is power in kilowatt (Kw), Q is discharge in m^3/sec, H is net head in metres (m) and η is the combined efficiency of turbine and generator. For fixing installed capacity, P has to be worked out for the design discharge Q. For fixing design discharge the annual water availability series at the project site of either 75 or 90% dependability is generally used. For a run of river SHP, normally 75% dependable flows are used for power studies. To estimate design discharge, from the annual 75% dependable flow series, a normal flow duration curve is used. The methods of developing the dependable annual flow series at the project site and the procedure to draw a flow duration curve are described in Chap. 7.

8.2 Flow Duration Curve (FDC)

The flow duration curve provides an estimate of the percentage of time a given stream flow is equalled or exceeded over a time period for which the discharge data is used. Thus FDC provides a single and comprehensive graphical view of the variability associated with the stream flow.

B. N. Asthana, *Planning and Design of Small Hydro Power Projects*,
https://doi.org/10.1007/978-3-031-96271-4_8

The shape of the FDC of a discharge site gives the idea of the flow characteristics of the river. The shapes of an FDC are classified as below

(i) Steep drop FDC.
(ii) Moderately flat FDC.
(iii) Flat FDC.

These are shown in Fig. 8.1. All three types represent similar nature of flow characteristics between the range of 50 and 100% of the time but between the range of 0 and 50% of the time these differ a lot. In steep drop FDC for a small change in percentage of time, the change in discharge is large. This means a large increase in discharge is available for a short period for energy generation which may not be economical. Such type of FDC generally represents a river that has large flood flows during monsoon and small flows in the rest of the year. Indian rivers fall in this category. In the case of flat FDC, for the same change in discharge, the change in %age of time will be large and so this increased discharge will be available for a longer time for energy generation. The flat FDC's are generally characteristics of European rivers where rainfall is more or less evenly distributed throughout the year. The behaviour of the moderately flat FDC is in between the two. Therefore, the FDC needs to be evaluated in terms of cost and energy to fix the design discharge.

Because of flat FDC, some projects in Europe (Mosonyi) are planned for a design discharge of 0 to 20% exceedance probability. Indian projects are designed for a design discharge of 20 to 50% exceedance probability because the river FDCs are steep to moderately flat. The percentage exceedance probabilities adopted for power projects of Uttarakhand (India) are as given in the following Table 8.1.

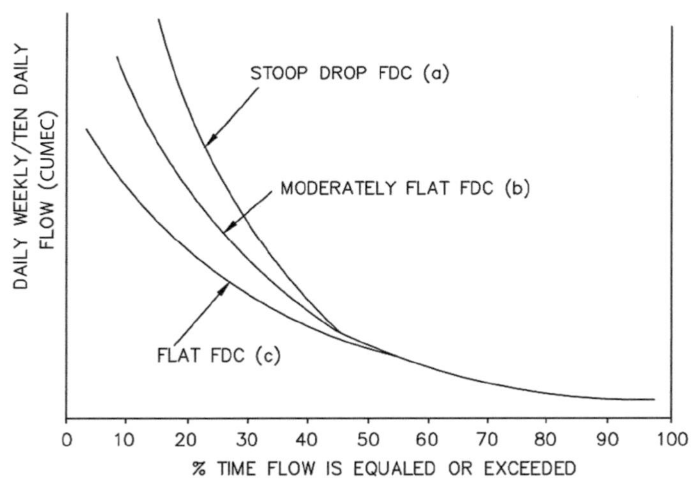

Fig. 8.1 % Time flow is equalled or exceeded

Table 8.1 Flows adopted for some projects

Sl. No	Name of project	Period of commissioning	Exceedance probability		Year of FDC
1	Maneri Bhali Hydro Project Stage I (3 × 30 MW) (on river Bhagirthi)	1970–80	46		Synthetic year flows of 90% dependability
2	Chilla Hydel Scheme (4 × 36 MW) (on river Ganga)	–do–	46 33	Before Tehri After Tehri	–do–
3	Maneri Bhali Hydro Project Stage II (4 × 76 MW) (on river Bhagirathi)	–do–	33		–do–
4	Srinagar Hydro Project (6 × 55 MW) (on river Alakhnanda)	Commissioned recently	21	Because of moderately flat FDC	90% dependable year flows

8.3 Approaches in Developing FDC and Energy Estimation

The FDC for a run-of-river scheme is generally developed using 10-daily discharges of a year of any particular dependability. These 10-daily discharges are arranged in descending order with rank 'm' which varies from 1 to n (where n is number of data which is generally 36 ten dailies in a year). Each rank is assigned the exceedance probability P by using the Weibull formula (Refer Chap. 7)

$$P = \frac{m}{n+1}. \tag{8.2}$$

There are various approaches to selecting the particular year to draw an FDC for energy estimation.

Nigam (1985) has recommended the use of synthetic year flows for developing the FDC. For the synthetic year, the 10-daily flows are taken as 90% or of any desired dependability from all the flows of that particular 10-daily for all the years of available hydrological data. Thus, the synthetic year flows do not correspond to any particular calendar year flows. This approach has been followed in the first three projects given in the above table.

Mosonyi recommends the estimation of the total energy of the stream at the site using the average flow duration curve of all the years of hydrological data. This will be the total area under the average year flow duration curve. Since harnessing this energy would not be techno-economically feasible, a reduction factor is applied which may vary from 75 to 80%. A high value of reduction factor may be applicable for flat FDC and a low value for steep FDC.

Mosonyi further suggested that technically utilisable power can possibly correspond to a discharge that is available 85 to 90% of the time. However, in many cases, the lowest cost of energy can be attained by utilising the flow available approximately 50% of the time in a year. In the interest of more effective utilisation of undeveloped hydro power potential, the trend has been to use discharge available for shorter duration, i.e. a discharge as high as Q^{34} (flow equalled to or exceeded 34% of time). This was considered feasible because of the integrated power grid system. Thus, according to Mosonyi, the FDC should be made for average year flow and the selection of design discharge for installed capacity should be in a range of Q^{34} and Q^{50} to be based on economic consideration, i.e. minimum cost of unit generation.

Gutherie Brown (1958) also suggested that the selection of design discharge should be based on economic considerations. He has suggested that mean or average discharge should be worked out. The maximum discharge that can be intercepted for power generation will be several times the mean discharge depending on the shape of the FDC. The economics should be worked out considering several multiples of the mean flow. In the U.K., this multiplying factor is generally 3.

From the above it is evident that there are different criteria for the selection of the year for the development of FDC. One is the use of a synthetic year of particular dependability, the other is the FDC of average year discharges and another criterion is to follow the FDC of dependable year flows. The AHEC Manual 1.3 recommends for an SHP scheme the average year discharges for the case when the discharge data available is for 10 or fewer years and 90% dependable year flow for the case when long period hydrological data is available for developing FDC to determine design discharge and the installed capacity.

8.4 Fixation of Installed Capacity

8.4.1 The Procedure that is Generally Followed for Fixing the Design Discharge and the Installed Capacity

(i) Determine the year for FDC. It may be average year (if data is for less than 10 years) and 90 or 75% dependable year for data more than 10 years. Generally, for SHP it is taken as a 75% dependable year. It can be determined either by arranging the yearly run-off in descending order or by using annual unrestricted energy in place of runoff.

(ii) Before preparing FDC, the mandatory environmental releases should be deducted from the discharges of the year selected for FDC.

(iii) Select a range of discharges from FDC. It may be between Q^{20} and Q^{50}. Alternatively, select a range of installed capacity considering the mean value corresponding to a discharge available for at least 8 months in a year.
(iv) Work out energy generation for 4 to 5 discharges or installed capacities in the selected range.
(v) Work out incremental energy per MW increase in the installed capacity and plot a curve.
(vi) Installed capacity is fixed at a value where the increase in incremental energy falls sharply.
(vii) The discharge corresponding to the selected installed capacity is the design discharge for the design of the scheme.

8.5 Selection of Size and Number of Units

8.5.1 The General Guidelines for the Selection of Size and Number of Units

(i) The number of units should be minimal as each extra unit increases the cost of civil works.
(ii) Single unit installation should be avoided because of a total loss of generation from the plant at the time of closure of the machine.
(iii) Large units have better efficiency.
(iv) Unit size may also be limited due to transport limitations, i.e. accessibility to the site.
(v) As far as possible, all the units in a power plant should be of the same size.
(vi) Considering the lean season discharge availability in schemes without pondage, the unit size selection should be such that atleast one unit is in operation at permissible part load which depends on the type of turbine.

8.6 Examples of Power Studies

Two case studies to illustrate the above procedure of fixing installed capacity and unit size and number of units are described in the following paras. One case is for a run-of-river scheme and the other is for a power plant at the toe of a dam.

8.6.1 *Case Study of a Run-of-River Scheme*

(i) The Scheme

The existing Karnah HEP (J & K) of 2 MW capacity is proposed to be remodelled for the increased installed capacity due to revised water availability. By shifting the site of power house more head is also expected. The power studies were carried out with the revised water availability and the head to fix installed capacity and the unit size and number of units.

(ii) Water Availability

The ten-daily average discharges for the years 2005–2006 to 2011–2012 were available. Since the data available was for less than 10 years, the average year discharges were considered for power studies. The year 2007–2008 was found to be the average year and its discharges were as given in Table 8.2.

As per guidelines, environmental releases equal to 20% of the average of the three months of lean season (December to February) were mandatory. These releases worked out as 0.57 cumec. This quantity was reduced from the available discharges. The net discharges available for power generation are shown in Table 8.2.

The FDC of discharge data (Fig. 8.2) showed that about 13 cumec discharge is available for eight to nine months of a year and it was considered as a mean value to select a range of installed capacities for further power studies.

(iii) Net Head

The gross head available with the changed site of the power house was found as 102.59 m. Head losses from the system were worked out for the mean discharge of 13 cumec and it was found as 4.01 m. Thus, the net head for power generation was worked out as 98.58 m. For this head, Francis turbine was selected and the combined efficiency of turbine and generator was taken as 92.12% for load varying from 100 to 70% and 80% for load varying from 70 to 50%.

(iv) Power Studies for Installed Capacity

The power potential corresponding to the mean discharge of 13 cumec and the net head of 98.58 m was worked out as 12 MW. The range of installed capacity selected for computing energy generation was adopted as 4 to 20 MW at an interval of 2 MW.

To fix the number of units, energy calculations were made for 2, 3 and 4 machine units for each installed capacity. The calculations for an installed capacity of 12 MW are given in Table 8.2 to illustrate the procedure.

Based on energy calculations, the incremental increase in energy per MW increase in installed capacity was worked out and the plot of this data against installed capacity is shown in Fig. 8.3. The plot revealed that the incremental increase in energy after an installed capacity of 14 MW is practically constant. Hence 14 MW was considered as installed capacity. The computations also revealed that power house would not be operative for some ten-dailies in a year with this installed capacity. Therefore, the installed capacity was fixed at 12 MW.

Table 8.2 Power studies for fixing installed capacity and number of units

Power potential studies for average year (2007–08)												
Water level in proposed Forebay						= 1516.85	Gross head				= 102.59 m	
Maximum T.W.S.						= 1414.26	Head loss				= 4.01 m	
Minimum T.W.S.						= 1413.26	Rated net head				= 98.58 m	
Riparian discharge						= 0.57 cumecs						
Assumptions												
Inc value = 2		Plant Capacity power =					12 MW		12 MW		12 MW	
		No. of machines =					2nos.		3nos.		4nos.	
		MW per Machine =					6.0 MW		4.0 MW		3.0 MW	
Month	Block	Days	Discharge in average year (Cumec)	Available discharge (less Riparian flow)	Power Gen (MW) (9.81*n*Q*H)/ 1000 (Unrestricted)	Uunrestricted Energy MU	Restricted to 6 MW per M/c (power MW)	Energy MU	Restricted to 4 MW per M/c (power MW)	Energy MU	Restricted to 3 MW per M/c (power MW)	Energy MU
Jun	I	10	23.93	23.36	20.81	5.00	12.00	2.88	12.00	2.88	12.00	2.88
	II	10	24.36	23.79	21.20	5.09	12.00	2.88	12.00	2.88	12.00	2.88
	III	10	24.72	24.15	21.52	5.16	12.00	2.88	12.00	2.88	12.00	2.88
Jul	I	10	18.96	18.39	16.39	3.93	12.00	2.88	12.00	2.88	12.00	2.88
	II	10	22.05	21.48	19.14	4.59	12.00	2.88	12.00	2.88	12.00	2.88
	III	11	22.92	22.35	19.91	5.26	12.00	3.17	12.00	3.17	12.00	3.17
Aug	I	10	15.36	14.79	13.18	3.16	12.00	2.88	12.00	2.88	12.00	2.88
	II	10	13.19	12.62	11.25	2.70	11.25	2.70	11.25	2.70	11.25	2.70
	III	11	12.22	11.64	10.38	2.74	10.38	2.74	10.38	2.74	10.38	2.74

(continued)

Table 8.2 (continued)

Sep	I	10	11.04	10.47	9.32	2.24	9.32	2.24	9.32	2.24	9.32	2.24
	II	10	9.24	8.68	7.74	1.86	7.74	1.86	7.74	1.86	7.74	1.86
	III	10	7.31	6.74	6.00	1.44	6.00	1.44	6.00	1.444	6.00	1.44
Oct	I	10	5.58	5.01	4.47	1.07	4.47	1.07	4.47	1.07	4.47	1.07
	II	10	6.01	5.44	4.85	1.16	4.85	1.16	4.85	1.16	4.85	1.16
	III	11	4.55	3.98	3.54	0.94	3.08	0.81	3.54	0.94	3.54	0.94
Nov	I	10	3.64	3.07	2.74	0.66	0.00	0.00	2.38	0.57	2.74	0.66
	II	10	3.30	2.72	2.43	0.58	0.00	0.00	2.38	0.57	2.74	0.66
	III	10	3.01	2.44	2.18	0.52	0.00	0.00	1.87	0.45	2.18	0.52
Dec	I	10	2.52	1.95	1.74	0.42	0.00	0.00	0.00	0.00	1.51	0.36
	II	10	2.61	2.03	1.81	0.44	0.00	0.00	0.00	0.00	1.57	0.38
	III	11	2.90	2.33	2.07	0.55	0.00	0.00	1.0	0.48	1.80	0.48
Jan	I	10	2.55	1.98	1.76	0.42	0.00	0.00	0.00	0.00	1.53	0.37
	II	10	2.54	1.97	1.75	0.42	0.00	0.00	0.00	0.00	1.52	0.37
	III	11	2.91	2.34	2.08	0.55	0.00	0.00	1.81	0.48	1.81	0.48
Feb	I	10	3.03	2.46	2.19	0.53	0.00	0.00	1.90	0.46	2.18	0.53
	II	10	3.12	2.55	2.27	0.55	0.00	0.00	1.98	0.47	2.27	0.55
	III	8	3.49	2.91	2.60	0.50	0.00	0.00	2.26	0.43	2.60	0.50
Mar	I	10	4.08	3.50	3.12	0.75	2.71	0.65	3.12	0.75	3.12	0.75
	II	10	0.41	3.84	3.42	0.82	2.97	0.97	3.42	0.82	3.42	0.82
	III	11	5.19	4.62	4.12	1.09	3.57	0.94	4.12	1.09	4.12	1.09
Apr	I	10	6.38	5.81	5.18	1.24	5.18	1.24	5.18	1.24	5.18	1.24

(continued)

Table 8.2 (continued)

	II	10	6.92	6.35	5.666	1.36	5.66	1.36	5.66	1.36	5.66	1.36
	III	10	8.90	8.32	7.42	1.98	7.42	1.78	7.42	1.78	7.42	1.78
May	I	10	11.98	11.41	10.16	2.44	10.16	2.44	10.16	2.44	10.16	2.44
	II	10	18.16	17.57	15.67	3.76	12.00	2.88	12.00	2.88	12.00	2.88
	III	11	20.92	2.35	18.13	4.79	12.00	3.17	12.00	3.17	12.00	3.17
			Gross energy generated MU			70.49		49.65		53.96		55.88
			Lean season energy generation (MU)					0.00		2.32		4.00
			Annual Design Energy (with 95% plant availability)					48.3		52.6		54.5
			Plant load factor (%)					47.2		51.3		53.2
			Utilization factor (%)					70.4		76.6		79.3
			Design discharge (cumecs)				13.47					

Combined T.G.Eff. above 70% load = 92.12%
Combined T.G.Eff. (50% to 70% load) = 80%
Plant availability = 95%
PLF has been considered as the ratio of gross generation (MU) to ideal generation (MU)
Ideal generation MU = installed capacity (MW) × 24 h × 365 days/1000

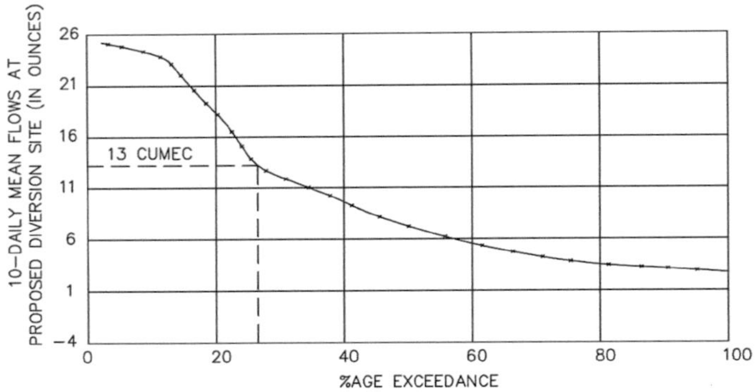

Fig. 8.2 Flow duration curve of average dependable year discharge (2007–08)

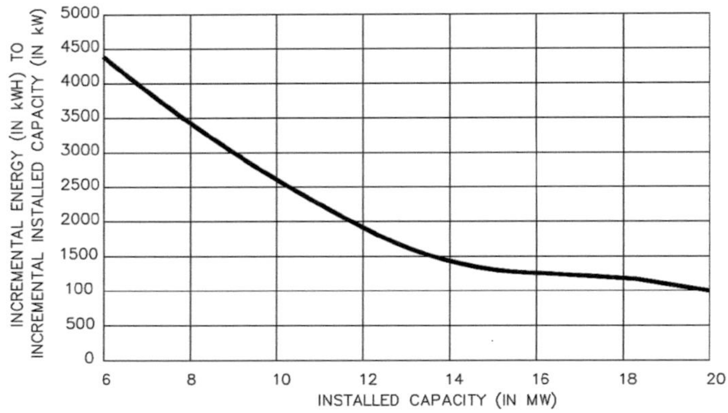

Fig. 8.3 Variation of I.E. to incremental I.C

The computations show that annual energy generation for installed capacity of 12 MW with 2, 3 and 4 units would be as below.

Units	Unit size (MW)	Annual energy (MUs)
2	6	48.32
3	4	52.64
4	3	54.55

To maximise energy generation 4 units each of 3 MW should be adopted. But the selection between 3 and 4 units should also be based on the cost of generation per unit of energy. Hence, 3 units were finally proposed.

8.6.2 *Example of Dam-Toe Scheme*

(i) The Scheme

A storage dam of medium height (about 20 m above the river bed) is under construction on river Betwa (District Jhansi, U.P.). The dam will store only a small portion of monsoon flows (storage is about 62 MCM, whereas the average annual run-off of river at site is around 3000 MCM) and non-monsoon flows (1st January to 30th June) of the river are to be released in the downstream. The proposed storage is to be used for the domestic water needs of the region and to feed 1.5 cumec in the tail of the existing canal system.

Since non-monsoon flows of the river are to be released from the dam uninterrupted, it was proposed to use these flows for power generation through a SHP at the toe of the dam. Feasibility stage power studies were carried out to fix the installed capacity. These are described below

(ii) Water Availability

The dam site is located about 70 km below the Paricha diversion weir. At this weir river flows are diverted into the Betwa canal. The dam site is also about 27 km upstream of the CWC discharge site at Mohana where the Betwa river meets river Dhasan. The study of available river flow data revealed that there are no water releases from Paricha during non-monsoon and there is no flow in river Dhasan at Mohana in this period.

Therefore, for water availability at the dam site, the 10 daily discharge data of the Mohana CWC site was considered for a period from 2000 to 2011. It was, for planning studies, assumed that the non-monsoon flows in the river from Paricha to Mohana are due to seepage from the groundwater and the seepage is at a uniform rate all along the length of the river. Thus, the Mohana discharges were reduced to 70% (in the ratio of distance) to arrive at the flows at the dam site and the same reduction was applied for monsoon flows. The 10 daily flows at the dam site were derived from Mohana site data on the above assumption and the annual run-off of each year was worked out.

The year for FDC was selected as a 75% dependable year which was worked out by arranging the yearly run-off of all the years, for which the data was available, in descending order and using Weibul's formula. It is shown in Table 8.3. The 75% dependable year was worked out as the year 2009. With the discharges of the year 2009, the FDC was drawn and it is shown in Fig. 8.4.

The FDC revealed that a discharge up to 36 cumecs (Q^{30}) would be available for about 8 months in a year. Hence this discharge was taken as a mean to select the range of installed capacity for power studies. The FDC also revealed that all time available flow is 4.5 cumec.

(iii) Net Head

The difference between the reservoir level and tail water level in the downstream of power house is the gross head. The maximum gross head is a difference between 156.5

Table 8.3 Computations for dependable year

Year	Annual run-off (Cumec-day)	Arranged in descending order		Rank No	Rank % (N*100)/13	Dependability (%)	Year corresponding to dependability
		Annual run-off	Year	N			
2000	80,251.95	80,251.95	2000	1	7.69		
2001	59,859.61	76,478.56	2003	2	15.38		
2002	27,753.51	71,224.95	2005	3	23.08		
2003	76,478.56	68,391.09	2008	4	30.77		
2004	36,805.37	60,787.88	2006	5	38.46		
2005	71,224.95	59,859.61	2001	6	46.15		
2006	60,787.88	36,805.37	2004	7	53.85	50	2004
2007	5542.90	27,753.51	2002	8	61.54		
2008	68,391.09	16,443.49	2011	9	69.23		
2009	13,556.65	13,556.65	2009	10	76.92	75	2009
2010	13,304.77	13,304.77	2010	11	85.00		
2011	16,443.49	5542.90	2007	12	92.31	90	2007

N = 12, N + 1 = 13

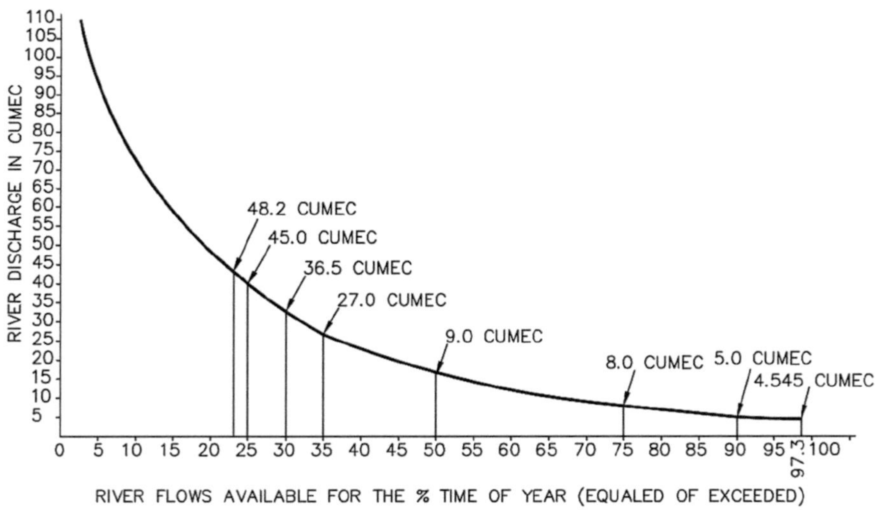

Fig. 8.4 Flow duration curve for the year 2009

(FRL) and 140.0 (minimum tail water level), i.e. 16.5 m but this will be available for a few days when the reservoir is at FRL and there is no water downstream. The FRL will be maintained during the monsoon months of July, August and September when there will be flow downstream also due to the spillway from the dam. FRL will start to go down from October and will reach 148 (MDDL) in June. Hence the reservoir level varies from 156.5 to 148 during October to June. The average reservoir level for head calculations from October to June is taken as (156.5–148) 2/3 + 148 = 153.6. The tail water level with machines running is taken as 140.3. Hence gross head from October to June is 153.6–140.3 = 13.3 m.

During monsoon when FRL is 156.5, the downstream level will depend on discharges from the dam. Because of several storage projects upstream of this site, an average downstream water level is assumed as 146.2 m on the assumption that large flows with higher downstream water levels will be for a few days in a year when power house would be closed. It was accounted for in forced outage in the power studies. Hence the gross head during monsoon months July to September was taken as 156.5–146.2 = 10.3 m.

Since penstock is of very small length, the head loss for a velocity of 4 to 5 m/sec in penstock will be small. It was thus assumed equal to 0.3 m. Thus, the net head during monsoon was taken as 10 m and 13 m during October to June.

(iv) Power Studies

The power potential corresponding to the mean discharge of 36 cumec and net head of 13 m worked out about 4 MW (8.3 × 13 × 36 Kw). Thus, energy computations were carried out for installed capacity ranging from 3.2 to 4.8 MW at an interval of 0.4 MW. The computations for 3.2 MW are shown in Table 8.4 for illustration and the results of other installed capacities are compiled in Table 8.5.

(v) Number of Units

The number of units was decided on the consideration that at least one unit may run at a permissible part load. The minimum flow in a 10 daily in 75% dependable year is 4.5 cumec. Considering 36 cumec as the design discharge of four units, one unit at full load would require 9 cumec. Being a low-head Kaplan turbine, it would be possible to run it at 50% discharge. Hence, four units were adopted.

(vi) Installed Capacity

The perusal of power study results of energy generation at different installed capacities revealed that beyond 4 MW the increase in energy had a sharp fall (Table 8.5). Hence installed capacity was fixed as 4 MW and the design discharge was fixed as 37 MW.

When for the installed capacity of 4 MW and design discharge of 37 cumec, the number of units was increased from 4 to 5 the energy generation had slightly increased from 15.74 MU but the cost of the power house and that of the non-overflow dam due to increasing in length increased significantly. Hence four units each of 1 MW capacity were proposed.

Table 8.4 Power studies for 3.2 MW installed capacity

Sl. No	Month	Ten daily blocks	No. of Days into a block	Discharge for the year 2009 (75% dependable year)	Water utilized to generate 3.2 MW	Gross head H	Head loss h	Net Head = H–h = H_d	Power P = 8.3 QHd1000	Ten daily generation (P × Days × 24) / 1000	No. of days all machines were running at full capacity
					cumec	m	m	m	MW	MU	
1	2	3	4	5	6	7	8	9	10	11	12
1	Jan	1–10	10	8.596	8.596	13.3	0.3	13.0	0.93	0.22	–
		11–20	10	8.274	8.274	13.3	0.3	13.0	0.89	0.21	–
		21–30	11	9.156	9.156	13.3	0.3	13.0	0.99	0.26	–
2	Feb	110	10	8.624	8.624	13.3	0.3	13.0	0.93	0.22	–
		1120	10	9.219	9.219	13.3	0.3	13.0	0.99	0.24	–
		2128	8	9.002	9.002	13.3	0.3	13.0	0.97	0.19	–
3	Mar	110	10	8.421	8.421	13.3	0.3	13.0	0.91	0.22	–
		1120	10	8.295	8.295	13.3	0.3	13.0	0.90	0.21	–
		2131	11	8.211	8.211	13.3	0.3	13.0	0.89	0.23	–
4	April	110	10	8.092	8.092	13.3	0.3	13.0	0.87	0.21	–
		1120	10	7.567	7.547	13.3	0.3	13.0	0.82	0.20	–
		2130	10	6.636	6.636	13.3	0.3	13.0	0.72	0.17	–
5	May	110	10	5.624	5.624	13.3	0.3	13.0	0.61	0.15	–
		1120	10	4.545	4.545	13.3	0.3	13.0	0.49	0.12	–
		2131	11	4.601	4.601	13.3	0.3	13.0	0.50	0.13	–
6	June	110	10	5.111	5.111	13.3	0.3	13.0	0.55	0.13	–
		1120	10	5.649	5.649	13.3	0.3	13.0	0.61	0.15	–
		2130	10	5.253	5.253	13.3	0.3	13.0	0.57	0.14	–
7	July	110	10	12.173	12.173	10.3	0.3	10.0	1.01	0.24	–
		1120	10	84.140	38.554	10.3	0.3	10.0	3.20	0.77	10
		2131	11	100.450	38.554	10.3	0.3	10.0	3.20	0.84	11

(continued)

Table 8.4 (continued)

Sl. No	Month	Ten daily blocks	No. of Days into a block	Discharge for the year 2009 (75% dependable year)	Water utilized to generate 3.2 MW	Gross head H	Head loss h	Net Head = H–h = H_d	Power P = 8.3 QHd1000	Ten daily generation (P × Days × 24) / 1000	No. of days all machines were running at full capacity
8	Aug	110	10	22.519	22.519	10.3	0.3	10.0	1.87	0.45	–
		1120	10	85.3540	38.554	10.3	0.3	10.0	3.20	0.77	10
		2131	11	52.717	38.554	10.3	0.3	10.0	3.20	0.84	11
9	Sept	110	10	86.840	38.554	10.3	0.3	10.0	3.20	0.77	10
		1120	10	255.710	38.554	10.3	0.3	10.0	3.20	0.77	10
		2130	10	26.803	26.803	10.3	0.3	10.0	2.22	0.53	–
10	Oct	110	10	165.760	29.657	13.3	0.3	13.0	3.20	0.77	10
		1120	10	109.620	29.657	13.3	0.3	13.0	3.20	0.77	10
		2130	11	46.963	29.657	13.3	0.3	13.0	3.20	0.84	11
11	Nov	110	10	38.188	29.657	13,3	0.3	13.0	3.20	0.84	11
		1120	10	34.937	29.657	13.3	0.3	13.0	3.20	0.77	10
		2130	10	40.229	29,657	13.3	0.3	13.0	3.20	0.77	10
12	Dec	110	10	25.382	25.382	13,3	0.3	13.0	2.74	0.66	–
		1120	10	9.051	9.051	13.3	0.3	13.0	0.98	0.23	–
		2131	11	8.484	8,484	13.3	0.3	13.0	0.92	0.23	–

[1] Installed capacity assumed for study = 4 machines × 0.8 MW = 3.2 MW
[2] Net head = H = 13.0 m, discharge required = 29.657 cumec (Jan–June, October–December)
[3] Net head = H = 10.0 m discharge required = 38.554 cumec (July–September)
Gross annual generation (MU) = 15.2
Forced outage 10% (MU) = 1.52
Net annual generation = 13.68
PLF = 0.49
Total 15.20

Table 8.5 Fixing installed capacity

MW	Cumec	MU (after 10% outage)	Increase in energy (MU)	PLF	Days of full capacity generation in a year
3.2	29.66	13.68	–	0.49	123
3.6	33.36	14.75	1.07	0.47	123
4.0	37.0	15.74	1.00	0.45	103
4.4	40.7	16.61	0.87	0.43	82
4.8	44.49	17.32	0.87	0.41	82

References

1. Gutherie B.: Hydroelectric Engineering Practice, Blackie and Sons Ltd, London (1958)
2. Guidelines for Small Hydro Development, Standard No. 1.3, Project Hydrology and Installed Capacity, AHEC, IIT, Roorkee (2013)
3. Emil, M.: Water Power Development, vol. 1, Publishing House of Hungarian Academy of Sciences, Hungary (1965)
4. Nigam, P.S.: Hand Book of Hydro Electric Engineering, Nem Chand and Bros, Roorkee (1985)
5. Project Reports (DPR) of the Schemes

Chapter 9
Sediment Management

9.1 General

The flow in natural rivers is always associated with sediment, which is contributed by the catchment of the river. The extent of sediment present in the flow during monsoon months is maximum and is directly proportional to the quantity of flow. When the sediment-laden water flows through the turbine and other underwater parts of the turbine assembly of a hydro plant, damage takes place on account of the erosive action of sediment. Studies have shown that excessive erosion reduces the efficiency of the turbine. It is seen that the parts that are affected by erosion in different types of turbines are as below:

(i) Francis turbine—Runner blades, guide vanes, bushes and seals, labyrinths and top cover.
(ii) Pelton wheel—Needles and nozzle trip liner, inside of buckets.
(iii) Kaplan turbine—Being low head turbines damage is not very significant. But where sediment content is large the periphery of blades is affected.

The other equipment which are affected are the seals of valves and the cooling water system. Studies have shown that the rate of abrasion may be expressed as below:

$$W \alpha S_1 \cdot S_2 \cdot S_3 \cdot S_4 \cdot M_r \cdot V^x \quad (9.1)$$

where, W is rate of erosion, S_1, S_2, S_3 and S_4 are coefficients of sediment concentration, hardness, shape and size, M_r is the coefficient for abrasion resistance of metal and V^x is relative velocity of flow with exponent 'x'. The value of 'x' based on experience is as below

$$x = 3 \text{ for Francis turbine}$$
$$= 2.5 \text{ for guide vanes}$$

B. N. Asthana, *Planning and Design of Small Hydro Power Projects*,
https://doi.org/10.1007/978-3-031-96271-4_9

=2.5 for Pelton wheel nozzles

=1.5 for Pelton wheel buckets.

This reveals that erosion damage is dependent on sediment characteristics and the type of turbine and its design. The sediment characteristics, that are responsible for erosion, are concentration, shape, size and hardness of particles. In this chapter aspects of managing sediment to minimise the damage due to sediment are dealt. For this purpose, at each project site, the data regarding the above sediment characteristics have to be collected and analysed to select an appropriate sediment management strategy.

9.2 Sediment Management of SHPs

SHPs are classified as (i) run-off-river schemes with diversion structures both with and without pondage, (ii) storage based schemes with a power plant at the toe of the dam or away from the dam and (iii) canal based SHP using canal falls either on existing canals or on new canal systems.

The problem of sediment damage is observed to be more severe in the case of a high head SHP of run-off-river type with weir/barrage as a diversion structure operating during monsoon and which is located on a stream in the Himalayas. In such cases, planning of a sediment management strategy needs careful study of sediment characteristics. The selection of suitable type of sediment mitigation measure should be examined for its efficiency against cost.

In a SHP, which is based on storage, the reservoir acts as a settling basin and most of the sediment settles and deposits in the reservoir. With time, the reservoir gets silted up to the crest of the spillway. Hence the invert of the crest of the intake structure is kept sufficiently above the crest level of spillway to reduce sediment entry into the intake. However, the invert of the crest of intake is to be placed below the MDDL. The sediment deposition in the reservoir often reduces the useful storage capacity of the reservoir. To counter this process of sedimentation, the practice is to provide low-level large sluices in the spillway, which may be operated to flush sediment during monsoon. In cases where the high-head power plant is operated away from the dam and harmful sediment is expected to enter the intake, some sediment extraction measures may be provided in the head reaches of the water carrier system.

In case of a SHP on canal fall, the problem of sediment management is generally taken care of in the design of the canal, as the canal is designed on non-silting, non-scouring criteria. Generally, it is assumed that the sediment coming into the canal does not cause significant damage to the turbine because of the low head. Hence sediment is no problem in a SHP on canal fall.

Since the impact of sediment is more predominant in the case of a run-off-river scheme especially when the concentration of angular or semi-angular particles having hardness exceeding 5 on Moh's scale and size larger than 0.25 exceeds 200 ppm, desilting measures are to be provided. This sediment erosion problem is observed to be very severe with projects on Himalayan streams where petrographic investigations on some streams have revealed a preponderance of quartz particles that have a hardness of 7 on Moh's scale and are sub angular in shape and the content varies up to 90%. This is not in all streams, hence detailed analysis of sediment samples is required for every case under consideration. The desilting measures generally adopted on run-of-river schemes are described below (Refer CBI & P, Pub. 301).

9.3 Desilting Measures

The desilting methods that are generally adopted on hydro projects are classified into two groups which are called preventive and curative. The preventive methods reduce the entry of sediment in the water conductor system and curative methods extract the sediment which enters the water conductor through intake before the flow reaches the turbine.

All these desilting methods are based on the simple concepts of hydraulic engineering which are stated below:

(i) Sediment concentration is higher in lower depths of the flow of a channel.
(ii) Coarser sediment particles move as the bed load along the bed of the channel.
(iii) The particles in suspension settle to lower depths of flow when flow velocity is reduced.
(iv) Sediment concentration is less on the outer curvature of flow.

The desilting methods adopted in different types of projects are described below.

9.3.1 In Run-of-River Projects

The run-of-river projects especially on Himalayan streams suffer due to sediment problems and all these are provided with some kind of desilting arrangement. The normal practices to manage the sediment are as below.

(A) **Scheme with a Weir/Barrage as Diversion Structure and a Power Channel**

In such a scheme the power house is located at the end of generally a long power channel. The two types of desilting methods that are generally adopted are as below.

(a) Preventive methods:

The following measures have been found effective in creating outer curvature of flow at the intake and to reduce sediment entry into the intake.

(i) By constricting the waterway of the diversion structure as shown in Fig. 9.1i.
(ii) By locating the intake at the curved branch of flow as shown in Fig. 9.1ii.
(iii) By providing the desired curvature in the guide bund as shown in Fig. 9.1iii.
(iv) Outer curvature at the intake can also be generated by the regulation of river flows.

The following practices of river flow regulation are found effective in reducing sediment entry.

- Still pond regulation: In this type of regulation the under sluice bays are kept closed and flood is passed through barrage bays. It is possible with the help of a divide wall. In this type of regulation, the canal draws the discharge from a still pond in the under sluice pocket.
- Semi-still pond regulation: In this type of regulation wedge type opening of under sluices (maximum opening of the farthest bay from the canal regulator) is adopted to pass the part flood through under sluices.
- In still pond regulation there is a need for frequent flushing of sediment from the undersluice pocket which is not often considered feasible on power channels due to the closure of the power plant during flushing. Hence semi-still pond regulation is generally preferred.

Observations on existing structures as well as model studies have shown that the inclination of intake with diversion structure and length of divide wall influence the entry of sediment in the canal. The adverse effect on flow at intake due to diversion at right angle is shown in Fig. 9.1iv. Studies have shown that an angle of 100 to 110° with barrage axis and divide wall of length up to 2/3rd width of canal intake are most effective in reducing sediment entry in the canal.
Other preventive methods generally adopted on the canal regulators are as follows. These depend on the principle that the bed load moves at a lower depth of flow and such low depth flow layers shall be prevented from entering the intake.

- The crest of the intake regulator is kept at an elevation higher than the crest of undersluice bays. Generally, the difference in the levels of the two should be 2 m.
- Provision of a few openings in the regulator crest below the crest elevation has in general been found helpful in removing the sediment that gets deposited by the side of the regulator crest.
- The provision of guide vanes in the form of submerged bars in front of the regulator crest may deflect the bed load away from the canal head regulator. Proper alignment of vanes is important.
- A cantilever platform projecting from the crest of the canal head regulator on the river side may separate out bottom layers of flow and prevent them from entering the canal. It is shown in Fig. 9.1vi.

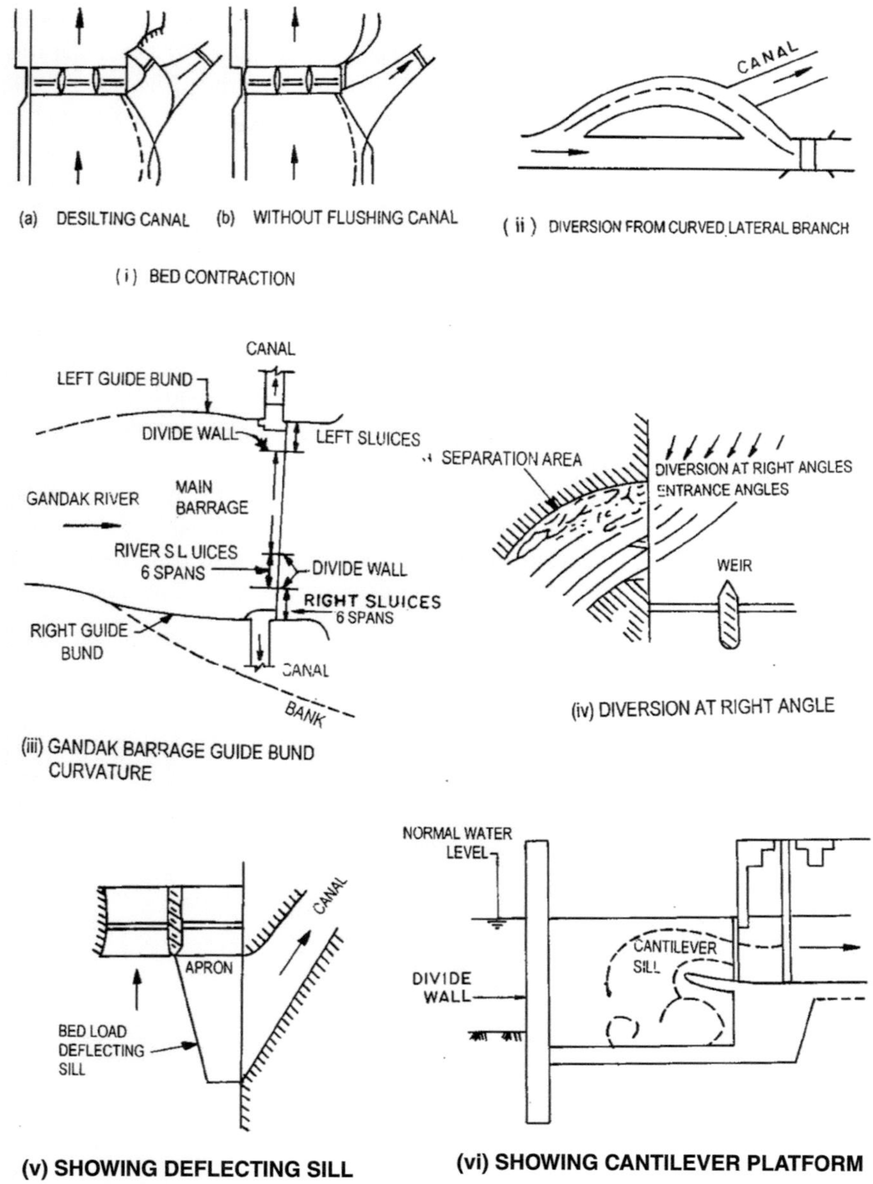

Fig. 9.1 Preventive measures at a barrage

- A combination of guide vanes and cantilever platform was later transformed into the shape of a tunnel type excluder and the first excluder was constructed in the pocket of Khanki headwork on river Chenab in 1934. Since then prototype observations on excluders and the model studies have helped in evolving its design parameters. A definition sketch of an excluder is shown in Fig. 9.2. Here vanes can be seen acting as walls and supporting the top slab of the excluder.

(b) Curative methods:

It is generally observed that despite preventive measures at the canal head, some sediment enters into the canal as a suspended load which is damaging to the turbine. Hence, some curative measures are adopted in the head reach of the canal to extract the harmful sediment before the flow reaches the power plant. These curative measures are described below.

(i) Sediment ejector

It is tunnel-type structure as shown in Fig. 9.3. It is similar to an excluder. It diverts lower layers of flow containing higher concentrations of coarser sediment particles into an escape channel which carries the diverted or ejected flow to some natural drainage. Thus,

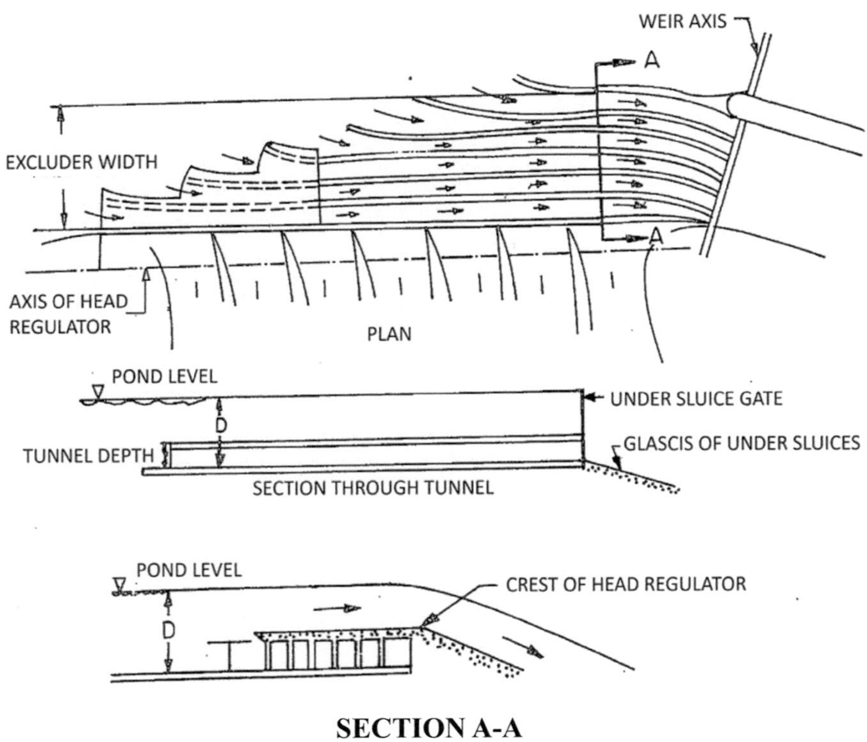

Fig. 9.2 Definition sketch of an excluder

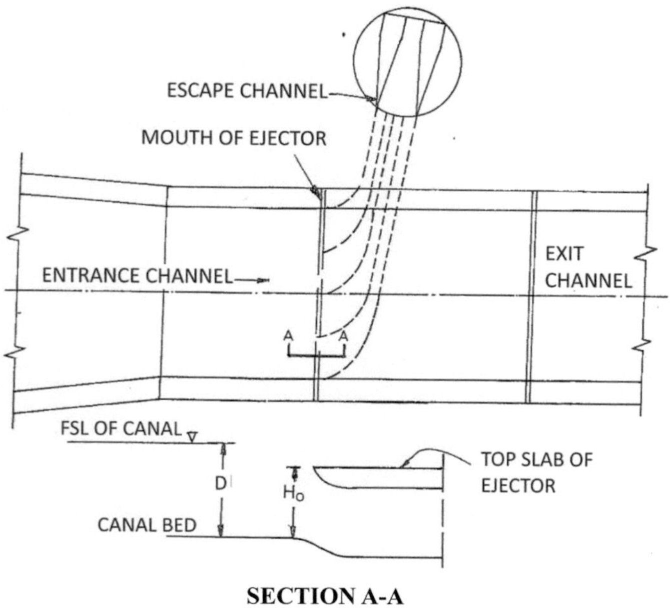

Fig. 9.3 Definition sketch of a silt ejector

the canal downstream of the ejector carries only the top layers of flow with lesser sediment. To achieve the desired effect, the top slab of ejector tunnels is generally kept at about 1/3rd to 1/4th of the water depth above the canal bed. The ejector is placed close to the canal head regulator but sufficiently away at a distance where the flow is expected to achieve the condition of uniform flow with coarser sediment moving as bed load.

(ii) Desilting Basin

In this type of desilting arrangement, the channel is widened and deepened to reduce the flow velocity so that the particle size to be extracted is settled to the bottom where some sediment removal or flushing arrangement is provided. Many hopper type desilting basins have been provided on hydro projects. A typical hopper-type desilting basin is shown in Fig. 9.4 which is provided at Maneri Bhali Project Stage-I, to exclude 0.3 mm size particles. Similar desilting basins have also been provided at the Ichari and Maneri Stage-II projects of Uttarakhand. The dimensions of the basin are dependent on the fall velocity of the particle size to be extracted and the flow-through velocity in the basin.

In view of high cost and time required to construct the hoppers and flushing conduits shown in Fig. 9.4, another arrangement has been developed in which the bottom slab of the basin is connected

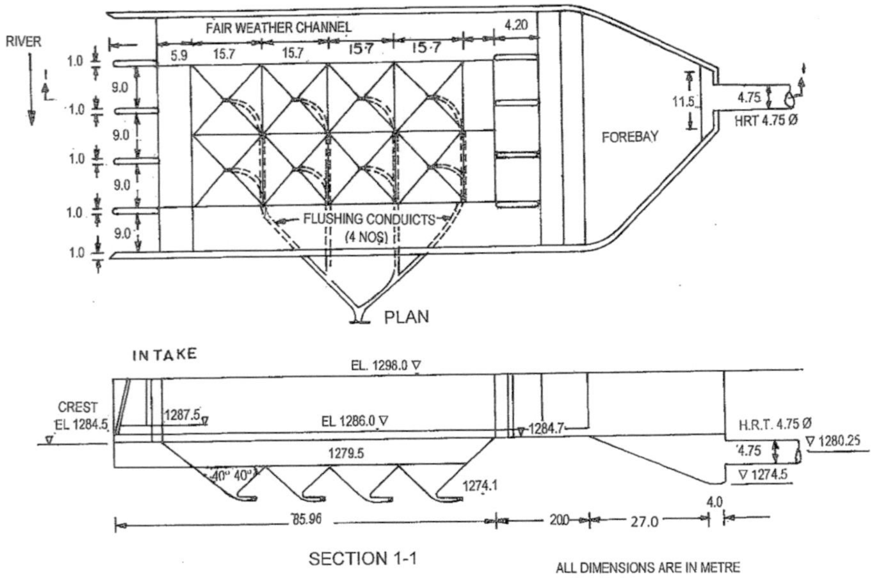

Fig. 9.4 Desilting chamber of Maneri Bhali stage I project

through a number of openings of varying diameter with a flushing duct made below the slab along the length of the basin. This arrangement is shown in Fig. 9.5 which is provided at Ranjit Hydro Project in Sikkim. Because of ease in construction, this has been adopted at many projects. The model studies have shown that both these types of desilting basins are practically of the same efficiency.

(iii) Vortex type desilting measures

The above-described types of desilting measures have been adopted on many SHPs. But in some cases, the available design discharge for power plant is small. In such a case, it becomes unviable to allow flushing discharge to the extent of 20 to a flushing discharge of 5 to 10% of the design discharge. These are 30% of canal discharge, which is required in a desilting basin to escape. The cost of these measures with respect to total project cost also becomes high. For such projects, the vortex type desilting measures have been evolved basically through hydraulic model studies and have been successfully adopted on some SHPs to extract + 0.2 mm size sediment particles. These are found to function efficiently with of two types as described below.

1. Vortex settling basin—It is a circular basin of a diameter of five times the width of the inflow channel with a hole at the bottom for flushing the sediment. The flow enters the basin tangentially and moves along the periphery of the basin. With the opening at the bottom, a vortex is formed and a radial flow is created. The bottom layer of flow which

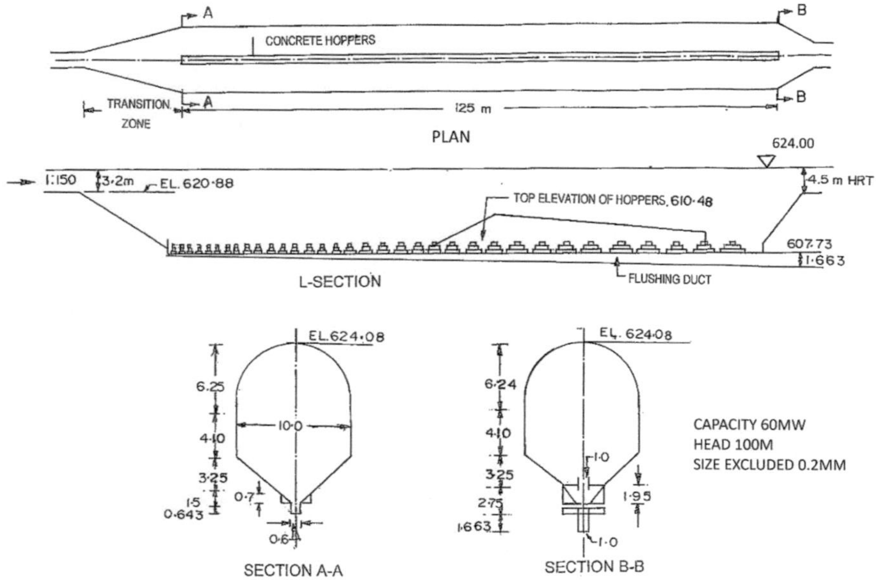

Fig. 9.5 Details of desilting basin at Ranjit dam (Sikkim)

contains a major amount of silt escapes through the opening and the top layer of flow enters the outflow channel over a crest at the other end of the basin. The vortex settling basin which has been adopted at Gaj project after model studies is shown in Fig. 9.6.

2. Vortex tube—A vortex tube is a tube of circular section made below the bed across the width of the channel with an opening at the bed level. The tube is joined on the downstream end with an escape channel with a smooth curve. The tube is placed either normally or at an angle to the direction of flow. To control the flow out of the tube, a valve is fitted at the downstream end of the tube. As water passes over the tube a shearing action across the open section of the tube sets up a vortex motion within the tube which sucks sediment from lower depths of flow into the tube. Extensive research and model studies have been carried out to fix the criteria to determine the parameters of this type of desilting measure. The vortex tube adopted at Ganvi Project is shown in Fig. 9.7.

(iv) Settling tanks:

Desilting measures described above can economically extract particles of size (+) 0.15 mm. The experience has shown that smaller size angular particles of quartz cause serious damage to turbine runners and other under water parts and sometimes need extraction. For this purpose settling tanks or basins have been provided. The low lying areas along the water conductor/canal are converted into settling tanks by constructing bunds

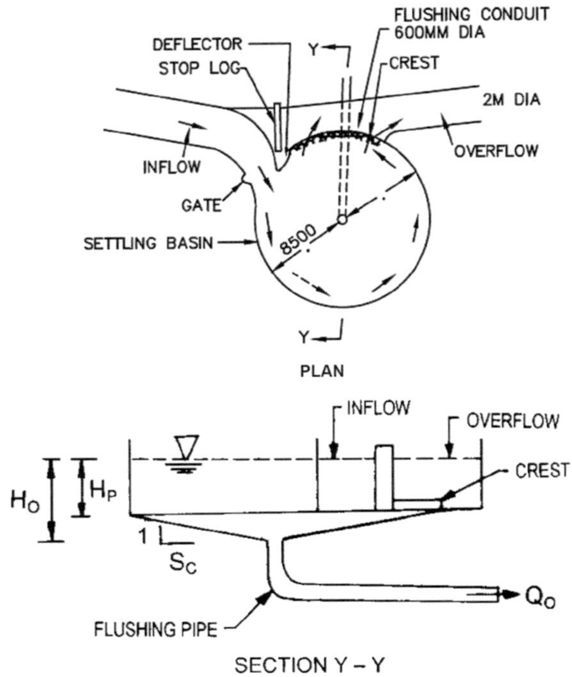

Fig. 9.6 Vortex settling basin at Gaj hydroelectric project

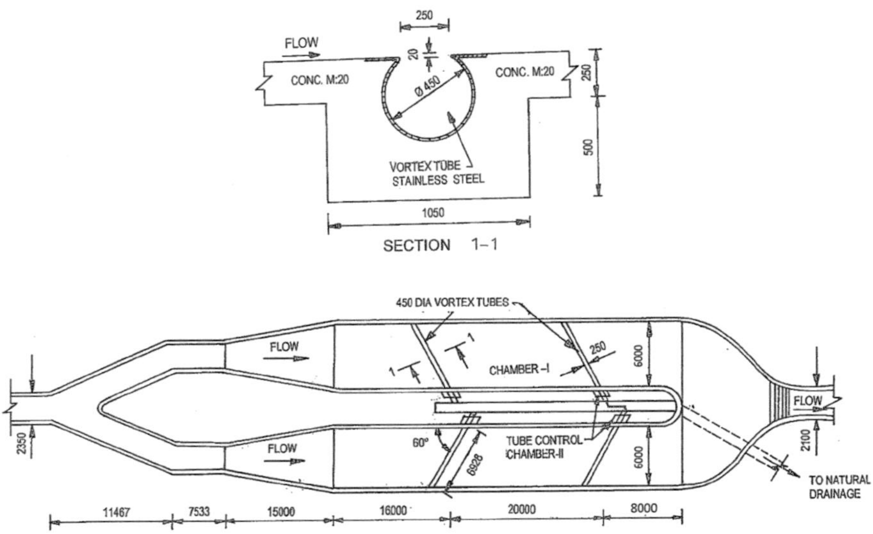

Fig. 9.7 Vortex tube at Ganvi hydroelectric project

along the periphery of the area. The flow enters from one side in the tank and the sediment drops at the bottom because the flow becomes practically stagnant due to the vast area of the tank. Thus, clean water flows out of the tank from the other end. The sediment that deposits in the tank is either manually or mechanically removed or flushed out depending on site conditions.

These settling tanks are often used as balancing reservoirs and also to store a few hours of inflow to run the power house for peaking purposes. These tanks if located close to the power house site can be used as forebay also (Refer Fig. 2.15).

A settling tank downstream of the silt ejector and upstream of the balancing reservoir was later introduced after model studies at CWPRS, Poona to remove particles (+) 0.075 mm of Trisuli Hydel Project (Nepal) which has an installed capacity of 18 MW operating under a head of 75 m with the head discharge of 31 cumec. It is shown in Fig. 9.8. Several such tanks have been provided at several projects where finer particles are required to be extracted.

(B) Schemes with Dam Projects

The dam projects are generally of two types. One is a storage dam project and the other is a diversion dam project. The SHPs with storage dams, which generally have large reservoir capacity, do not face any sediment problem. The reservoir acts as a large settling basin and clear water enters the water conductor through the intake which is placed much above the expected level of silting after 50 to 100 years of operation.

In the case of diversion dams, the reservoirs are generally small because these only divert the flows in the water conductor at the required pool level. These are often used for storing diurnal variations of flow to be used for operating the power house for a few hours of peaking. Since reservoirs of diversion dams are small a lot of sediment enters the intake of the water conductor system when power house is operated during monsoon. The hydropower scheme with

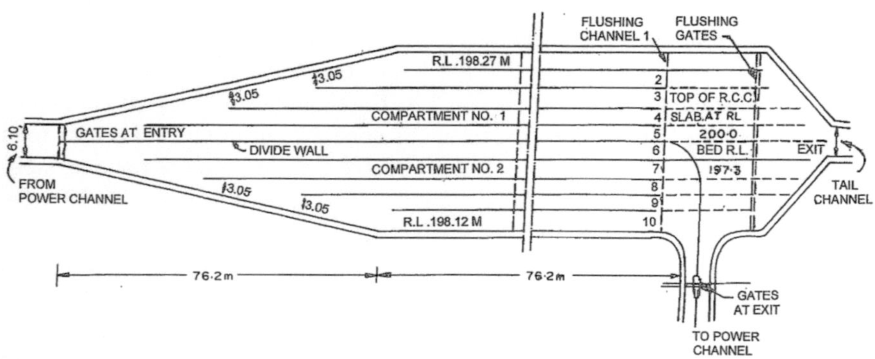

Fig. 9.8 Settling cum-flushing tank on Trisuli power channel

a diversion dam face a lot of sediment problem and need desilting measures. Both types of dam schemes have power plants either at the toe of the dam or away from the dam. The problem of sediment management is more serious in the case of plants away from the dam because these operate at relatively higher heads. The preventive and curative measures generally adopted for dam projects are discussed below.

(a) **Storage Dam Projects**

(i) Preventive measures:

The reservoir planning is done in such a way that the crest of the intake is sufficiently above the expected level of silting and below the MDDL.

(ii) Curative measures

Generally, no curative measure is required because only very fine suspended sediment enters the intake.

(b) **Diversion Dam Projects**

In these projects, the reservoirs are of insignificant storage capacity and small length and these get silted up after a few years of operation generally up to the spillway crest resulting in the entry of sediment-laden flow in the water conductor through the intake causing damage to the turbine and other under water parts. Hence both preventive and curative measures are taken on such projects.

(i) Preventive measures

(1) In order to reduce sediment entry in the intake, the crest of intake is placed sufficiently above the spillway crest.

(2) Sediment deposition in the reservoir reduces the storage capacity provided for peaking and sometimes a lot of sediment deposits along the crest of intake. This tendency of sediment deposition is significantly arrested by flushing the reservoir. Large size openings fitted with gates are provided at or close to river bed level in the dam which may have a capacity of passing design flood or a significant proportion of design flood to flush the reservoir. Flushing with full drawdown condition has been found most effective. In this condition, the reservoir level is drawn down to bed or close to bed elevation by making the gates fully open. A sketch showing the impact of such flushing on sediment is shown in Fig. 9.9. The figure also shows the conditions of silting in case the flushing is done with insufficient drawdown. In case of flushing with full drawdown the power house remains inoperative during flushing operation. So it is feasible in cases where the reservoir may get filled up to FRL in a few hours after flushing.

(3) In some cases where reservoirs are of short length, sediment bypassing has been successfully adopted. Sediment bypassing in such cases requires the construction of either a diversion dam or a weir upstream of the reservoir to bypass the flood in the river downstream

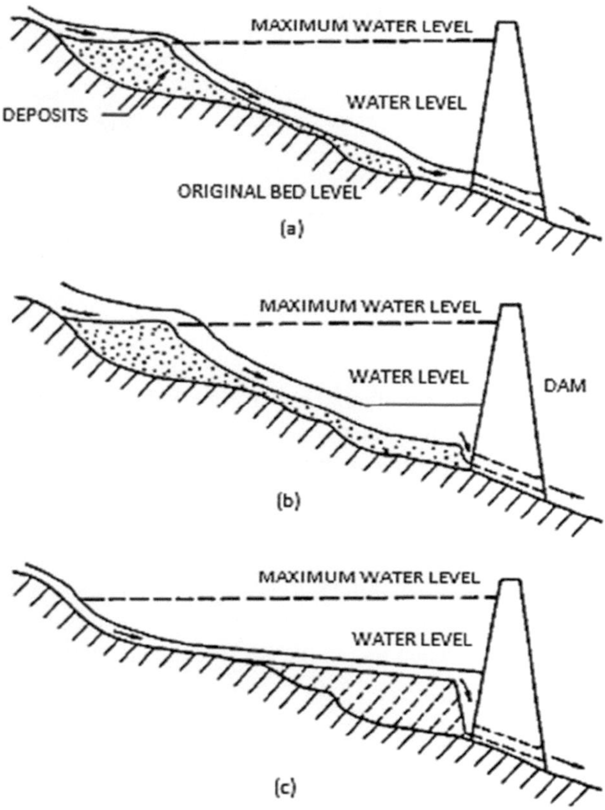

Fig. 9.9 Showing condition after **a** flushing with full drawdown **b** flushing with insufficient drawdown and **c** long period flushing with insufficient drawdown

of the main dam through a tunnel or channel aligned around the reservoir. The selection of the structures depends on reservoir length and configuration, the slope, geology, river hydrology, economics and the operation and maintenance problems. This has been adopted through canals and tunnels on small reservoirs in many countries such as China, Japan, Switzerland, Italy, Austria, and South Africa. A typical layout of bypassing the Amsted reservoir in Switzerland is shown in Fig. 9.10. Because of the high cost, the sediment bypassing method is adopted in cases where other methods are not effective to the desired extent.

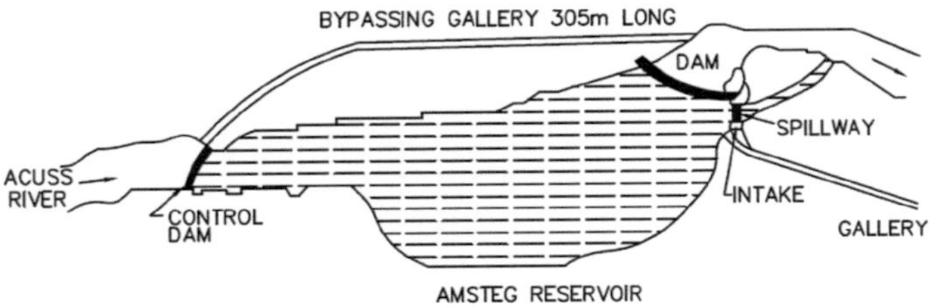

Fig. 9.10 Bypassing of Amsteg reservoir in Switzerland (After Reed 1931). *Source* Batuca et al. 2000

(ii) Curative methods

It has been observed that in run-of-river SHPs with diversion dams, lot of sediment enters the water conductor through intake despite the preventive measures and causes damage to turbines. In such situations, curative measures to extract the harmful particle sizes are provided in the water conductor in the upstream of powerhouse and close to intake. These are the same as described above. In many cases, more than one method is adopted. Generally, desilting basins are provided but sometimes a combination of desilting basin and vortex tube or desilting basin and settling tank is provided depending on the sediment characteristics and the head.

9.3.2 *In Other Types of SHP*

The other types of SHPs are those that are located either at the toe of a storage dam or at canal falls. As discussed above (para 9.2) such types of SHPs do not suffer significant damage and hence do not need desilting measures.

9.3.3 *Effectiveness of Desilting Measures*

Generally, the effectiveness of the desilting measures provided at the hydro project is gauged from the extent of damage observed on turbine runners/buckets. In case the damage is reasonably within the limits of annual maintenance of the machines, the provision of the desilting measure is considered satisfactory. In case the damage is significant then additional desilting measures should be planned and adopted if they are feasible and cost effective. Alternatively, the water conductor may be closed from the intake when the sediment concentration in river flow is above a certain limit. This limit is generally decided by the operational experience of the project. The days for which the power house will be shut during a season and the resulting loss in power generation shall be taken into consideration before finalising the water conductor closure alternative.

References

1. Asthana B.N.: Sediment Management in Water Resources Projects, Pub. No.301, CBI&P, New Delhi (2007)
2. I.S. 6518.: Code of Practice for Control of Sediment in Reservoirs", B.I.S. Manak Bhawan, New Delhi (1992)
3. Sediment Control at Intakes—A Design Guide, Edited by P. Avery, Pub. BHRA, The Fluid Engineering Centre, Bedford, England (1989)

Chapter 10
Scope of Design of Structures

10.1 General

A small hydro project consists of the following civil structures:

1. Diversion/storage structure.
2. Intake structure.
3. Water conductor system.
4. Desilting measures.
5. Balancing reservoir.
6. Forebay/surge tank.
7. Penstock.
8. Power house.
9. Tail race channel.

The above structures are normally required in a run-of-river scheme, but all these are not required for an SHP at the toe of a storage dam and at a canal fall.

A SHP at the toe of a storage dam requires an intake structure, a short length penstock, power house and a small tail race channel. A canal fall SHP requires a by-pass channel, power house and an escape channel.

The layout details of these structures are described in Chapter 2. Before proceeding with the design of these structures their functional requirements are required to be understood besides their layout. These are briefly described below.

An arrangement for the regulation of the flow of water through these structures is also required. Hence, all the civil structures are provided with gates or valves as per their functional requirement. The design aspects of these structures are dealt in separate chapters.

B. N. Asthana, *Planning and Design of Small Hydro Power Projects*,
https://doi.org/10.1007/978-3-031-96271-4_10

10.2 Functional Requirements

The functional requirements of above-mentioned civil structures are briefly described below.

1. Diversion structures—A diversion structure is an important component of a small hydro project of run-of-river type. Its functional requirement is to divert the required flow into the water conductor system. These are of two types: one which diverts the flow as and when available and the second is that which has some storage to meet the peaking requirement of the power house. Weirs are the first type and the barrages are the second type of diversion structure. Both these are provided across the perennial streams.
2. Storage structures like dams are provided on the streams which are normally dry during non-monsoon periods. The monsoon flows are stored in dams which are released to run the SHP during non-monsoon periods. These are generally planned and operated as multipurpose projects. These are normally provided on the streams of South India.
3. Intake structures—This is a structure normally located close to the diversion structure in its upstream on the banks. It allows the required flow in the water conductor. If the water conductor is a canal, it is generally called canal head works. The intake structure is generally fitted with regulating gates which help in feeding the regulated discharge to the water conductor. It is also fitted with a trash rack to prevent the entry of trash into the water conductor. It is aligned and designed in such a way that minimum sediment enters through the intake.
4. Water conductor system—This structure helps to carry water to the power house. It can be either an open channel, a tunnel or a combination of the two. The choice of a water conductor depends on topography, geology, economy and ease of construction and maintenance. An open channel has to be aligned at a required slope so normally its length is more as compared to a tunnel. The channel generally has to cross a number of drainages requiring cross-drainage works. Tunnels need geologically strong rocky strata. Tunnel needs several adits for construction to complete tunnelling within the time schedule if its length is large. These adits are generally plugged after construction.
5. Open channels are designed as lined channels using the continuity equation and Manning's equation. Free flow tunnels are designed in the same manner as an open channel. A pressure tunnel running full is designed on the consideration of permissible velocity.
6. Desilting Measures—When the sediment entering the intake is considered harmful to the turbine, the desilting measures are provided in the water conductor to extract the sediment particles of size that may damage the turbine. These are of various types and their choice depends on site-specific conditions. These are described in detail in Chapter 9. These are placed close to the intake structure.

7. Balancing Reservoir—It is provided only when a big natural pond as a depression is available in the alignment of the water conductor. It can serve as a settling basin to extract fine sediment and as a supplementary storage for peaking purposes. It can act as a fore-bay if it is close to the location of the power house.
8. Fore-bay/surge tank—Fore-bay is provided at the end of the open channel. It has a small storage capacity to store water which may be supplied to the turbine, when it is started after a closure, for a short period during which the flow in the channel is accelerated. It also protects the canal from the surges which are generated by the sudden closure of the turbine. In case of a long tunnel running full, the purpose of the fore-bay is performed by a surge tank. It is a circular shaft of size bigger than a tunnel made at the downstream end of the tunnel.
9. Penstock—It is generally a steel pipe that carries water from the fore-bay or surge tank to the turbine. It runs fully under the water head. It is generally laid along the slope of the hill. It is laid either exposed to the surface or buried below natural ground. The velocity in the penstock is decided based on sediment characteristics and permissible head loss. The velocity in the penstock is generally limited to 4 to 5 m/sec.
10. Power house—It is generally a structure to house the generating equipment i.e. turbine and generator along with other associated equipment required to control the operation of power generation. A valve at the power house end of the penstock is provided for closing the water supply to the turbine when it is required. It is located upstream of the turbine in the power house. The power house building is a framed structure of reinforced cement concrete. In line with the generating units, sufficient space is provided for the erection and maintenance of generating equipment. Generally, a closed space is provided for the control room. Outside the power house building and close to the erection bay some space is provided to place the transformers. The switchyard which transmits power to the load centre is located near the transformers.
11. Tail Race Channel—It is a channel which is provided at the downstream of powerhouse to carry the flow after power generation to a natural drainage. It is normally a small length open channel. Sometimes the escape channel from the forebay meets the tail race channel.

10.3 Scope of Design

The hydraulic and structural design aspects of all these structures are described structure-wise in separate chapters. One separate chapter deals with the aspects of gates and valves that are required to regulate the flow through these structures.

Chapter 11
Diversion Structures

11.1 General

Diversion structures are classified as weirs or barrages. Various types of weirs have been used for SHPs. These are structures of small height (say up to 4–5 m) above the river bed to divert the flow into the water conductor and have no storage capacity. These have an ungated crest at the top to pass river flow. These are generally classified by the name of the material used to construct them.

Barrages are gated structures with a slightly raised crest above the river bed. Gates raise the water level to feed the water conductor at the required level. In barrages fitted with high head gates (15–16 m high above crest), some storage is provided to operate the power house for peaking purposes.

The design aspects of these diversion structures are discussed in the following paras.

11.2 Raised Weirs

Weirs of different materials have been constructed for many projects. With the type of material used in construction, weirs are often classified. The use of locally available construction materials makes the project economical. However, the selection of a weir type should be made on other considerations, such as the importance and size of SHP, foundation conditions, the height to which the minimum water level of the river is to be raised, the design flood to be passed and the river geometry, i.e. the river cross section and the longitudinal slope. The details of the weirs which are given below are based on publications of CBI&P and AHEC.

B. N. Asthana, *Planning and Design of Small Hydro Power Projects*,
https://doi.org/10.1007/978-3-031-96271-4_11

11.2.1 Solid Boulder Bar

This type of raised crest is formed of boulders available locally in the river bed. The boulders are placed in wire crates (gabions) of galvanised iron (GI) wires. Generally, crates are made of size 2 m × 2 m × 1 m with opening size of 3–4 inches (7.5–10 cm). The boulders to be placed in crates should be bigger than the size of the opening. The crates are placed one over the other as shown in Fig. 2.1e forming steps on the downstream face. The steps help in energy dissipation. The upstream and downstream faces of the boulders bar are covered with 10 cm thick reinforced concrete to make the bar water tight. These are made of low height (up to 4 m) with per metre length discharge up to 3 cumec. Generally, these are made on a strong rocky foundation. The base width of the boulder bar is 1.5 times the height for stability. As a measure against scour in the downstream a few (1 or 2) rows of boulder-filled wire crates are placed on the river bed.

It has been sometimes observed that downstream steps of the boulder bar get damaged due to the impact of falling boulders during floods. To avoid such damage in the streams carrying big size boulders, the steps of the boulder bar are made on the upstream side as shown in Fig. 11.1 (AHEC Memo 2.2–2.3). As shown in Fig. 11.1, it is generally difficult to economically procure earth for backfilling the bar in boulder reaches of hill streams. This earth backfilling slope may also need rip rap protection. Instead of earth, the backfilling may be made of boulders. The top of the bar and downstream slope should be covered with 10 cm thick reinforced concrete. In order to dissipate energy due to vertical drop a water pool should be provided by depressing the river bed downstream of the bar. The bed of the pool should be of concrete. Beyond the pool, a few rows of boulder wire crates or CC blocks should be provided as shown in Fig. 11.1 to safeguard against scour.

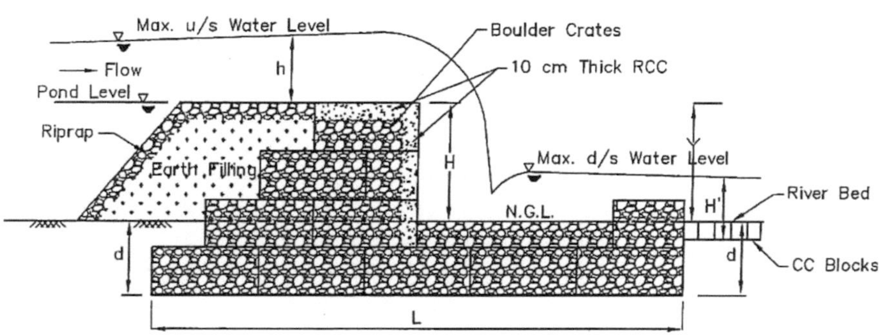

Fig. 11.1 Bed and water surface profile of Boulder weir

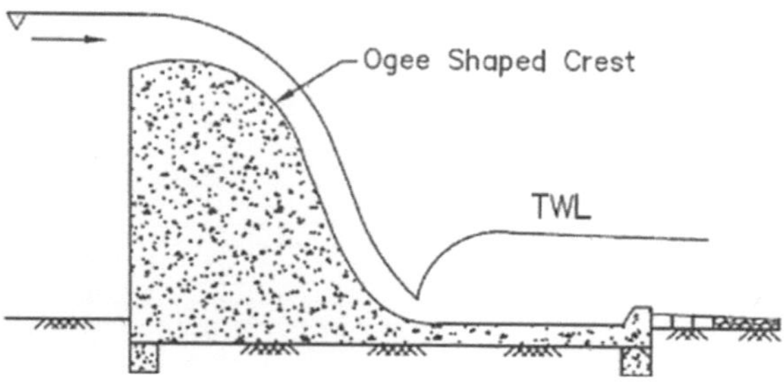

Fig. 11.2 Concrete dam

11.2.2 Raised Crest Type Weir

These are made of concrete or stone masonry for small heights (4–5 m) on strong rocky foundations on small hilly streams preferably where boulder bar is not considered feasible. The base width is made 1–1.5 times the height. In that case, no analysis for stability is required. The crest is made either ogee-shaped or broad-crested. The downstream protection against scour should be provided in the form of a few rows of boulder crates (gabions) or CC blocks. These are shown in Fig. 11.2.

11.2.3 Crib Dams

Crib dams are raised crest weirs of small height made of locally available ballies, bamboo and timber. It is economically feasible where timber is available in abundance. These are also made on hard rocky strata. These are made of 10–15 cm diameter tree trunks placed at a spacing of 600–900 mm. These trunks are nailed together to form cages which are filled with boulders/pebbles, etc. The upstream and downstream faces are covered with wooden planks to prevent seepage. The downstream face is made in steps to dissipate energy (shown in Fig. 11.3a). It is also made sloping at a slope of 3H:1V as shown in Fig. 11.3b. In both cases, the river bed beyond the downstream slope is protected either by gabions or CC block.

Sometimes frame-type crib dams are made as shown in Fig. 11.3c. Its upstream sloping face is planked to prevent seepage. The water load on the upstream slope gives stability to the structure. On the downstream, there will be a vertical drop of water so a water pool with a concrete base should be provided as shown in Fig. 11.3c for the dissipation of energy. The river bed downstream of the pool should be protected by gabions or CC blocks.

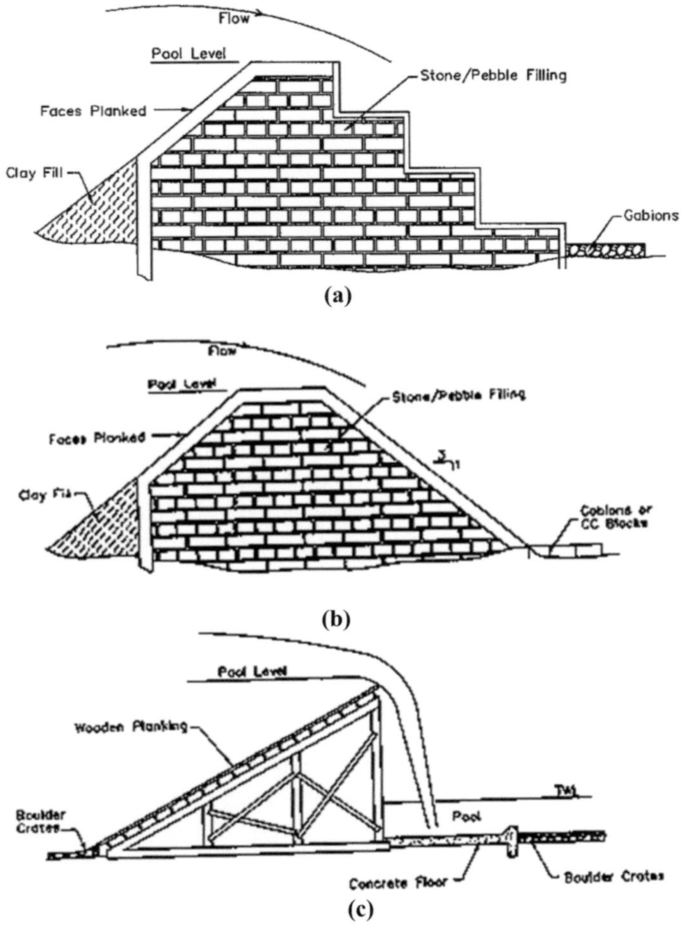

Fig. 11.3 Various crib dams

11.2.4 Hydraulic Design Aspect

The raised crest weirs described above have the crest at the required pond level. The flood is passed over the crest. So the crest length should be adequate to pass the design flood with 2–3 m head of water over the crest. The length of the crest can be worked out by the following relation:

$$Q = C.L.H.^{3/2}$$

where

Q Design flood discharge (cumec).

C Coefficient of discharge (for ogee-shaped crest it is 2.0 and for broad crested weir it is 1.7 when the downstream tail water level is below crest level). If the crest is drowned a reduction factor should be applied.
L Length of weir in metres.
H Water depth over the crest in metres.

Thus, *L* depends on the value of *H* which is decided from the considerations of submergence created upstream and the channel width available at the site of the weir. If the streams are not very wide the crest may be made in the entire width of the stream section. But the crest may be limited to the required length in the central portion of the main channel if the stream width is large. The channel length beyond the crest may be made as the non-overflow section with the top adequately above the maximum water level (crest level + H). Depending on the channel section, the non-overflow section may be provided either on one side of the spilling section or on both sides. Generally, it is made of the same material that is used in making the spilling section, but other types of construction material may be used depending on the availability.

11.2.5 Structural Safety Considerations

The weirs described above are of small height (4–5 m) and these rest on strong rock foundations. When base width more than the height is provided the weir can be considered reasonably stable and structurally safe. Measures for water tightness to prevent seepage through the weir are also provided both on upstream and downstream faces. Some measures as described above are also taken for energy dissipation in the downstream along with river bed protection. Because of the small water head and strong rock foundation excessive seepage through the foundation is not expected.

Another important aspect of weir safety is its junction with the abutments. It should be ensured that the weir is not bypassed during floods and that excessive seepage does not take place through the contact of the weir with the abutment. Hence, the weir crest should be taken inside the abutments for a certain length. At the contact with the abutment the area of the weir section should be increased. To increase the seepage path, the stable slopes of abutment both upstream and downstream of the weir should be made watertight for a certain length and up to a height adequately above the maximum flood level. For this purpose either the shot-creting should be done or the concrete facing of the weir should be extended on both the upstream and downstream slopes of abutments. If abutment slopes are unstable a retaining wall of concrete or masonry may be made. The joint of the weir with the retaining wall should be provided with water stops to check seepage through the contact.

11.2.6 Weir on Pervious/Alluvial Foundation

If a weir is to be constructed on a pervious foundation it should be made safe against downstream scour and piping through the foundation. To prevent damage due to scour a cut-off at the downstream end should be provided. The bottom of the cut-off should be at least up to a depth of 1.5 times the scour depth below the design flood tail water level. The scour depth is worked out as below

$$R = 1.35\left(\frac{q^2}{f}\right)^{1/3}$$

where

- R scour depth in metres.
- q discharge intensity in cumecs (design flood/length of weir).
- f silt factor $= 1.76\sqrt{d_{50}}$, $d50$ is size of the foundation material in millimetres. It should be determined for finer particles in the foundation (after removing boulders and pebbles).

In order to safeguard against piping through the foundation, the base width of the weir should be about 10–12 times the difference between the pond level and the minimum tail water level (Bligh's concept). However, it should also be checked that this floor length of the weir along with the downstream cut-off ensures a safe exit gradient for the type of foundation (as per Khosla theory).

In addition to the above measures the other design aspects described in the preceding paras of the hydraulic design (Sect. 11.2.4) and structural safety (Sect. 11.2.5) should also be adopted for the complete design of a weir on pervious foundation. A typical section of a boulder/rock fill weir on a pervious foundation is shown in Fig. 11.4. As shown in Fig. 11.4, normally a 3–4 m deep concrete open cut-off can be provided downstream. If a deep cut-off is required, then other methods of construction would be required. A 1–2 m deep cut-off is also provided in the upstream. River bed protection in the form of C.C blocks and gabions is provided in the downstream. The length of 1 m thick gabions is made equal to twice the depth of scour below the stream bed. In the upstream of upstream cut off a few rows of C.C. blocks or gabions are placed as river bed protection. The weir is placed on a firm foundation after removing loose river bed material which normally is 1–2 m deep. On the firm foundation base, a 150 mm thick concrete mat is placed and then boulder rock fill is placed to complete the weir. Since the base width is large the concrete or masonry partition walls are made as shown in Fig. 11.4.

A concrete weir can also be constructed on an alluvial foundation. A typical section of such a weir is shown in Fig. 11.5. The depth of cut-off and floor length and thickness, etc. are designed according to Khosla theory which is discussed later with the design of barrage.

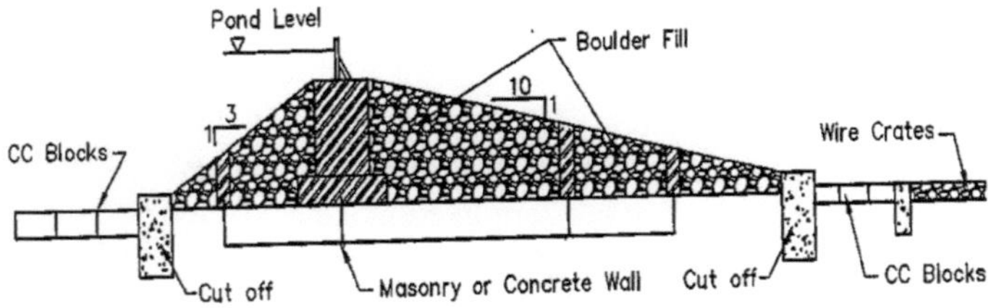

Fig. 11.4 Rock fill weir on alluvial foundation

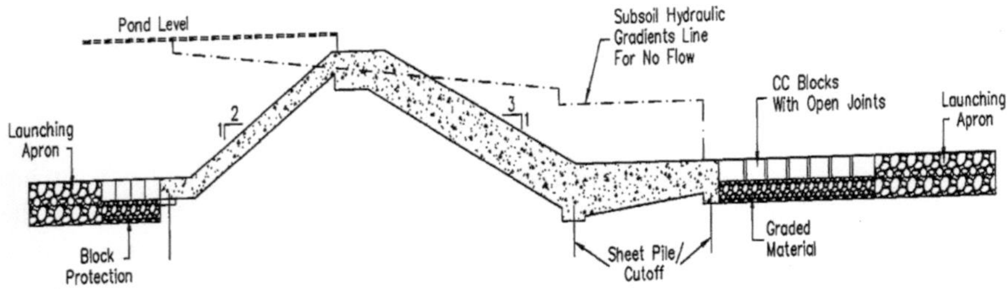

Fig. 11.5 L-section of weir on permeable foundation

11.3 Trench Weir

The trench weir also known as bottom intake or Tyrolean has been extensively used in small hydro schemes on small size steep hilly streams. These are adopted for diverting discharges normally in a range of 0.5–25 cumec for power generation. The trench weir has the following advantages:

(i) It can be constructed on any type of foundation.
(ii) Since it has no structure above the river bed, it is not affected by the action of rolling boulders.
(iii) It does not affect the regime of the stream.
(iv) It does not obstruct the movement of aquatic life.

The trench weir consists of a rectangular/trapezoidal channel made of concrete or stone masonry below the stream bed. The trench is constructed across the flow of the stream and generally in its full width. The top of the trench is normally kept at the river bed and the entire top width of the trench at the river bed is covered by a trash rack. The trash rack prevents entry of trash, boulder and pebbles in the trench.

11.3.1 Trash Rack

Trash rack is provided with a slope generally of 1:10 (V:H) in the direction of flow so that boulders and pebbles may roll down instead of settling on the rack. But the rack needs regular cleaning during and after monsoon. The trash rack is made of bars which are either round bars or rectangular strips with depth more than width. The strips are normally 10–12 mm thick. Clear spacing between the bars is generally kept approximately equal to the width of the bars or the bars are placed with centre-to-centre spacing of 20–30 mm. The bars are fixed in a steel frame and the steel frame is rigidly fixed with the two concrete side walls of the trench on which the trash rack is supported. The trash rack should be made strong enough to withstand the load and impact of rolling boulders. The trash rack is constructed in parts so that a part requiring repair may be easily removed and repaired.

11.3.2 Trench

The rectangular/trapezoidal trench of adequate width is constructed below the stream bed across the flow of the stream in full channel width ending in an intake well on the bank which has the intake structure for the water conductor. The trench is provided with a bed slope between 1 in 10 and 1 in 25 (V:H) in the direction towards the intake well so that sufficient velocity is generated to carry away the sediment that may find entry through the trash rack. The normally recommended slope is 1:12.5 (V:H). The bed profile can be evaluated as described in the hydraulic design of the trench.

The river bed upstream and downstream of the trench is protected against scour. The protection consists of gabions of size 1.5 × 1.5 × 1.0 m covered with 100 mm thick M_{15} concrete on top to protect wire mesh of gabions from abrasion due to rolling boulders. Instead of gabions cc block protection may be provided. The length of the protection should be designed on scour considerations. To safeguard against dislodging of blocks cut-off walls of concrete one on the upstream and two on the downstream should be provided. The depth of cut-off should be 1.5 times the scour depth. The bottom of the trench should also be below scour depth. The typical X-section and L-section of a trench weir are shown in Fig. 11.6 (CBI&P, Pub. No. 280).

11.3.3 Hydraulic Design of Trench

(a) **Width of Trash Rack**

The flow over the trash rack of the trench weir is a gradually varied flow (discharge decreases) and the solution of the flow equation is quite complex. An approximate solution has been obtained on the following assumptions (Singh et al. 1985; Kumar 2000).

- The flow through trash rack is considered a flow-through slotted or round orifice.
- Specific energy is constant along the flow and rack bars are horizontal.
- The flow over the rack is regarded as a flow over a broad crested weir.

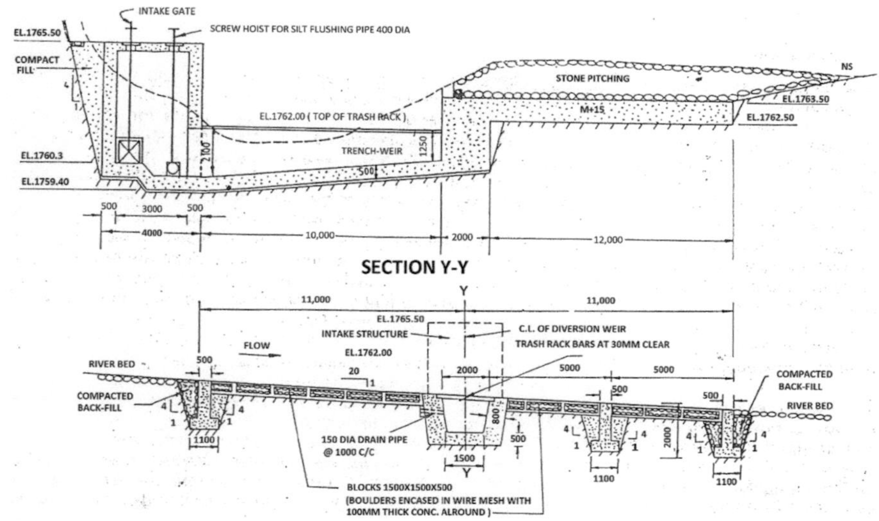

Fig. 11.6 Typical section of trench weir and intake

The width of the trash rack L_o is expressed as

$$L_o = \frac{Q_1}{\varepsilon_0\ \varepsilon_c\ C_d\ b\sqrt{2gE}}$$

where

- L_o Width of trash rack (m).
- Q_1 Discharge diverted (design discharge Q + flushing discharge). $Q1$ is taken as 1.2 Q (cumec).
- $\in_o$ Opening ratio $= \frac{\text{Clear area of trash rack opening}}{\text{Total area of trash rack}}$.
- $\in_c$ Clogging ratio (taken as 0.5).
- C_d Coefficient of discharge (it depends on shape of rack bars which may be trapezoidal, rectangular or round and the spacing of bars). As an average value it is taken as 0.46.
- b Width of stream over which the trash rack is placed.
- E Specific energy $= \left(\frac{Q_1}{C_d'\cdot b}\right)^{2/3}$, $C_d' = 1.7$ (for broad crested weir).

In practice the racks are provided in a width twice the value of L_o worked out from above relation to take care of excessive clogging (higher value of $\in_c$) and reduced effective length of trash rack because after monsoon the channel flow width may be much smaller. The recommended values of trash rack width (CBI & P) for various discharges are as below.

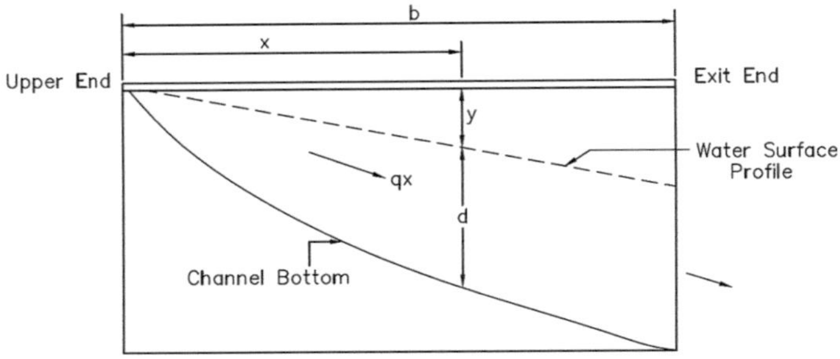

Fig. 11.7 Water surface and bed profile inside the trench weir

S.No.	Discharge (Cumec)	Width of trash rack $L_o(m)$
1	Upto 0.5	1.5
2	1.0	2.0
3	5.0	2.5
4	12.5	3.0

(b) **Size of Trench and Bed Profile**

For determining the bed profile and size of trench, the hydraulic design principle of a side channel spillway is used with simplified assumptions. It is assumed that flow in the trench is uniform in the entire length of the waterway. The discharge at any section at a distance x from the upper end of trench (see Fig. 11.7) is taken as $q \cdot x$ where q is discharge per unit length which is equal to Q_1/b.

The velocity at any distance is expressed as

$$V = a \cdot x^n$$

where 'a' and 'n' are constants and V is average velocity at distance x. The value of n is taken as 0.5 for a straight line water surface profile and a concave bed in the trench. The ordinate y for the water surface measured downwards from the trash rack level is expressed as

$y = \frac{n+1}{n} \cdot h_v$ where h_v is the velocity head at end of trench.

y at the end of the trench is given by

$y = \frac{A}{2T}$ where A is area of trench and T is top width of the water surface

The two expressions give

$$h_v = \frac{n}{n+1} \cdot \frac{A}{2T}.$$

If d is water depth at the end then, h_v can be given by

$$Q_1 = A\sqrt{2gh_v}.$$

Thus, water depth and velocity at the end can be worked out which on substitution to the expression of velocity will give the value of 'a'. Thereafter, y and d can be worked out for any value of x.

The following explains the above procedure

Assume rectangular section with $T = 2$ m for $Q = 3$ cumec and $b = 15$ m

$A = 2\text{d}$, d is depth of water at the end

$$h_v = \frac{0.5}{1 + 0.5} \cdot \frac{2d}{2.2} = \frac{d}{6}$$

$$3 = 2d\sqrt{2 \times 9.8 \cdot \frac{d}{6}}$$

$$\text{or} \quad 3 = 3.6\ d^{3/2}$$

$$\text{or} \quad \text{d} = 0.85 \text{ m}$$

$$\text{y at the end} = \frac{2d}{2.2} = \frac{d}{2} = 0.425$$

$$\text{Total depth at end} = 0.85 + 0.425 = 1.275 \text{ m}$$

$$\text{Velocity at end} = \frac{3}{2 \times 0.85} = 1.76 \text{ m/s}$$

$$\text{Hence} \quad 1.76 = \text{a}.15^{0.5}$$

$$\text{Thus} \quad \text{a} = 0.456$$

Now Q at any distance x can be worked out. Depth can be worked out after working out velocity at that distance which will be equal to $0.456\,x^{0.5}$. Thereafter, depth y can be worked out equal to $\frac{n+1}{n}\frac{v^2}{2g}$. The bed level of trench will be trash rack level—$(d + y)$ at any distance x.

The hydraulic design of a trench weir which is enclosed as Appendix 1 illustrates the procedure of hydraulic design. The computations for the water surface and bed profile levels at different distances from the upper end of trench are given in the enclosed table and their plots are shown in the enclosed figure. This design was also tested on a hydraulic model constructed on a scale of 1:10 at IRI Roorkee. The observed water surface profile is also shown in the figure. The observed velocities have also been compared with the computed values in the enclosed table. The model results are found comparable with the design values within the limits of observational errors. This confirms the applicability of the above design procedure.

In practice, it is experienced that it is not practically possible to construct a curved bed profile. Hence in practice, the required bed profile is approximated with at least two straight lines of different slopes, one near the upstream end and the other in the rest of the length of the trench as shown in the figure enclosed in Appendix 1.

11.3.4 Structural Design Considerations

The trench is an RCC structure made in M_{20} concrete. For its design, the loads and forces corresponding to HFL condition should be considered. The bottom slab may be designed as a slab fixed with the side walls. The side walls may be designed as cantilever walls.

11.3.5 Intake Well

The intake structure is generally a vertical rectangular well located at the end of the trench on the river bank. It has three openings. One is at the end of trench through which the flow from the trench enters the well. The other is to feed the power channel so its location should match with the power channel. The third is at the bottom of the well to flush the sediment. All three openings are controlled through gates operated from the top of the well which is sufficiently above the HFL. The bottom of the well is kept 0.5 to 1m below the bed level of trench at the end so that sediment may settle to the bottom of well and may be flushed through the flushing conduit which should discharge in the river downstream of the trench weir above the HFL.

The size of the well should be adequate to create stilling of the flow entering through the trench so that sediment may settle before the flow reaches the power channel outlet.

The well is generally an RCC structure and the walls of the well should be designed for the water pressure from inside assuming no backfill as well as for the pressure from the outside back fill when well is empty. The top of well should be covered with a slab. It should have a manhole. The typical details of a trench weir intake structure are shown in Fig. 11.6.

11.4 Other Types of Weirs

Other types of weirs which are used for SHPs are briefly described below. However, their experience in Indian conditions is very limited.

11.4.1 Conda Weir

It is an ogee-shaped concrete weir in which a self-cleaning screen (known as Conda-effect screen) is fitted on the downstream slope of the weir between the crest and horizontal floor for energy dissipation. The screen filters sediment from the flow and comparatively sediment-free water drops into the hollow space made in the weir under the crest and sediment flows down the screen to the river channel. Clean water from the hollow space is taken out through a pipe to the power channel. The arrangement is shown in Fig. 2.4. It has been used for SHPs in the USA, Canada, etc.

The observations have shown that the screen opening of 1mm can exclude 95% of 0.5 mm size sediment particles whereas a screen with an opening of 0.2 mm has an efficiency of 90% for particle size 0.1 mm. The head loss in these screens is very high (1–1.5 m). The one major disadvantage of this type of weir is that total non-monsoon flow cannot be diverted for power generation through the Conda screen.

11.4.2 Rubber Dams

These are the weirs of small height (up to 5 m) and are made of inflated rubber or canvas bags. These work like falling shutters on a weir. The rubber bags are anchored to a concrete base across the stream which is to be dammed. In the inflated condition, the rubber dam holds water upto the proposed pond level like a gate. In deflated conditions, the rubber bag falls on the base slab to pass the river flow just like a falling shutter.

The rubber dams are fixed on the concrete base by either single line anchoring or double line anchoring as shown in Fig. 2.5. Single line anchoring is provided in case of a small head of water and double line for a higher head of water. Strong anchoring is very important for stability. Strong J bolts are embedded in concrete base upto sufficient depth to provide anchorage. Anchorages are also carried on the side, which may be inclined or vertical. The rubber bags are inflated either with air or water. Water-filled dams are more stable and free from pulsations and vibrations.

Rubber dams are made of canvas of nylon fibre with protection layers of rubber or neoprene hot pressed with synthetic rubber. The rubber sheets are available in rolls and can be installed without an intermediate structure up to a length of 150 m but a cost-effective span is about 40 m. The rubber dam in inflated and deflated condition is shown in Figs. 2.5 and 2.6.

If the river carries heavy bed load of large size particles the rubber dams are prone to heavy damage and would need frequent maintenance and repair. But any repair underwater is difficult. The experience of such rubber dams in India is still very limited.

11.5 Barrages

The barrages are also used as diversion structures for SHPs. The barrages are low-crested weirs with regulation gates over them. The gates can regulate the passage of river flow which is not possible with a weir. A barrage is generally provided when river flow is large and correspondingly the diversion discharge is also large, the river

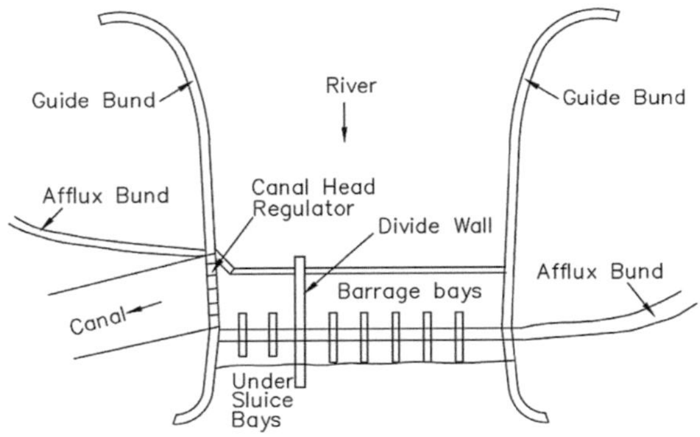

Fig. 11.8 Typical plan of a barrage

slope is flat (say 1 in 100) and an extra head is required to be created through high gates. The provision of high gates helps in creating storage required for peaking purposes which is not feasible with a weir. A typical plan of a barrage showing its various components is shown in Fig. 11.8.

Generally, in the case of barrages in wide alluvial reaches, the waterway provided is much less than the river section. In that case, the river flow is guided to the barrage through the guide bunds as shown in Fig. 11.8. The waterway is also divided into barrage bays and the undersluices and the two are separated through a divide wall which generally extends up to the upstream end of the canal intake. The undersluices and divide wall help in proper flow regulation in diverting water with less sediment into the canal intake. Undersluices have a crest lower than the crest of barrage bays and have the capacity to pass about 20% of the design flood or twice the canal discharge whichever is more. For the movement of fish in the river across the barrage a fish ladder is provided generally by the side of the divided wall.

The design of a barrage has practically been standardised and for details, any book on Irrigation (Singh 2005; Varshney et al. 1993) may be referred. For the design of various components of a barrage, I.S. codes are also available at BIS, New Delhi. The data required for design and the design concepts are briefly given in the following sections.

11.5.1 Data Required for Design

The following data is normally required for the design of a barrage:

- Contour plan of barrage site, river cross section and L-section of river.
- Borehole log chart and soil investigations.
- River discharge data to estimate water availability and design flood.
- Stage discharge curve at barrage site (can be developed theoretically using Manning's equation, if observed data is not available) which is essential for the design of energy dissipators.

- Details of floodplain, submergence and existing structures that may be affected due to ponding.
- Pond level: It is canal full supply level + working head. Sometimes upstream conditions limit pond level. In that case pond level minus working head will be canal FSL. A high pond level even above HFL may be adopted to increase head for power generation.
- Afflux: It is an increase in HFL in the upstream due to reduction in waterway. Generally, in the upper reaches of a river, it is taken as 1.0 m but in lower reaches of a river (plains), it is limited to 0.3 m. It also depends on river conditions upstream.
- Retrgression: It is the degrading of the river bed downstream due to the construction of a barrage. It lowers the river stage downstream. A value of about 1.0 m is adopted in alluvial reaches at the low river stage and 0.3–0.5 m at design flood.

11.5.2 Design Flood

For the waterway and design of other components, a design flood of 50 year return period is adopted. Depending on the importance and risk it may be increased to a 100-year return period or to a standard project flood (IS: 11223). For freeboard, the design flood should be a flood of 500 years return period.

11.5.3 Waterway

In the case of a meandering alluvial river, a waterway is provided using a looseness factor to Lacey's perimeter which is worked out by

$$P = 4.83\ Q^{1/2}$$

where Q is design flood (cumec) and P is Lacey's perimeter (m)

Silt factor	Looseness factor	Water way
<1	1.2–1	1.2–1 times P
1–1.5	1–0.6	1–0.6 times P

Silt factor $f = 1.76\ d_{50}^{1/2}$ where d_{50} is average particle size of river bed material in mm.

In boulder reach of a river where the channel is confined between the banks, the waterway is provided equal to the channel width at design flood. In many cases, it ranges from 0.25 to 0.9 times Lacey's P.

The waterway is divided into bays separated by piers. Bay size depends on the size of gate. The maximum span of the gate used so far is 18 m. The under-sluice bays cater for 15–20% of the design flood. Barrage bays and undersluice bays are separated by a divide wall.

11.5.4 Crest Level

In the case of undersluices the crest level is kept as near to the deepest level of the channel as possible. The barrage bay crest level is kept 1–1.5 m above that of the undersluices. The crest level of both the barrage bays and the undersluice can be determined by using the broad crested formula given below in which the coefficient of discharge may be taken as 1.7.

$$Q = 1.7\ (\mathrm{L} - 0.1\ \mathrm{nH}) \cdot H^{3/2}$$

where

- L Total clear water way in m.
- n Number of end contractions.
- H Head over crest in m.
- Q Design discharge in cumec.

In case of submergence of crest C (coefficient of discharge) may reduce to 1.18. For the value of C at different submergence ratios, the curve given in Fig. 11.9 may be used.

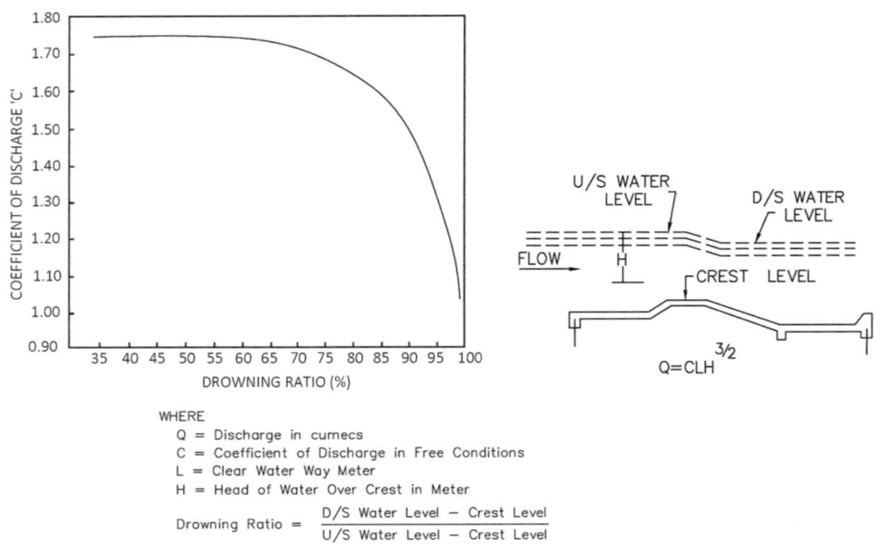

Fig. 11.9 Drowning ratio versus coefficient of discharge curve (in MKS Units) (based on Malikpur Curve) applicable in broad crested barrages

11.5.5 Shape of Barrage Crest

Generally upstream floor level is at the average bed level and the downstream floor level is determined from the energy dissipation consideration. The crest is provided flat with a width of about 2 m or more depending on the gate and stop log requirement. The top of the crest is joined with the upstream floor by a slope of 2:1 and with the downstream floor by a slope of 3:1 (H:V). Steeper slopes may be provided if the barrage is on a rocky foundation.

A typical section of barrage on the alluvial foundation is similar to that of a weir on the alluvial foundation (refer Fig. 11.5) except that the crest is kept at or near the river bed and the pond level is maintained by gates provided between the piers instead of shutters.

11.5.6 Floor Length

The total imperious floor length is worked out from exit gradient considerations in case the barrage is on the alluvial foundation

$$G_E = \frac{H}{d}\frac{1}{\pi\sqrt{\lambda}}$$

$$\lambda = \left(1 + \sqrt{1+\alpha^2}\right)/2 \text{ and}$$

$$\alpha = \frac{b}{d} \quad \text{(For values of } \alpha, \lambda \text{ curves (Khosla curves) are available)}$$

where H is maximum static head (pond level—minimum tail water level), b is total floor length, d is depth of d/s cut-off and G_E is safe existing gradient which varies with the type of river bed material. The recommended values of G_E for different river bed materials are as below

Shingle	1/4 to 1/5
Coarse sand	1/5 to 1/6
Fine sand	1/6 to 1/7

Knowing H and d, b can be worked out for a safe exit gradient.

The downstream floor beyond the glacis end is provided as per the jump basin requirement for energy dissipation. The rest of the required length of the floor is provided in the upstream. The jump basin is normally designed in accordance with I.S. 4997.

11.5.7 Divide Wall

A divide wall is provided between undersluice bays and barrage bays to check cross flow which is caused by the difference in crest levels of undersluice and barrage bays. It also helps in maintaining a still pocket in front of the head regulator which is required for canal operation from sediment entry considerations. Its length in upstream generally covers 2/3rd or the full length of the canal regulator. In downstream it extends up to the end of the floor.

11.5.8 Cut-Off

Cut-offs are provided at the end of the floor both in its upstream and downstream. In alluvial reach, cut-offs are provided through sheet piles but in boulder reaches these are made of concrete. The depth of cut-offs is based on the depth of scour R which is worked out by

$$R = 0.473\left(\frac{Q}{f}\right)^{1/3} \quad \text{where looseness factor is more than 1}$$

$$R = 1.35\left(\frac{q^2}{f}\right)^{1/3} \quad \text{where looseness factor is less than 1}$$

where Q is the design flood and q is discharge intensity per metre width and f is silt factor.

The upstream cut-off is designed for 1.0 R and the downstream cut-off for 1.25 R. These values are subtracted from HFL to get the bottom level of cut-offs. The depth of downstream cut-off may be increased from safe exit gradient consideration to adopt an economical b/d ratio.

In case of barrages on rocky foundations, it is not possible to estimate the scour depth so cut-offs are normally provided of depth say 1–3 m below floor level. The floor length requirement from exit gradient consideration in this case generally becomes longer than usual. Hence the length beyond the downstream jump basin requirement is provided in the upstream of the crest. The river bed protection longer than required from scour consideration (described below) is also provided on the downstream to take care of the river bed scouring.

11.5.9 Protection Works

In upstream and downstream of the impervious floor, river bed protection works as shown in Fig. 11.5 are provided. These comprise CC blocks of adequate size so that they are not dislodged during floods. Generally, blocks of size $1.5 \times 1.5 \times 1.0$ m in a length approximately equal to D and 1.5D (D is the depth of scour below the floor level corresponding to 1.5 R and 2.0 R) are provided upstream and downstream of the floor respectively. The downstream blocks are placed over a graded inverted filter conforming to the following filter criteria:

$$\frac{D_{15}\text{ of filter}}{D_{15}\text{ of foundation}} \geq 4 \geq \frac{D_{15}\text{ of filter}}{D_{85}\text{ of foundation}}\ \text{(Filter may be laid in two or three layers).}$$

The downstream blocks shall be laid with open joints to be filled with shingle. Toe walls shall be made at the end of block protection.

Beyond the block protection, in the case of barrages/weirs located on an alluvial foundation, a launching apron made of loose boulders or stones is placed. The stone or boulder shall not be less than 300 mm in size and no stone shall weigh less than 40 kg. The quantity of stone shall be adequate to cover the slope of the scour hole which may develop due to scouring in the river bed. Its slope shall be assumed not steeper than 2:1 nor flatter than 3:1 (H:V). The latter should be adopted for rivers with fine sand. The thickness of the stone covering the slope of the scour hole shall generally be 0.6–1.0 m. More thickness shall be provided in finer river bed material. The quantity of stone required shall be placed on the river bed in a length of 1.5 D with increasing thickness towards the outer edge (For details refer IS: 6966).

11.5.10 Hydraulic Design of Floor

The floor shall be designed for both surface and subsurface flow. For surface flow a stilling basin in the downstream with adequate length with appurtenances is generally provided, from energy dissipation considerations. The type of stilling basin required to be provided depends on the Froude number $\left(\frac{v}{\sqrt{gd}}\right)$ which is generally low in the case of barrages (around 2.5). IS: 4997 shall be followed for the design of the stilling basin. The design shall be carried for a discharge intensity with a 20% increase for concentration.

The subsurface flow through the foundation in ponded conditions causes uplift force at the bottom of the downstream floor. In the case of barrages on permeable foundations of infinite depth, the uplift at various locations is worked out by Khosla's theory taking into effect the cut-offs, thickness of floor and slope of glacis, etc. Any standard book (Singh 2005; Varshney et al. 1993, etc.) on irrigation structures may be referred to for details. In case the depth of the permeable foundation is limited, the uplift pressure may be computed by FEM or electrical analogy model or by the method given in CBI & P, Pub. No. 179.

11.5.11 Structural Design

(a) **Floor**

The upstream floor is not subjected to any differential force, so it is generally laid in 10 m × 10 m panels of 0.5 m thick plain concrete. Waterstops are provided between panels to make the floor impervious.

The downstream floor is designed for uplift pressures. There are two ways to design it. One is to make it a gravity floor of plain concrete with a thickness adequate to counterbalance the uplift force. The other is to make the floor a RCC raft. The floor as a raft shall be assumed supported on piers, which are extended up to the end of the floor in this case, and may be designed as a continuous beam on several piers loaded with uplift pressure from the bottom. Depending on the spacing of piers alternate barrage piers may be extended in specific cases. The continuous beam span is also limited because of loading, the depth of the beam and the amount of reinforcement required. Normally a raft of 2 m thickness is provided with 1% steel. The downstream floor is normally covered at the top with 0.3 m thick rich concrete to resist abrasion during floods. Nominal temperature reinforcement is provided at the top of this concrete.

In case of barrages on rocky foundation, the downstream floor may be made 1.0 m thick of plain concrete which shall be adequately anchored to the foundation rock through rock bolts. The anchorage depth should normally be determined on the basis of pull-out tests. Generally, 25 m Ø bars are used as anchors. The spacing is generally half the embedment.

(b) **Divide Wall**

It is generally a RCC wall about 1.5–2 m thick. It is designed as a cantilever wall for differential head which is caused by the difference in water level on its two sides. It is maximum when the barrage bay adjoining the divide wall is open and the undersluice bay is closed.

The upstream and downstream noses of the divide wall should be protected against scour by placing CC blocks and launching apron as it is done for the floor. The nose should also be protected from abrasion. In alluvial reaches where the bearing capacity of the foundation is low, the divide wall is placed on the well foundation. In other cases, a raft foundation below the floor may be provided for the divide wall.

(c) **Piers**

The piers are of RCC and these provide side support to gates. Since these contain gate groves, their thickness is normally 2 m or more. Water load through gates is transferred to piers. The water load acts as a horizontal force along the length of the pier and from the other direction the pier is acted upon by the differential water pressure which is caused when one side gate is open and the other side gate is closed. It is also acted upon by the forces transferred by the bridge which is supported on the top of piers. The pier transfers all the vertical and horizontal forces to the foundation.

The pier is designed as a cantilever column fixed at the base. For detail design I.S. 13551 may be referred.

If the bearing capacity of the foundation strata is good, the piers may have independent footings below the crest. The other alternative is to fix the pier with the concrete of the crest. In that case, the crest is suitably reinforced at the top to distribute the loads that are transferred through the piers.

11.5.12 Special Precautions in Boulder Stage

Some of the precautions that shall be taken in case of a barrage in boulder stage river are given below:

- High crest shall be avoided.
- Chute blocks and friction blocks in the stilling basin shall be avoided.
- The floor shall be protected against abrasion due to rolling boulders either by providing high-performance concrete on top or by anchoring rails on top of the floor.
- The pier noses shall be protected by steel cladding against the damage due to rolling boulders.

Annexure 1

Data

Design discharge = 2.8 cumec

Width of stream and trash rack = 14 m

Opening ratio $\in_o = 0.49$ (rack is made of 25 mm ϕ bars @ 30 mm)

Clogging ratio $\in_C = 0.5$

Slope of the rack = 1 in 12(same as the slope of river bed)

$$C_d = 0.47$$

Width of Rack

$$\text{Specific energy } E = \left[\frac{Q}{C'L}\right]^{2/3} = \left[\frac{2.8}{1.7 \times 14}\right]^{2/3}$$
$$= 0.25 \text{ m}$$

$$\text{Width of rack } L = Q/\in_0 \cdot \in_C \cdot C_d \cdot b\sqrt{2gE} = 0.8 \text{ m}$$
$$\text{The width of trash rack adopted} = 2.0 \text{ m}$$

Bed and Water Surface Profile in Trench

Top width of trench = 2.0 m (assume rectangular section)

$$\text{Length of trench} = 14.0 \text{ m}$$
$$\text{Design Discharge} = 2.8 \text{ m}$$
$$\text{Discharge intensity} = 2.8/14 = 0.2 \text{ cumec/m}$$
$$\text{Area A} = \text{Td} \quad \text{where, T} = 2 \text{ m and d is depth of water}$$
$$\text{Velocity head } h_v \text{ at end of trench is worked out by}$$

$$\begin{aligned} h_v &= (n/n+1)(\text{A}/2\,\text{T}) \\ &= (0.5/0.5+1)(2\text{d}/2.2) = \text{d}/6 \text{ (assuming } n = 0.5) \\ \text{Q} &= A\sqrt{2\text{gh}_\text{v}} \\ 2.8 &= 2.\text{d}\sqrt{2 \times 9.81 \times (\text{d}/6)} = 3.617\text{d}^{3/2} \\ d &= 0.843 \text{ m} \end{aligned}$$

Velocity at exit and = (2.8)/(2 × .843) = 1.66 m/s

$$V = a \cdot X^n$$
$$1.66 = \text{a}.14^{0.5}$$

Therefore, a = 0.444 Hence, V at any distance $X = 0.444\, X^{0.5}$
Y at any distance X will be

$$Y = \frac{n+1}{n} \cdot h_v = (1.5/0.5)\ h_v = \left(3\text{V}^2/2\text{g}\right) = 0.153\ \text{V}^2$$

For different values of X, Q and V, the values of d and y can be worked out to plot the water surface and bed profile of the trench. The computations of y and d at one distance, i.e. $X = 9$ m are given below

$$\begin{aligned} X &= 9 \text{ m}, \ Q = 28 \times 9/14 = 1.8 \text{ cumec} \\ V &= 0.44\ \sqrt{9} = 1.33 \text{ m/s} \\ Y &= 0.153 \times 1.33^2 = 0.26 \text{ m} \\ A &= \frac{Q}{\text{V}} = \frac{1.8}{1.33} = 1.35 \text{ m}^2 \\ A &= 2.\text{d} = 1.35 \\ d &= 0.68 \text{ m} \end{aligned}$$

This y and d will give water and bed profile at $X = 9$.

The plots of water surface and bed profiles are shown in the enclosed figure. The design was tested on a 1:10 scale model for water profile and velocities. The observed values are compared with theoretical values. The variations are within the limit of observational errors. This establishes the adequacy of the theoretical design concept.

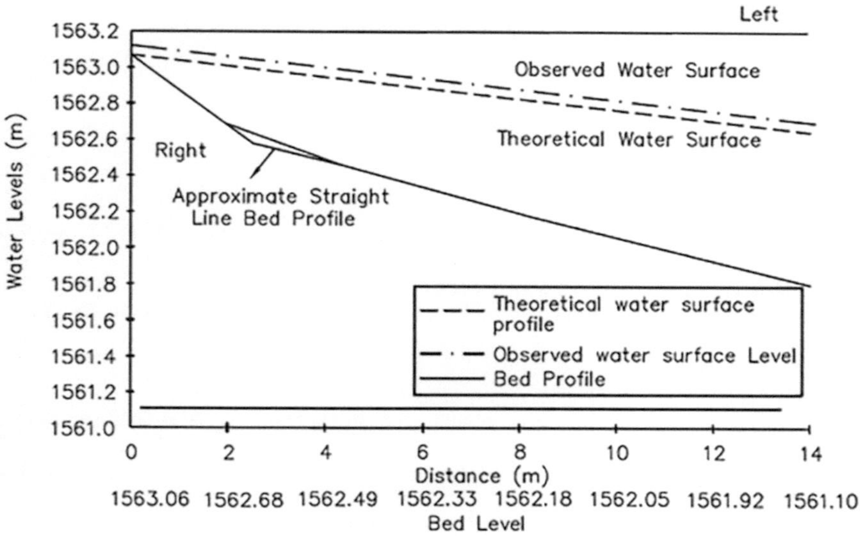

Bed and Water Surface Profile of Banu Trench Weir

Velocities in Banu Trench Weir at Design Discharge 2.8 cumecs

Distance along the trench in metres (right to left)	Particulars	
	Velocity in m/s (observed)	Velocity in m/s (theoretical)
1	–	0.444
2	1.0	0.627
3	0.885	0.769
4	0.885	0.888
5	0.885	0.993
6	0.885	1.09
7	0.885	1.17
8	0.90	1.26
9	1.084	1.33
10	1.252	1.40
11	1.328	1.47
12	1.468	1.54

(continued)

(continued)

Distance along the trench in metres (right to left)	Particulars	
	Velocity in m/s (observed)	Velocity in m/s (theoretical)
13	1.77	1.60
14	–	1.66

References

CBI&P.: Manual on barrages and weirs on permeable foundation. Pub. No. 179 Vol-1 CBIP, New Delhi (1985)

CBI&P.: Manual on planning and design of SHP. Pub. No. 280, CBI&P, New Delhi (2001)

Guidelines for Small Hydro Development, Standard No. 2.2 and 2.3.: Civil Works—Hydraulics and Structural Design. AHEC, I.I.T., Roorkee (2013)

I.S. 11223.: Guidelines for fixing spillway capacity. BIS, New Delhi (1985)

I.S. 13551.: Structural design of spillway pier and crest. BIS, New Delhi (1992)

I.S. 4997.: Criteria for design of hydraulic jump type stilling basin with horizontal and sloping apron. BIS, New Delhi (1995)

I.S. 6966 (Part-I).: Guidelines for hydraulic design of barrages and Weirs-Alluvial reaches. BIS, New Delhi (2005)

I.S. 11130.: Structural design of barrages and weirs. BIS, New Delhi (2005)

Kumar, R.: A study of diversion structures with special reference to trench weir. M.E. Dissertation, WRDTC. University of Roorkee, Roorkee (2000)

Singh, R., Mahajan, S.K.: Hydraulic Design of a Trench Weir. Published in 52nd R and D Session, CBI&P, New Delhi (1985)

Singh, B.: Fundamental of Irrigation Engineering. New Chand and Sons, Roorkee (2005)

Varshney, R.S., Gupta, S.C., Gupta, R.L.: Theory and Design of Irrigation Structures, vol. II (Canal and Storage Structures). Nem Chand and Sons, Roorkee (1993)

Chapter 12
Design of Dams

12.1 General

The art of designing and constructing dams is as old as the practice of harnessing river flows for domestic purposes and for irrigating crops. The evidence and records show that till the twentieth century small-height earth embankments and masonry walls were constructed to store monsoon flows and their design approaches were empirical and ad hoc. In the twentieth century, the design and construction technology for dams developed fast and a large number of high concrete and embankment dams of different types have been constructed in many countries of the world. The design procedures have been rationalised and largely standardised and construction has been mechanised. In India, dam construction has been taken up on a large scale after independence and many high- and medium-height dams of concrete gravity type and embankment type have been constructed for irrigation and hydro power development. A large amount of literature is available regarding all aspects of planning, design and construction for various types of dams in the form of books, manuals, journals and codes. Some of these are given in reference and may be referred to for details.

For small hydro projects, dams are required either for diversion purposes or for storing monsoon flows to be used during non-monsoon periods when rivers carry insignificant flows, especially in South India. The diversion dams generally provide an extra head for power generation and some storage for the peaking operation of the power house. The choice for a diversion dam depends on topography and the geology of the site and where other types of diversion structures are not feasible. Commonly, the diversion dams are concrete/stone masonry gravity dams. Arch dams may be constructed if abutment rock is competent to take the water thrust which is transferred to abutments due to arching action. Other kinds of dam have also been used in SHP projects.

Storage dams, to store monsoon flows, are generally provided on rivers of South India where non-monsoon river flows are insignificant. These rivers are generally wide and their banks are not very steep. In such cases, the storage dams are generally

B. N. Asthana, *Planning and Design of Small Hydro Power Projects*,
https://doi.org/10.1007/978-3-031-96271-4_12

embankment type or composite type. The embankment dams are either earthfill or rockfill type covering the entire river section with a spillway either on one side on an abutment or separate from the dam. In a composite dam, the spillway is provided in the centre of the main river channel and an embankment is provided on both sides of the spillway to cover the wide river section. The spillway is of the same type as is provided in a gravity dam and the embankment on its sides is either an earthfill or rockfill dam.

The different types of concrete and embankment type's diversion dams and storage dams described above which are normally of low height (up to 30 m) are shown in Figs. 12.1 and 12.2, respectively.

The layout and design aspects of the above type of dams which are normally used for SHPs are briefly described in the following sections. For a detailed study, the references given in the bibliography may be referred. The figures and other information used in this chapter are generally based on the reference of 'Dam Engineering'.

12.2 Concrete/Masonry Gravity Dams

The section of a gravity dam is designed in such a way that the entire section is under compression under all load combinations. Hence, the compressive strength of construction material is important. Concrete can be designed for any strength so concrete gravity dams can be constructed for large heights. The highest concrete gravity dam is 284 m high (Grande Dixence in Switzerland). In India, the highest concrete gravity dam is Bhakra (226 m). The strength of stone masonry dams depends on the type of rock used in construction and so these are limited to a maximum height of around 100 m. Since the diversion dam of an SHP is generally of a height of less than 30 m, it can be constructed of both concrete and stone masonry. The merits and demerits of the two are given below.

The advantages of masonry dam over concrete dam are as follows:

(i) No elaborate construction equipment is required.
(ii) No arrangement of aggregate cooling is required.
(iii) No requirement of shuttering.
(iv) Joints are less.

The demerits of masonry dams are as follows:

(i) These are labour extensive.
(ii) These need longer construction period.
(iii) These are less impermeable than concrete.

The advantages of concrete dams are their impermeability, smooth finish for the surfaces of high-velocity flow and quicker construction.

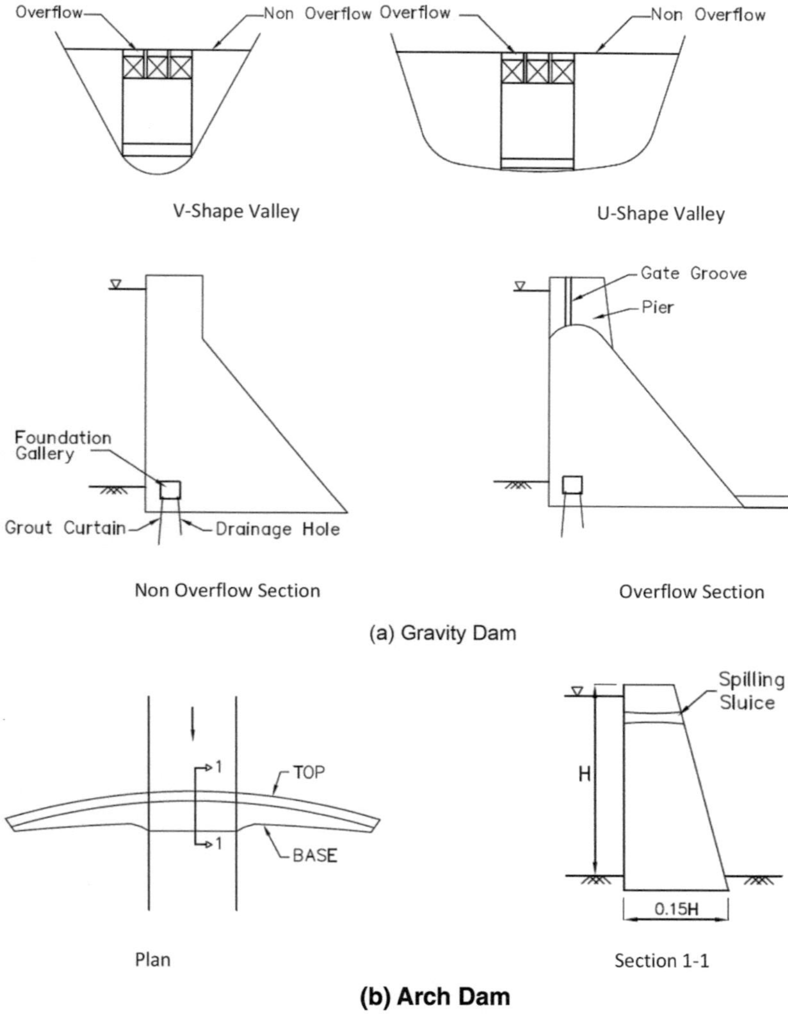

Fig. 12.1 Concrete dams

The selection of the type of dam for a project depends on the consideration of the above merits and demerits as well as on the availability of construction material, resources (men and machine) for construction, time for construction and the economy. Generally, concrete gravity dams are preferred in the Himalayas where the rivers flow in narrow sections within high abutments.

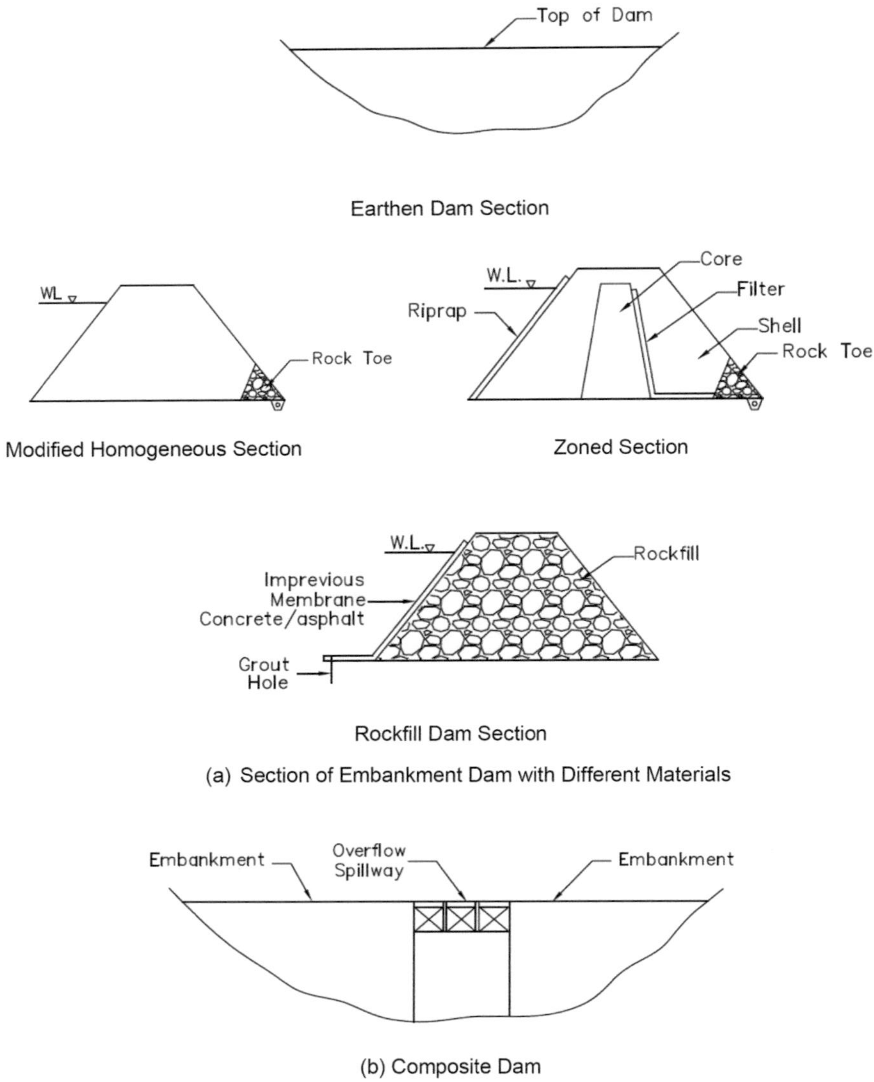

Fig. 12.2 Embankment dams

12.2.1 Selection of Dam Site

The following are the considerations for a good dam site:

(i) The river section at dam site should be narrow for minimum length of the dam and the valley should be wide in the upstream of the site.
(ii) The foundation rock should be strong and of less permeability.
(iii) The abutment should be high enough and the slopes should be stable.

(iv) Construction material such as sand and coarse aggregate of desired quality and in required quantity should be available at a close distance for a concrete dam and rock for a masonry dam.

(v) Availability of infrastructural facilities such as the approach to site, communication, power supply and nearness to the township is important because these aspects reduce project cost and time requirement to complete the work.

(vi) The submergence should not affect important structures of historical, religious and high economic value such as mines and factories.

(vii) Reservoir rim should be geologically stable.

12.2.2 Planning of a Dam

For proper planning and selection of a dam site, topographical surveys, hydrological data collection and analysis, geological mapping and geotechnical investigations and material surveys are required. These are briefly described below and for details relevant I.S. codes may be referred (some are given in references). Readers may also refer Chaps. 6 and 7.

(i) Topographical surveys: Contour plan of the reservoir area up to 5 m above FRL/MWL should be prepared. It shall give details of area of submergence. The submergence survey is helpful in finalising the FRL and in computing storage capacity up to FRL.
River L-sections and X-sections should also be prepared. These can be used to develop gauge discharge curve. The X-section at dam site will be helpful in dimensioning the dam.
A reference bench mark shall be established at the dam site which shall be used for layout of works during construction.

(ii) Hydrological data: The long-term discharge data at the dam site is required to work out the water availability for utilisation, design flood and diversion discharge.
If the long-term observed discharge data is not available, it should be computed from the available discharge data of other sites either on the same river or from adjoining catchment. Rainfall and discharge data can be obtained from IMD. It is advisable that the gauge discharge site for observations should be established at the proposed site at the first opportunity if it does not exist. In addition to discharge data, sediment data should also be measured or collected from other sources.

(iii) Geological mapping: The geological map which is prepared from the visual inspection of the area and the available information shall give the idea of the overall geology of the area and the geology of the site with the details of the expected rock type and its extent as well as the presence of geological structures such as major faults and shear zones.

(iv) Geotechnical investigations: These are carried out through bore holes on the dam axis and in the area of the spillway and the intake structure. The borehole data will give the details of the extent of over burden on rock, rock type, its soundness on the basis of N values and recovery data and the permeability based on water tests. The shear strength parameters of rock are obtained from laboratory tests on rock samples.

(v) Material surveys: The natural river bed material near the dam site should be investigated for the required quantity of fine and coarse aggregate and their quality. If natural material is not available, then the rock of desired quality should be searched near the dam site and the quarry areas should be identified. The quarried rock should be properly crushed and screened before use.

Since quarried material is costly, it should be used as minimum as required for mixing with natural material to make up for the deficiency in quantity and quality of the natural material. The quality of the aggregate should be tested as per standard procedures conforming to the relevant I.S. codes.

12.2.3 Layout of Dam

(i) **Dam Axis**

The dam axis is commonly defined as the line of the upstream face of the dam if it is vertical, otherwise the centre line of the road bridge on the top of the dam is taken as the dam axis. It may be made slightly curved towards upstream to take advantage of topography to shorten the length or to build the dam on geologically sound rock. The axis shall be selected with the following considerations:

(i) Entire dam base is on one type of rock.
(ii) Foundation excavation is minimum.
(iii) Foundation treatment is minimum.
(iv) Foundation rock does not slope in downstream direction.
(v) Abutments are stable with minimum overburden.

(ii) **Dam Top**

The top of the dam is fixed after adding the freeboard to the FRL if the MWL and FRL are the same. The freeboard required is a function of the wind velocity and the fetch (reservoir length perpendicular to the dam axis). The criteria and methodology for calculating the freeboard are given in I.S. 6512. The wave height for dams above 30 m height may be worked out by the following expression:

$$h_w = 0.0032\ \mathrm{FV}$$

where h_w is wave height (m), F is fetch (km) and V is wind velocity (km/h). The recommended freeboard is 1.33 times the wave height corresponding to the design reservoir elevation plus the wind setup which is a rise in still water against the u/s face and is worked out by

$$S = \frac{V^2F}{62000\,D}$$

where S is wind setup (m) and D is average depth of reservoir (m). S is generally small in small reservoirs. However, the minimum freeboard above MWL should be 1.0 m. The I.S. code also prescribes that if the design flood is not the PMF, then top of dam shall not be lower than the reservoir level corresponding to PMF. It also recommends that a 1.0 m high solid parapet wall shall be provided in all concrete gravity dams and it should not be considered as part of freeboard.

(iii) **Dam Section**

The section of a dam, cut by a plane perpendicular to the dam axis, is a solid triangular section. Theoretically, the thickness at top is zero, but it is widened to accommodate the road bridge and other requirements of operation and construction. Generally, the minimum top width is 3.0 m. This width is provided either up to the FRL or up to a depth equal to the width. Thereafter generally upstream face is vertical up to the base and in downstream the top is connected to the base width required from stability consideration giving a slope that commonly ranges between 0.70 and 0.85: 1 (H:V) (refer Fig. 12.1a). Sometimes a slope is provided in the upstream face of the dam either in part of the height or at full height of dam. The weight of water on the sloping portion upstream adds to the stability of dam. This slope varies normally from 0.1 to 0.4 H:1V.

(iv) **Length of Dam**

The length of the dam is the distance between the two abutments at the top level of dam. The dam length is divided into a number of blocks which are separated by contraction joints. Each block behaves under water pressure as an independent cantilever. The joints are provided with water-stops (I.S. 12200) on the upstream to prevent seepage. The blocks are made 12–18 m wide to prevent cracking due to thermal stresses. These are shown in a typical plan of a concrete gravity dam in Fig. 12.3.

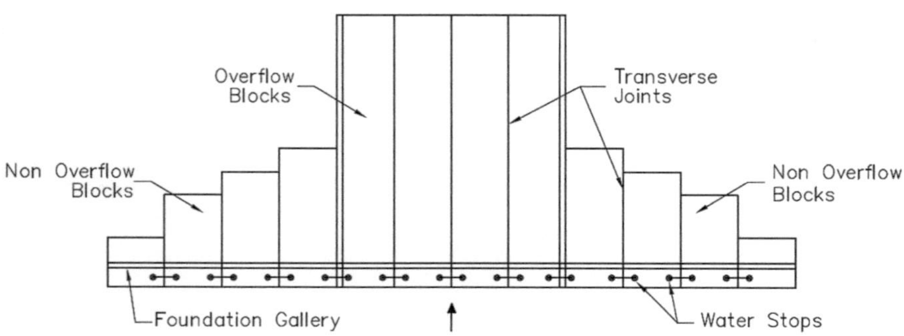

Fig. 12.3 Typical plan of gravity dam

The length of a concrete gravity dam is generally divided into overflow and non-overflow parts. The overflow portion is also called the spillway which discharges excess river flow downstream and is normally located in the centre of the main river channel. The non-overflow part is located on either side of the spillway extending up to the abutments. Some details of these two portions are described below.

(v) **Overflow Section (Spillway)**

(a) *Shape*

The typical section of the overflow section is shown in Fig. 12.1a. The overflow section is generally an ogee-shaped crest conforming the nappe of a free-falling jet over a sharp crested weir to avoid the creation of excessive negative pressures on the concrete surface. The shape of the ogee from top of crest on the downstream side is a parabola and it is tangentially joined with the downstream edge of the base width. The shape of the ogee upstream of the crest is a circular curve either of a single radius or a compound curve. The definition sketch is shown in Fig. 12.4 where the downstream portion is defined by the equation

$$\frac{y}{H_o} = -K\left(\frac{x}{H_o}\right)^n$$

where K and n are constants whose values depend on slope of the upstream face and on the velocity of the approach. The values of K and n can be obtained from the curves given in Design of Small Dams—USBR (shown in Fig. 12.4) or from any other book on irrigation structures. The velocity of approach is defined as

$$V_a = \frac{q}{P + h_o}$$

where q is discharge per metre length, P is height of crest above river bed and h_o is head of water above the crest upstream of the crest. If V_a is of small value say about 1–2 m/s and upstream face is vertical, then $K = 0.5$ and $n = 1.85$.

The elements of upstream curve can be obtained from the available curves. When the upstream face is vertical and the velocity of approach is small $x_c = 0.3$ and $y_c =$ 0.15 of design head. A suitable circular curve can be fitted for the upstream portion.

(b) *Length*

The length of the spillway portion and the crest level (below FRL) are determined from hydraulic, economic and geological considerations to pass the design flood. The discharge over ogee crest is given by

$$Q = \text{CL}\ \text{H}_0^{3/2}$$

where $Q =$ design flood discharge, $C =$ discharge coefficient, L is effective length of crest and H_o is head over crest including velocity head. The discharge coefficient has

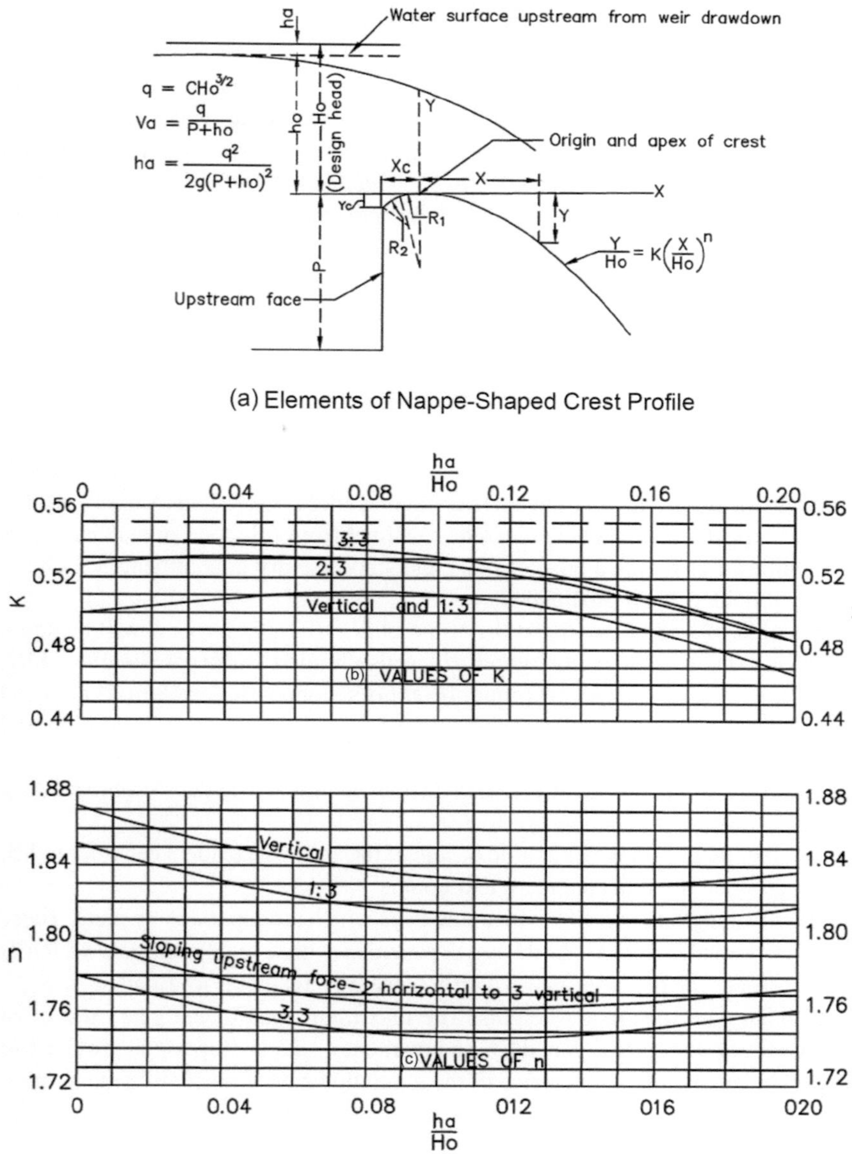

Fig. 12.4 Ogee spillway (U.S.B.R.)

a maximum value of 2.21 for an ogee crest when there is a free fall (tail water level is below crest level) and the upstream face is vertical. For various values of the ratio of dam height/head, submergence of crest reduces. If dam height/head ratio is less than 1, the value of C reduces sharply below 2.21. The effect of various parameters on the value of C can be estimated from the curves given in standard texts and USBR—Design of Small Dams.

Q, the design flood is worked out from the observed/computed discharge data of the dam site. The flood frequency analysis is carried out to work out floods of a certain return period. The SPF and PMF, if required, are worked out using the unit hydrograph method. For the procedure of such an analysis, refer Chap. 7. For more details, any standard book on Applied Hydrology may be referred. The spillway capacity should be fixed conforming to I.S. 11223—'Guidelines for fixing spillway capacity'.

The effective length L is less than the length of crest due to the effect of side contractions caused by the shape of piers and abutments. It is expressed by

$$L = L' - 2(\mathrm{nk}_p + k_a)H_o$$

where L' is length of crest, n is number of piers, and k_p and k_a are contraction coefficients of piers and abutments. The value of these coefficients varies with the shape of piers and abutments and can be obtained from I.S. code or standard texts on hydraulics. For rounded nose of piers and smooth entry at abutments which are normally provided $K_p = 0.01$ and $K_a = 0.1$.

The length of crest depends on the value of H_o. If H_o is large, L will be small and the vice versa will also hold good. H_o is also limited to the gate height if FRL and MWL are the same. The maximum height of the gate used is 18 m. If the head is large then the discharge intensity (discharge for metre length) will be high and in that case, a deeper stilling basin for energy dissipation will be required. Hence a combination of length and head over crest should be judiciously decided considering hydraulic efficiency and economy.

For further details on the hydraulic design of the spillway, readers can refer to I.S. code no. 6934 or any other standard text book.

Generally, length of crest is divided into bays which are separated by piers. These are generally made of RCC. Each bay is provided with a regulation gate which is either a vertical fixed wheel gate which is generally made up to a height of 8–10 m or a radial gate for greater heights. The maximum span of the gate used is 18 m. Upstream of these gates generally, an emergency gate or stop log is provided to facilitate the repair and maintenance of the regulation gate. Considering the number of bays one or two sets of stop logs are provided. The piers accommodate the bridge and the gate hoists at the top.

In some situations, uncontrolled (ungated) spillways are provided. In that case, the crest of the spillway is at FRL. The flood passes over the crest creating a head over it, so the MWL will be crest level plus the head. The top of dam is fixed at MWL plus the freeboard. So, in such cases, the height of the dam becomes more as compared to a gated spillway. With an ungated spillway, the flood cannot be moderated as is possible with a gated spillway. Hence, if other factors permit, generally a gated spillway is adopted.

(c) *Structural Design Considerations*

The piers are structurally designed for differential water head created due to one side bay of a pier being open and the other side bay being closed and the loads which are transferred from the bridge and the gate and the gate hoist. The pier is considered fixed with the concrete of the ogee crest. The main reinforcement of the pier is therefore carried inside the ogee for sufficient depth as required for fixity. The crest is also provided with reinforcement under the pier to transfer the stresses caused due to reactions at the base of the pier. For the design of the pier, I.S. 13551 may be referred.

The base width of the ogee spillway is worked out from stability considerations for the water load as well as dead load and the vertical loads transferred from the piers. The water load is transferred through gates to the piers and then to the ogee spillway crest. For stability analysis generally, all the loads transferred through the pier are assumed to be uniformly distributed throughout the bay. The stability analysis procedure is described in Sect. 12.2.4. The downstream surface of the spillway is subjected to high-velocity flow. Therefore, the surface should be made smooth and of high strength concrete.

(d) *Energy Dissipators*

The flow at the toe of the overflow section in supercritical (Froude number $V/\sqrt{gd}$ is greater than one) and the velocities are high which may scour the river bed downstream. This scouring may damage the structure. Hence, energy dissipation arrangements are provided at the toe of the spillway. Generally, a hydraulic jump basin is provided at the toe of the spillway.

A hydraulic jump is formed in the basin when supercritical flow meets the subcritical flow in the basin where the water depth will be greater. For a supercritical flow depth 'd_1' at the toe of the spillway a conjugate depth 'd_2' in jump basin is required for the formation of the jump. The energy is lost in the jump. Hence, to form a good jump, the water depth (tail water level for the corresponding discharge) in the basin should be close to the required 'd_2'.

The requirement of the length of the basin and dimensions of other appurtenances such as chute blocks, basin blocks and end sill depend on the Froude number. The design details of hydraulic jump basin, whose typical sketch is given in Fig. 12.5, can be worked out according to I.S. code no. 4997.

A good jump forms when Froude number is 4 or more. If it is less than 4 a weak jump will form and loss of energy in the jump will be less.

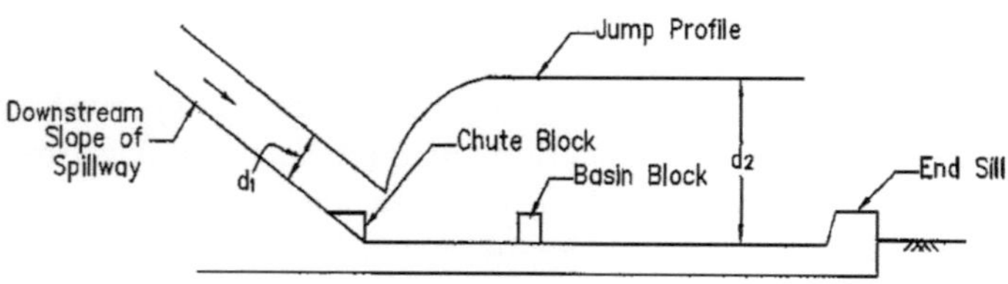

Fig. 12.5 Typical sketch of jump basin

If the tail water depth is more than conjugate depth, then a submerged jump will form with the pulsating flow which may travel long distances downstream causing damage to the river bed. In such cases instead of a jump basin a roller bucket is provided. Similarly, if the tail water depth is less than the conjugate depth the high-velocity jet will travel a long distance causing scouring in the river bed. In this situation, a ski jump bucket is provided. For the design of buckets I.S. code 7365 may be referred.

(vi) **Non-over Flow Section**

The non-over flow section on either side of the spillway is basically a triangular section with a base width as required from stability considerations. The top is widened to accommodate the road bridge. A typical non-overflow section is shown in Fig. 12.1a.

A training wall separates the overflow and non-overflow sections. The height of the training wall depends on the depth of flow along the overflow section and in the jump basin. The top of the training wall is generally 1.0 m above the water surface. One or two non-over flow blocks adjoining the overflow blocks are normally widened beyond the road to accommodate a control room for gates of the spillway, site office, elevator shaft, stair case, etc. Sometimes stop logs and their operating crane are also parked in this space.

Normally in a gravity dam, intake structures for the outlets/penstocks for the release of regulated flow for power generation, etc. are located in one or more non-overflow blocks. The trash racks for intakes are placed on the upstream face. These are generally of two types: (i) straight type to be used with small-height dams and (ii) semi-circular type for high dams. Normally the gates of intake are operated from the dam top. These are preferred when the power house is at the toe of the dam. In other cases, the intake may be placed at a suitable location separate from the dam. These are discussed in detail in the chapter on Intakes.

(vii) **Drainage Arrangement**

There is always a possibility of seepage of reservoir water either through the foundation or the body of the dam or from both despite all the measures and precautions taken during design and construction. The seepage is therefore collected as close to the upstream face as possible in a foundation gallery and is pumped out in the river downstream of the dam.

The foundation gallery is provided throughout the length of dam and is constructed near the upstream face of the dam. It is shown in the plan in Fig. 12.3 and in section in Fig. 12.1. The drainage holes are made in foundation to collect seepage downstream of the grout curtain.

The seepage through transverse joints, which may take place despite the provision of water-stops, is collected through a form drain provided just downstream of the water-stops. These form drains discharge in the foundation gallery.

Seepage may also take place through upstream face concrete of dam from the cracks which may occur due to temperature variation. To arrest this seepage, form drains are provided at about 3 m from upstream face at a spacing of 3–4 m. These form drains also discharge in the foundation gallery. The arrangement is shown in Fig. 12.6.

A drain in the foundation gallery carries all the seepage collected in the foundation gallery to a sump which is provided at the lowest point of the gallery from where the seepage water is pumped out by an automatically operated submersible pump. The sump may be provided at other suitable locations also.

(viii) **Galleries and Openings**

A number of galleries and openings are provided for different purposes in a gravity dam. Generally following galleries are provided.

- Foundation gallery for grouting and drainage runs along the dam axis at full length. Minimum size is 2.25 m high and 1.8 m wide. It is made at a minimum distance of 1.5 m from the upstream face and about 1.5–2 m above the foundation rock.
- Sometimes another gallery at mid-height parallel to the foundation gallery is provided for inspection and instrumentation in high dams.
- Adits normal to the dam axis are provided to approach the foundation gallery.
- Vertical shafts are provided for intake gates, and for housing the elevator shaft and stair case.

For layout and design of openings, I.S. code 12966 Part 1 and Part 2 may be referred.

12.2.4 Design of Dam

(a) **Forces acting on the dam**

The forces acting in a gravity dam are as follows:

1. Dead load.
2. Water pressure.
3. Silt pressure.
4. Uplift pressure.
5. Wave pressure.
6. Wind pressure.
7. Earthquake forces.
8. Ice load.

These forces are estimated for the unit length of the dam and are explained below:

1. Dead Load: It includes the weight of the concrete section which is assumed for analysis plus the weight of water on the upstream slope if any, and the weight of other structures supported on the dam. The weight of concrete is taken as the product of the area of the section and the density of concrete which is taken as $2.4^{T}/m^{3}$. It is assumed to act at the centre of gravity of the section. For the convenience of estimation, the dam section is considered in parts as shown in Fig. 12.7. For masonry dam, the density is taken as $2.3^{T}/m^{3}$.

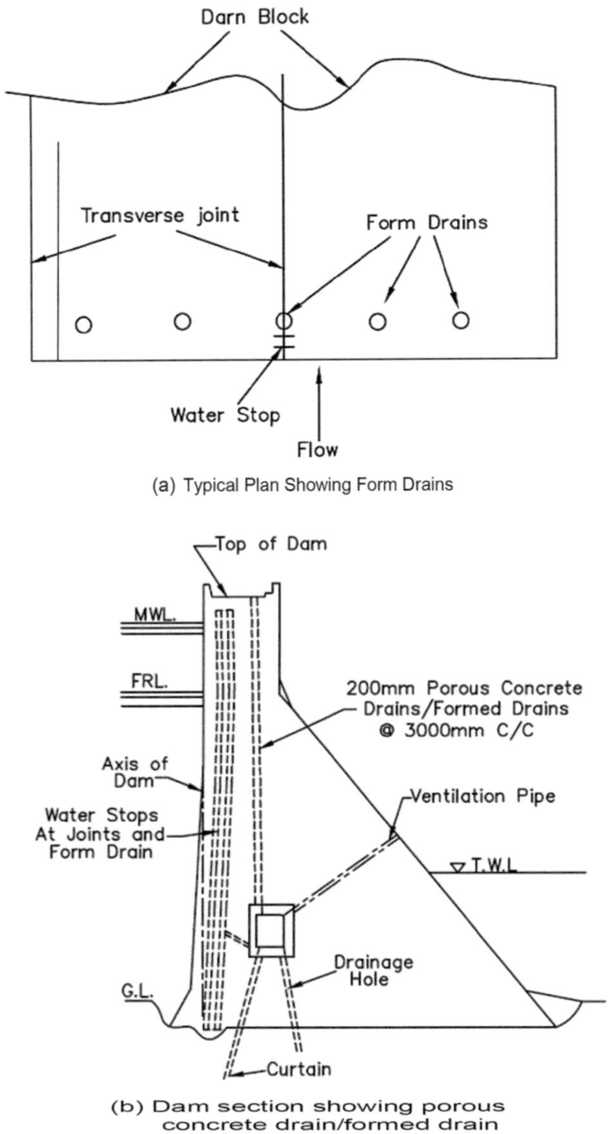

(a) Typical Plan Showing Form Drains

(b) Dam section showing porous concrete drain/formed drain

Fig. 12.6 Details of form drains

2. Water Pressure: The water pressure at any level of the dam section is equal to the hydraulic head at that level. At the base of the section, the water pressure is equal to reservoir depth, i.e. wh (see Fig. 12.7) where w is the unit weight of water and is taken as $1^T/m^3$. The water pressure diagram is triangular and the total water force is equal to $\frac{1}{2}$ wh^2 acting at $h/3$ above base. If the dam section

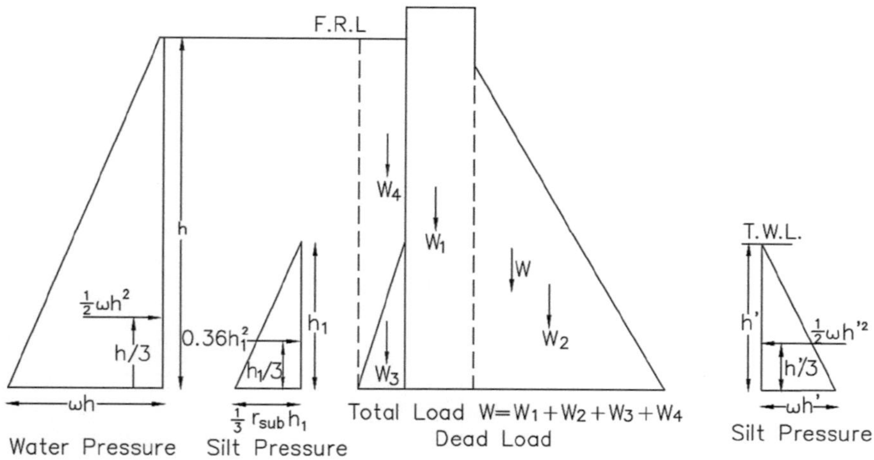

Fig. 12.7 Dead load water and silt pressure on dam

has water in the downstream the water force from downstream corresponding to the tail water level is worked out in the same way and is shown in Fig. 12.7.

3. Silt pressure: Silt pressure is assumed to act on the upstream face corresponding to the level up to which sediment is expected to get deposited. The horizontal force due to triangular silt pressure is taken as 0.36 h_1^2 acting at $h_1/3$ where h_1 is the height of deposited silt. The density for vertical load estimation of silt deposited on the upstream slope is taken as $1.9^T/m^3$. It is shown in Fig. 12.7.
4. Uplift pressure: It is caused due to seepage through the foundation. The uplift pressure is assumed to act on the entire base equal to the reservoir depth at the base level. It acts upwards opposite to dead weight and tries to destabilise the dam. It is maximum at the upstream end of the base and its distribution from upstream to downstream is a straight line. It is shown in Fig. 12.8. The uplift pressure is assumed to be reduced at the location of the drainage holes in the foundation gallery as shown in Fig. 12.8. The uplift force U is the area of the uplift diagram shown in Fig. 12.8.
5. Wave pressure: It is a pressure which is exerted by the waves on the upstream face of dam. The waves are created due to wind on the reservoir surface. The wave height depends on wind velocity and fetch of reservoir (refer Sect. 12.2.3). If the wave height is h_w then horizontal force P_w is taken as 2 wh_w^2 (w is unit weight of water) and it is assumed to be acting at 0.375 h_w above still water level (FRL) as is shown in Fig. 12.9.

Fig. 12.8 Uplift diagram

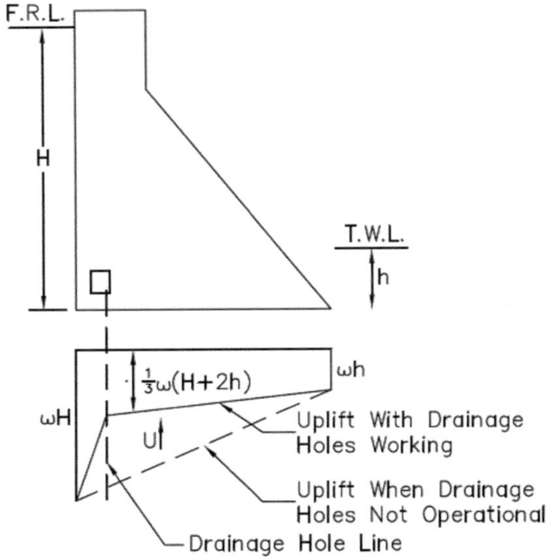

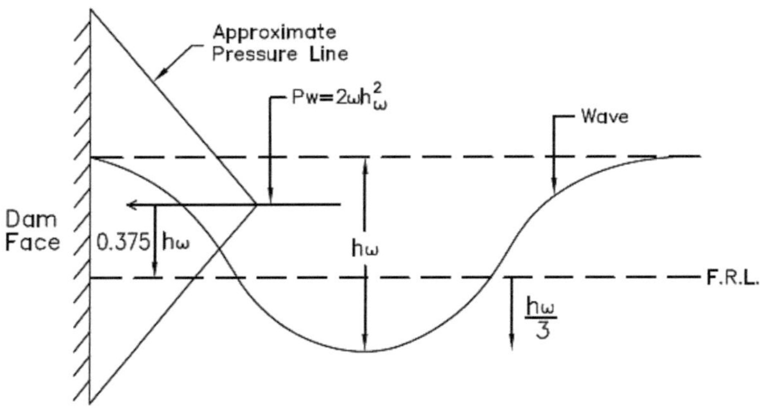

Fig. 12.9 Wave pressure diagram

6. Wind pressure: It is the pressure exerted by the wind when it strikes the exposed upstream face of the dam. It will be significant when the reservoir level is low.
7. Earthquake force: It is considered when the dam is located in a seismic area. India is divided into five seismic zones. Zones IV and V are in the Himalayan belt and are prone to severe earthquakes. Usually for small-height dams pseudo-static approach of analysis is followed. In this approach, static inertial forces for gravity loads combined with the hydrodynamic force of reservoir water are considered for analysis. The two are assumed to be acting in the same direction.

An earthquake causes acceleration which is expressed as a percentage of acceleration due to gravity (g), i.e. $\propto \cdot g$ where $\propto$ is called seismic coefficient. The value of $\propto$ depends on the seismic zone in which the dam is located. The basic horizontal seismic coefficient $\propto_{ho}$ of each zone is given below (I.S. 1893).

Zone	I	II	III	IV	V
$\propto_{ho}$	0.01	0.02	0.04	0.05	0.08

Horizontal seismic coefficient $\propto_h = \beta.I. \propto_{ho}$ where β depends on foundation type and is 1.0 for rock and hard strata and I is an importance factor which is taken as 3.0 for dams. So normally for dams in the Himalayas $\propto_h$ is taken as 0.24 and $\propto_v$ is taken as 0.75 of $\propto_h$.

Inertial force due to the dead weight of dam in a horizontal direction is taken as $\propto_h$ W acting at the centroid of the dam section and in a vertical direction it is $\propto_v$ W. In stability consideration, the vertical inertial force is considered acting upwards.

Hydrodynamic pressure due to the seismic acceleration effect on reservoir water is assumed to act horizontally in the direction of inertial horizontal force and is computed by the Zangers method which assumes that the hydrodynamic pressure diagram is parabolic and it is shown in Fig. 12.10.

Pressure at any depth is given by pe $=$ C $\propto_h$ w.y.h where w is unit weight of water:

$$C = \frac{C_m}{2}\left[\frac{y}{h}\left(2 - \frac{y}{h}\right) + \sqrt{\frac{y}{h}\left(2 - \frac{y}{h}\right)}\right]$$

C_m is maximum value of coefficient C and its value for $y = h$ and the vertical face of dam is 0.73.

$$\text{The total force } P_E = 0.726\ C_m \propto_h\ \text{wh}^2.$$

$$\text{Point of application of this force} = \frac{4\text{h}}{3\pi}.$$

$$\text{Moment of this force} = 0.299\ C_m \propto_h\ \text{wh}^3.$$

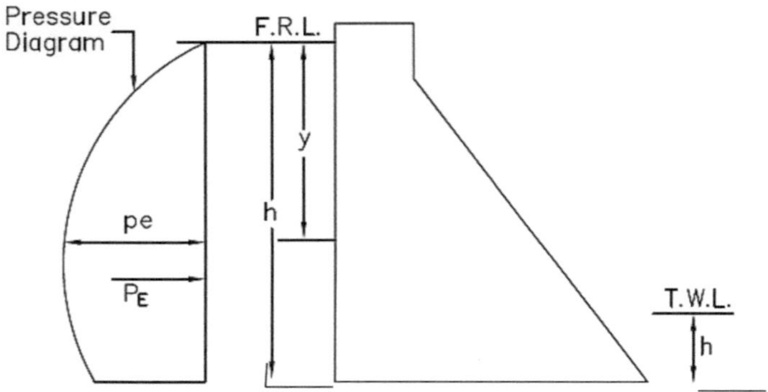

Fig. 12.10 Hydrodynamic pressure

The value of C for various slopes of the upstream face and different values of $\frac{y}{h}$ can be obtained from the curve given in 'Design of Small Dams'—USBR.

For high dams or for dams on unusual foundation conditions or for those located in regions of very severe seismicity, dynamic analysis with FEM using the response spectrum method should be used. For the details of this method, any text on concrete dam (Design of Gravity Dams, USBR) may be referred and experts shall be consulted.

8. Ice Load: This load is considered where the dam is in cold region and the reservoir is expected to freeze. The magnitude of ice pressure has been estimated ranging from 0 to 80 tonne per linear metre of the contact with the upstream face of the dam. A value of 25 t/m is widely accepted for analysis.

(b) **Load Combinations**

The following load combinations are analysed to ensure a safe and economical dam section:

(i) A—(construction stage)—dam completed but reservoir not filled.
(ii) B—(normal operating conditions)—reservoir at FRL, normal tail water level and normal uplift (drains operative).
(iii) C—(flood condition)—reservoir at MWL, maximum TWL and normal uplift.
(iv) D—A + earthquake.
(v) E—B + earthquake.
(vi) F—C with full uplift (drains inoperative).
(vii) G—E with full uplift.

The above load combinations are grouped as follows:

- Combinations A and B are usual or normal conditions of loading.
- Combination C is the unusual condition of loading.
- Combinations D to G are extreme conditions of loading.

These load combinations have different values of permissible stresses and factors of safety and these are indicated in the following paras of stability analysis.

(c) **Stability Analysis**

(i) The stability analysis is carried out with the following assumptions:

(1) Dam is divided into blocks and each block behaves as an independent cantilever fixed at the foundation.
(2) Vertical stress varies linearly at the base or at any other horizontal plane from upstream to the downstream end. This assumes that the foundation is rigid.

(ii) The requirements for the stability analysis are as follows:

(1) The stresses in all load combinations at all elevations and at the base should be within permissible values of compression and tension in concrete. Normally in normal load combinations, the entire section should be under

compression and the resultant of all forces must be within the middle third of the base.

(2) The stresses which are transferred to the rock foundation should be within the permissible bearing capacity of the rock which should normally be worked out by testing. For preliminary studies, the permissible bearing capacity may be assumed as given below:

Compact sand – gravel mix	50 T/m^2.
Shattered and broken bed rock	100 T/m^2.
Rock in sound condition like slate, schist	350 T/m^2.

(3) The section should be safe against overturning, i.e. the ratio of stabilising moment and overturning moment should be more than one in all loading conditions.

(4) The dam should be safe is sliding at the base and at the planes at other elevations. The criteria for checking the safety against sliding are as follows:

(a) Frictional resistance between rock and concrete resists sliding due to horizontal forces. The coefficient of friction μ (tanϕ) for normally encountered rock is taken as 0.65–0.75 for normal loading conditions and 0.85 for extreme loading conditions. The factor of safety (FOS) is worked by

$$\text{FOS} = \frac{\mu \Sigma V}{\Sigma H}, \text{ where } \Sigma V = \text{total vertical loads and}$$

$$\Sigma H = \text{total horizontal loads}$$

The FOS for normal loading conditions should be >1.5 and not less than 1.0 for extreme loading conditions.

(b) For important and high dams on unusual foundations, the sliding safety is evaluated considering both frictional and cohesive strengths of foundation material as given below:

Shear friction factor (SFF) $= \frac{\mu \Sigma V + CA}{\Sigma H} > 1.5$ for normal loading conditions and not less than 1.0 for extreme loading conditions. The following values of μ and C may be used:

	μ	C (kg/cm^2)
Sound massive rack	0.8	30
Fractured and jointed rock	0.6	7
Compacted gravel sand mix	0.5	0

To check sliding at any elevation of dam section, the shear friction factor given above should be evaluated using shear resistance between two concrete surfaces. The value of ϕ is normally taken as 45° and the value of C is taken as 0.05 fc, where fc is the compressive strength of concrete.

(iii) Procedure for stability analysis

A dam section is assumed for stability analysis on the basis of similar existing dams. A safe and economical section is worked out by trial satisfying the above requirements. The vertical and horizontal loads and their moments about the toe are worked out for all A–G load combinations. Thus, the stresses at the heel and toe of the dam section are worked out for all load combinations using the following expression:

$$\sigma_{ZU}, \sigma_{Zd} = \frac{\Sigma V}{b}\left(1 \pm \frac{6e}{b}\right)$$

where σ_{ZU} and σ_{Zd} are normal vertical stresses on upstream and downstream ends; b is base width of section and e is eccentricity, the distance of the point at which the resultant force cuts the base from the mid-point as shown in Fig. 12.11.

$$e = \bar{y} - \frac{b}{2}$$

$$\bar{y} = \frac{\Sigma M_{toe}}{\Sigma V}$$

As stated above, the normal vertical stresses should be within the permissible compressive strength of concrete. This shall be taken as the 1-year strength of concrete (fc) divided by 3, 2 and 1 for normal, unusual and extreme loading conditions, respectively. Normally fc ranges from 150 to 300 kg/cm^2 in dams. The compressive stress in no part of dam should exceed 70 kg/cm^2 in normal loading conditions and 100 kg/cm^2 in extreme loading conditions.

If $e > \frac{b}{6}$ tension will occur. In normal loading conditions, i.e. combination A and B no tension should be allowed. A tension of 5 kg/cm^2 may be allowed in extreme loading conditions.

The normal vertical stresses are transferred to the foundation and shall be within the permissible bearing capacity of the rock. This may be evaluated by dividing the compressive strength of the rock by 4.0, 2.7 and 1.3 for normal, unusual and extreme loading conditions if the rock has been tested. Otherwise, if the permissible bearing capacity given in para ii (b) above are used, these may be increased by 50% for extreme loading condition.

The safety against sliding as discussed above should also be checked for all load combinations.

(d) Principal Stress

The concrete design shall be based on the principal stresses and shear stresses that will develop in concrete. These can be worked out in terms of normal vertical stresses

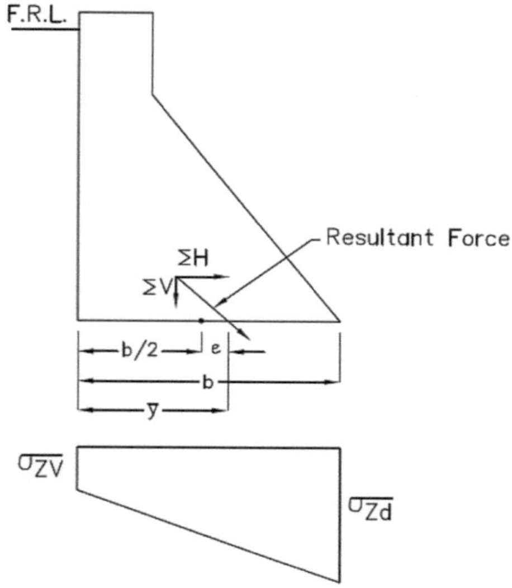

Fig. 12.11 Normal vertical stresses

which are evaluated as described above. These are maximum at the upstream and downstream ends at the base.

The plane at which there is no shear stress is called the principal plane and the stress acting normally at the principal plane is called principal stress. At every point in concrete there will be two principal planes perpendicular to each other and the stresses on these two planes are called major and minor principal stresses.

(i) Downstream face: The stresses on an element of the downstream face at the base are shown in Fig. 12.12a. Since no shear force acts on d/s face, it becomes a principal plane and the other principal plane will be perpendicular to it. There is no stress on d/s face so the $\sigma p_1 d$ will be a major principal stress and τ_{zyd} is the shear stress. σ_{zd} is normal vertical stress and is known from stability analysis. σ_{pld} and τ_{zyd} in terms of σ_{zd} are given by

$$\sigma_{pld} = \sigma_{zd} \sec^2 \phi_d$$
$$\tau_{zyd} = \sigma_{zd} \tan \phi_d$$

(ii) Upstream face: The stresses on an element of u/s face at the base are shown in Fig. 12.12b. Similar to d/s face, the u/s face is one principal plane with water pressure on it and no shear stress. The stresses in terms of σ_{zu} are as follows:

Fig. 12.12 Principal stresses

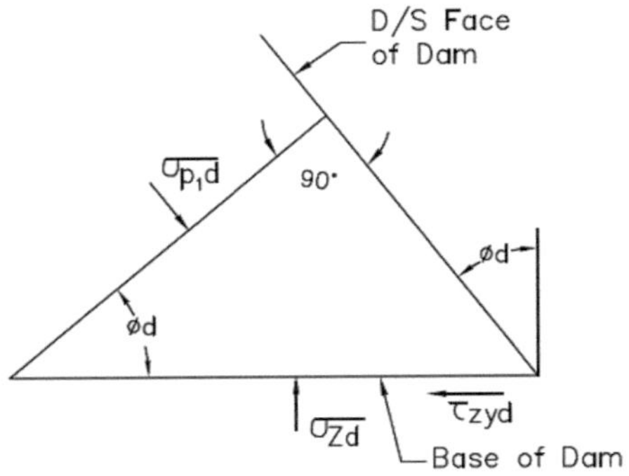

(a) Principal Stress in D/S Face

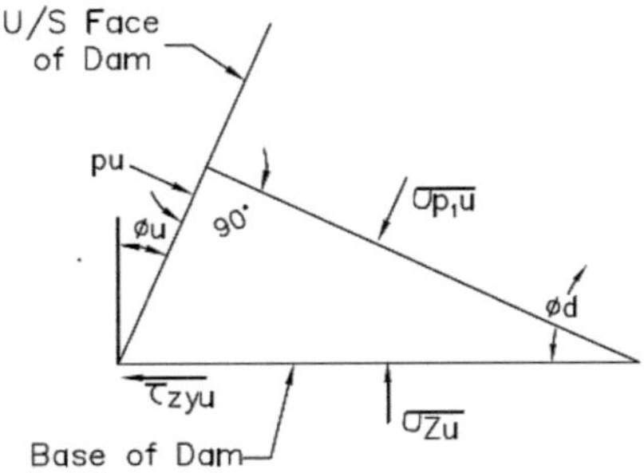

(b) Principal Stress on U/S Face

$$\sigma_{plu} = \sigma_{zu} \sec^2 \phi_u - P_u \tan^2 \phi u$$
$$\tau_{zyu} = -[\sigma_{zu} - P_u] \tan \phi_u$$

For derivation of above expressions, any text on concrete gravity dam may be referred.

12.2.5 Other Aspects of Gravity Dams

Besides planning and investigations, layout and design aspects described above there are several other important aspects that require careful consideration and these are briefly described below.

(i) **Thermal stresses and control**

Thermal stresses in a concrete dam develop due to restraint to volume change. This may be caused by temperature changes in concrete mass and the restraint due to the bond between concrete and rock. When these stresses, which are tensile in nature, exceed the tensile strength of concrete then cracks may develop and if these are not prevented then structural stability and water tightness of dams, etc. get affected. Thermal stresses due to restraint in volume change are given by

$$f = \text{R.C. E}_{\text{ff}}\Delta\text{T}$$

where f is tensile stress, R is restraint factor which for rock is taken as 1.0, C is the coefficient of thermal expansion of concrete,

$$E_{\text{ff}} = \frac{\text{Ec}}{1 + 0.4\left(\frac{\text{Ec}}{\text{Er}}\right)}$$

where Ec and Er are modulus of elasticity of concrete and rock and $\Delta T = \text{tp} + t_r - t_f$, where t_p is placement temperature, t_r is temperature rise due to heat of hydration of cement and t_f is final stable temperature of concrete mass which depends on the atmospheric temperature at dam site.

It can be seen that except ΔT other factors responsible for thermal stresses are the properties of concrete and rock. Hence f, the tensile stress can be controlled by managing ΔT which means by reducing tp and t_r as t_f depends on site conditions. ΔT should usually be limited to 22 °C. To achieve it the placement temperature should be kept around 10–12 °C. It is generally achieved by pre-cooling of aggregate and using ice flakes in mixing water and by avoiding placement of concrete during day time in summer when temperatures are high.

The temperature rise (t_r) due to heat of hydration of cement is normally 20 °C for a lift of 1.5 m which is placed 72 h after the placement of the previous lift. This can be reduced by reducing the quantity of cement. The requirement of cement can be reduced by replacing it with 15% pozzolana which is generally fly ash.

The curing of exposed surfaces after removal of the form work is also a measure of controlling cracking.

Fig. 12.13 Dental treatment

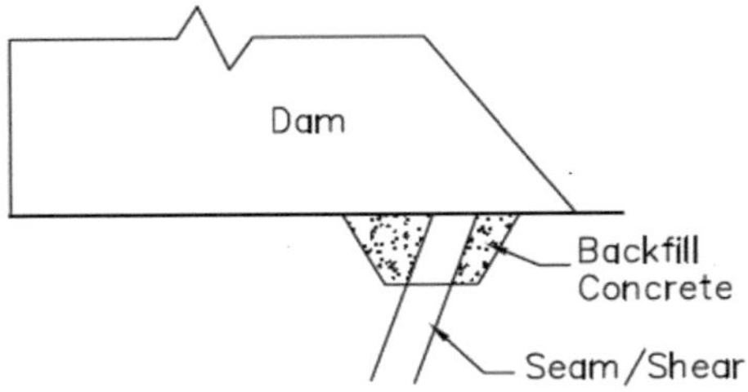

(ii) Foundation Treatment

Concrete gravity dams are normally built on rock foundations. But it is very rare to get a foundation of strong rock which is water tight and free from geological weaknesses such as shears, soft rock areas, and stratification. Hence, practically all concrete gravity dam foundations need treatment which is called normal treatment and some having major geological weaknesses in the foundation need site-specific special treatment.

(a) **Normal treatment**: It comprises dental treatment, consolidation grouting, curtain grouting and drainage.

If at the foundation level the bed rock has thin seams or shears containing soft/crushed rock, these should be excavated up to a depth say 1.5 m and backfilled with concrete. It is called dental treatment. It is shown in Fig. 12.13. It is advisable either to anchor the backfill concrete with the rock or grout the surface between the backfill concrete and the rock.

Consolidation grouting is done under the entire base of the dam to improve foundation bed rock which gets weakened due to excavation and removal of overburden for which heavy equipment and blasting techniques are normally used. It is a low-pressure shallow grouting. Normally, it is 6–8-m-deep holes grouted at a pressure of 2–3 kg/cm^2. Initially, holes are drilled in a grid of 6 m which is later reduced to a 3 m grid. Holes are grouted in one stage in full depth.

Curtain grouting is done when the permeability of the rock is more than 3 Lugeons (one Lugeon is water loss of one litre/minute/metre of hole). It is done to create a seepage barrier in the foundation. It is carried out through a foundation gallery. In small-height dams, it can be done through a concrete pad projected from the upstream face of the dam at the base. Generally, a single-row grout curtain is provided. For grouting, generally vertical NX (75 mm) size holes are made. These can be made slightly inclined upstream. Initially, holes are made at a spacing of 6 m which may be reduced as required up to 1.5 m to satisfy the condition that permeability reduces to less than 3 Lugeons. As per IS: 11293, the suggested depth of the grout curtain is

$$\frac{2.H}{3} + 8, \text{ where } H \text{ is height of dam.}$$

The depth of grout curtain should also be subject to the water test result in the holes. The holes should be drilled up to a depth at which the permeability is less than 10 Lugeons.

Generally grouting is done with water cement mix (5:1). The general rule for determining the grout pressure at the collar of the hole is that maximum pressure should not exceed 0.23 kg/cm^2 per metre depth of rock. So, as the depth of grouting increases the grout pressure increases.

In good rock where the hole is not expected to collapse the hole is drilled up to the required depth and grouting is done from the deepest portion in stages. Each stage may be of 3–6 m depth. In poor rock, the holes are drilled in 2–3 stages of equal depth. Each stage is drilled and grouted before the second stage is drilled.

Drainage: It is experienced that grout curtain generally does not provide a leak-proof seepage barrier. So to arrest the leakage through the grout curtain, drainage holes just downstream of the grout curtain are provided. These are also made through the foundation gallery as shown in Fig. 12.14. These also ensure reduction in the uplift pressure on the dam base downstream of drainage holes. The NX size holes are drilled generally vertical or slightly inclined towards downstream up to a depth of about 70% of grout curtain at a spacing of 1.5 m. For detailed design refer I.S. 10135. A drain in the foundation gallery is provided. Seepage from the holes is collected through the drain in a sump located at the lowest point of the gallery and is pumped out.

(b) **Site-Specific Treatment**: These are required where foundation rock is not geologically sound.

Geological weaknesses are commonly of two types: (i) existence of soft rock pockets of low strength and (ii) smooth bedding planes or thin clay seams having orientation conducive to sliding failure. The measurers generally adopted are as follows.

To treat the weak zones, the general practice is to remove these up to sufficient depth and to backfill it with concrete. If it is not feasible then an RCC mat of sufficient

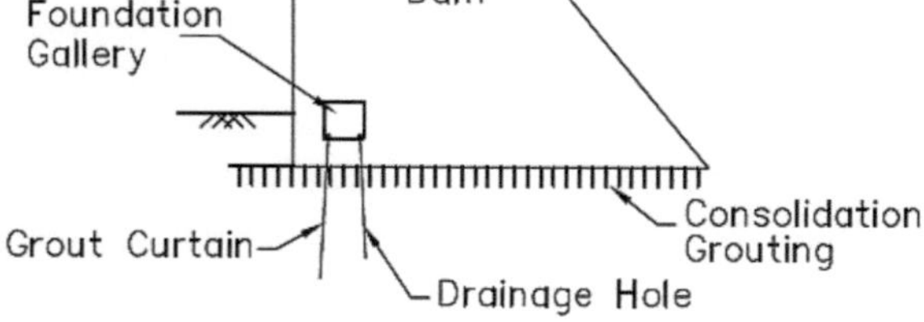

Fig. 12.14 Grouting and drainage of foundation

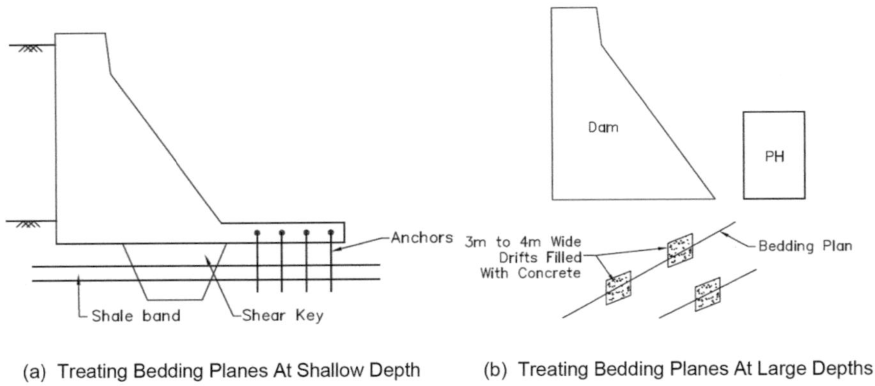

Fig. 12.15 Treatment of weak zones

thickness is placed to cap the weak zone so that load is transferred to adjoining strong rock.

To treat the bedding planes or clay seams, shear keys and anchors are used if the bedding planes are at a shallow depth. If the depth of such planes is large then underground shear keys across the planes are constructed. They are sketched in Fig. 12.15.

(iii) **River Diversion**

Before starting the construction of a dam, the river flow has to be diverted, so that the construction area could be isolated from the river flow. The diversion scheme and the structures required for diversion of flow depend on river discharge, topography and the geology of the area.

The diversion method to be adopted is decided on the basis of the river section and its width and the maximum flow required to be diverted. It can be a single stage diversion in which the river flow is diverted either through a tunnel or a channel located in the abutment. It will require a coffer dam both upstream and downstream of the construction area. The arrangement is shown in Fig. 12.16. The structures are planned to serve the purpose for the entire construction period.

If the river is wide, a multi-stage river diversion method may be adopted. The non-monsoon river discharge may be allowed to flow through a part of the river section and the rest of the river section may be isolated by constructing a dyke or masonry wall for dam construction during the working season. In the next year, flood may be allowed to pass on an incomplete concrete gravity dam and in the working season the flow is passed through the openings in the constructed dam and the remaining area may be isolated for construction. The scheme is sketched in Fig. 12.17. Depending on river width two or more stages of diversion are adopted.

The diversion discharge for a concrete dam generally varies from 1 in 10 to 1 in 25 years return period flood and sometimes a discharge corresponding to a non-monsoon peak flow of 1 in 25 year return period, because monsoon flood can be passed over

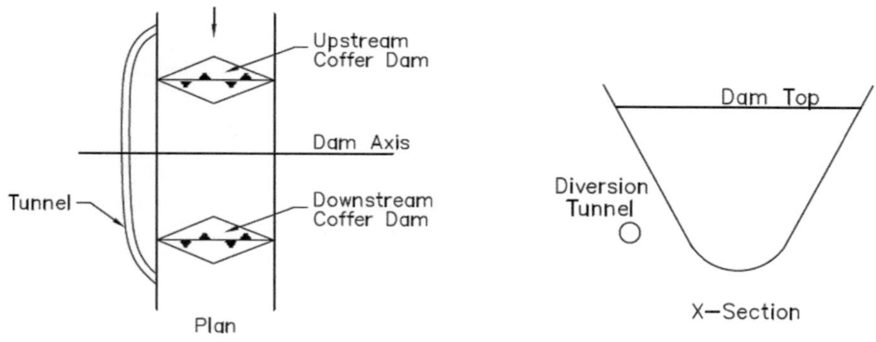

Fig. 12.16 Single stage diversion scheme

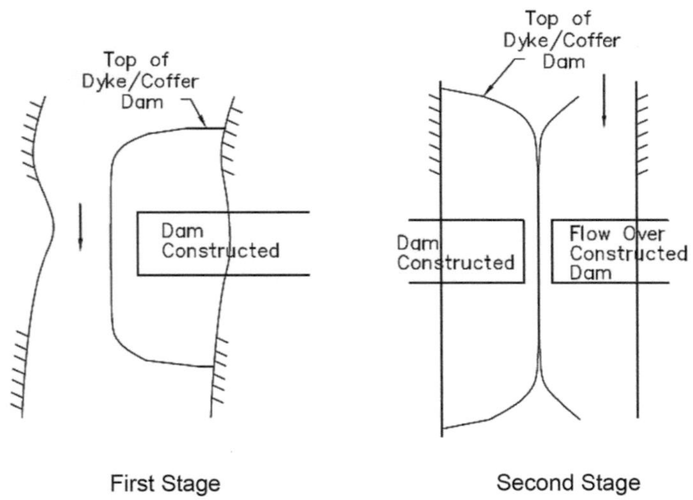

Fig. 12.17 Multi-stage diversion scheme

an incomplete dam without any appreciable damage. In small projects, the diversion discharge is limited to a frequency of 5–10 times the construction period.

Generally, a diversion tunnel is designed for running full condition under the head created by the coffer dam. An optimal size of the tunnel and height of the coffer dam should be adopted. Higher velocities up to 15 m/s may be allowed in the tunnel which is generally lined with M_{20} grade concrete. The minimum size of the tunnel is limited to 3 m. Coffer dam is constructed after the non-monsoon flow is diverted through the tunnel for which a low-height dyke is made across the river. The river closure requires special attention during the construction of dyke. The coffer dam may be an embankment or a masonry/concrete dam with adequate provisions against seepage through it and its foundation.

(iv) **Sediment Management**

When a dam is constructed across a river, sediment will get deposited in the reservoir. In a small reservoir of SHP, the sediment deposits up to the crest of dam spillway and gets its entry into the intake. To manage minimum sediment entry into the intake generally the crest of intake is placed sufficiently above the crest of spillway and the big size low level sluices are provided in the spillway to flush the sediment. More details are given in a separate chapter on Sediment Management.

(v) **Instrumentation**

In a concrete dam, instruments are installed for following type of measurements:

- Uplift pressure at the base of dam in at least three transverse sections.
- Seepage through the foundation and dam can be measured at certain locations on the drain in the foundation gallery.
- Temperature of the interior of dam—Thermometers may be embedded in the body of dam and also on or near the concrete surface at a few elevations.
- Suspended plumb bob—Generally, it is placed in the deepest block to measure the tilt of the block during the filling and emptying of the reservoir.
- Geodetic measurement—Surface markers are placed on the top and downstream face of dam. Their observations give a displacement of dam with respect to abutments.

I.S. 7436 (part-2) may be referred for the guidelines for measuring instruments which are supplied and installed by specialised agencies.

12.3 Roller Compacted Concrete Dams (RCC Dams)

12.3.1 General

The roller compacted concrete dam is basically a concrete gravity dam described in Sect. 12.2. Both are similar in shape, size, planning, layout and structural and hydraulic design. The difference is in the concrete mix and its transportation and placement techniques. The concept of roller compacted concrete, which is a no-slump concrete with less cementitious content and which can be transported and compacted by using heavy earth-moving equipment which is used in the construction of embankment dams, was first used for the construction of a gravity dam in Japan in 1978. This concept gained acceptance the world over because of reduced cement requirement, savings in construction time and cost of the dam as compared to the conventional concrete gravity dam. Since 1980 many countries have constructed high RCC dams. This concept of design and construction of gravity dams has been adopted in India also and the first such small dam was constructed in 2003 (12 m high as part of Ghatghar, pumped storage project). After this, a few more RCC dams have been constructed. The middle Vaitarna RCC dam which is 102 m high was completed in 2 years in 2012. The Testa IV low dam project which is about 30 m above the river bed was completed as an RCC dam in 2015.

The main advantages of RCC dams over conventional concrete gravity dams are as follows:

- Low cement content reduces heat of hydration thereby reducing the need for temperature control measures like pre-cooling of aggregates, and the use of ice flakes in mixing water.
- Reduction in the number of transverse joints, efforts required in the preparation of lift joints and requirement of form work on dam faces.
- Reduction in construction time due to easy and speedy transportation and placement of concrete.
- Reduction in overall cost of the project which is estimated as about 15%.

Roller compacted concrete is a low cement and low workability concrete which is able to support heavy earth-moving equipment. It is transported from the batching and mixing plant to the placement site in large trucks, spread by dozers and compacted by vibratory rollers.

12.3.2 Types of RCC Dams

Different countries have slightly different concepts of design and construction based on type of mix:

- Lean RCC dam: The RCC mix is made of low cementitious content, say up to 100 kg/m^3, with pozzolana content up to 40%. To check seepage through lift joints bedding mixes are to be used which are of 5-cm-thick rich conventional concrete. A 20 mm thick bonding mortar layer made of 1 part cement and $2\frac{1}{2}$ part sand can also be used.
- Medium paste RCC dams: The mix is made of cementitious content ranging from 100 to 150 kg/m^3 with 40–60% pozzolana.
- High paste RCC dams: The mix has a cementitious content of more than 150 kg/m^3 with high pozzolana content (70–80%).

The medium and high paste RCC dams are preferred for storage purposes because of their low cement content, high density and strength and better bonding properties. Lean RCC dams may be used as low dams or dams for flood control where storage is for a short period.

The layout includes all the provisions such as spillway, intake, galleries and other openings and the design of the RCC dam is the same as of a conventional concrete dam. However, efforts should be made to minimise the provision of galleries and other openings for speedy construction. Attempts should also be made to locate all the openings in a group on one side of dam.

12.3.3 Design of RCC Mix

Mix design for an RCC dam needs special attention for a specific project and should be evolved through laboratory tests. Before laboratory trials, the following should be initially decided:

- Desired strength and workability (Vebe consistency time of 15–20 s).
- The type of paste $(C + P)$ content. P is the Pozzolana which is generally fly ash.
- Maximum size and volume of coarse aggregate. The maximum size generally used is 75 mm, sand content is taken as 30–35% of total volume of aggregates. The rest will be coarse aggregate.
- Initially, C/P is taken as unity.
- Initially, $W/(C + P)$ is taken as 0.7–1.0.

Then laboratory tests are carried out on parametric trial mixes to work out the most economical mix of pozzolana, cement and water conforming to the requirement of strength, workability and durability. Generally, admixtures are not used for mix design. For mix design procedure USBR publication on Roller compacted concrete, 2005 and AHEC manual 2.2 and 2.3 which has reference of ASCE Committee 207 on RCC-August 1980 may be referred. The final design mix should also be tested on the test section of the dam at site.

12.3.4 Construction Approaches

The construction approach for an RCC dam differs from country to country. In the Japanese approach, generally, RCC is used in the central portion of dam and u/s and d/s faces are provided with 2–3-m-thick conventional concrete. In the American approach, the use of formwork is minimised and transverse joints are reduced. Other countries slightly vary the construction technique to suit their conditions. Generally, the glacis of the ogee spillway is covered with conventional concrete as shown in Fig. 12.18a. In the case of an ungated small-height spillway, the energy dissipation is achieved by stepping the RCC profile but steps are covered by conventional rich mix concrete for protection against erosion. In non-overflow section, the formwork may be replaced by concrete curbs or facing units (precast concrete panels) of conventional concrete 60–75 cm high. RCC can be placed generally in layers of 300 mm thickness after the curbs are hard enough as shown in Fig. 12.18b. In many projects, upstream

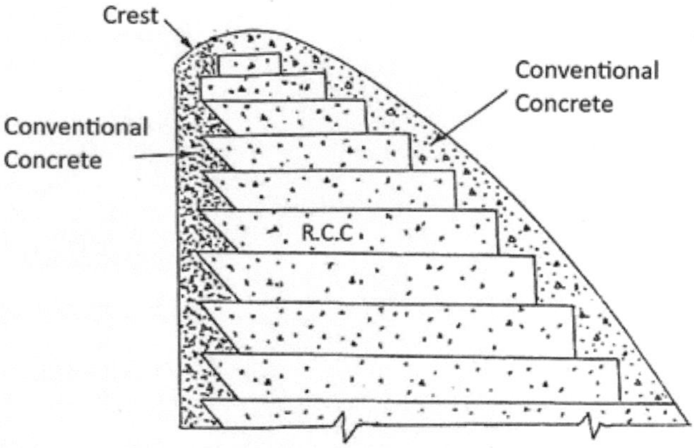

(a) Typical design of spillway for small and medium RCC dam

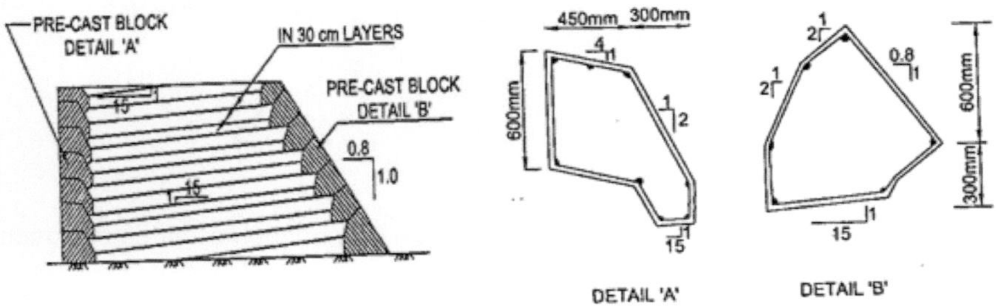

(b) Roller Compacted concrete dam section

Fig. 12.18 Details of RCC dam (AHEC-2.2)

face is made of 2–3-m-thick conventional concrete. The RCC lift is immediately placed after the conventional concrete and the two are compacted together. The provision of conventional concrete provides water tightness, convenience in placing water-stops at transverse joints and the construction of foundation gallery. Such a dam section is shown in Fig. 12.19.

12.4 Embankment Dams

12.4.1 General

Embankment dams are structures that are constructed of naturally available material (soil or rock) without any binding material other than those inherent in natural material. The materials used are generally those which are available at or near the dam site. These materials form a relatively flexible structure that can slightly deform to

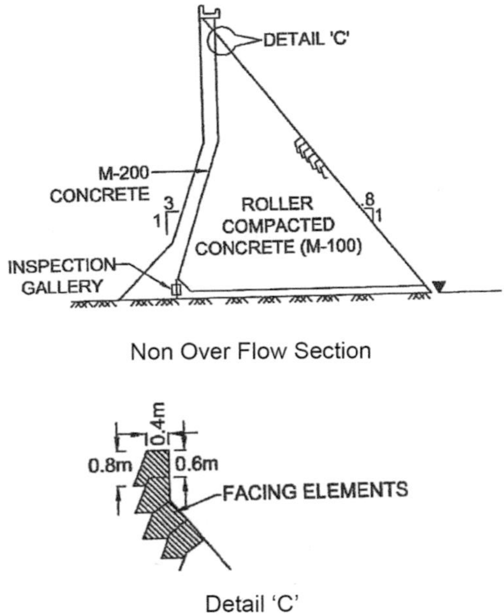

Fig. 12.19 Typical section of RCC dam (AHEC-2.2)

conform to foundation deflection. This makes the construction of an embankment dam possible on any type of foundation.

The construction of embankment dams has been practiced in many countries since the early days of civilization. The design remained ad hoc and practice-based. The dam height was generally less than 30 m till 1930. Higher dams based on scientific design procedures came into existence after rapid advances in soil and rock mechanics and the construction technology using heavy earth-moving machinery during the second half of the twentieth century.

12.4.2 Types of Embankment Dams

In SHP, the embankment dam is provided either to cover the river section at full width or as dykes on both sides of a centrally placed ogee-shaped spillway in the river channel as in the case of a composite dam (refer Fig. 12.2). In both the cases, depending on the topography, geology, foundation conditions and material availability, any of the following type of embankment dams may be provided.

There are basically two types of embankment dams:

(a) Earthfill dams.
(b) Rockfill dams.

(a) **Earthfill Dams**: These are made of soil and are generally constructed on an alluvial foundation. These are further divided into the following types:

(i) Homogeneous.
(ii) Modified homogeneous.
(iii) Zoned.

(i) A homogeneous dam section is made of one type of soil which is clay or impervious soil. The height is limited to 6–8 m. These are seldom used for storing water and are generally used as flood control embankments. Despite the impermeability of soil, seepage emerges on the downstream slope approximately at a height of 1/3 of the water depth in the reservoir as shown in Fig. 12.20a. Hence homogeneous dam sections are normally provided with some kind of drainage arrangement made of pervious material to protect the downstream slope from seepage and to provide safe exit to seepage.
(ii) Modified homogeneous dam section is the homogeneous section with some drainage arrangement in the downstream. Different drainage arrangements such as rock toe, horizontal blanket and chimney drain are shown in Fig. 12.20b along with the modified seepage line (called phreatic line).
(iii) Zoned embankment: In this type of dam, a seepage barrier in the form of a clay core is provided in the body of the dam. A shell of pervious material is provided on both sides of the core. A filter is provided along the slope of the core between the clay core and the shell to safeguard piping through the core which is subjected to high water pressure. Normally a vertical core is provided in the centre of the dam section. Cores sloping upstream are also provided. The zoned sections with different core locations are shown in Fig. 12.21.

The constructional advantage of an inclined core is the independent placing of the D/S shell and the grouting below the core, if required, can be done while the D/S shell is being constructed. The top and base widths of the core are normally provided as given below:

Top width	6–12 m (minimum 3 m)
Base width	(i) 30–50% of water head (for any type of soil)
	(ii) 15–20% of water head (with proper filter)
	(iii) Should not be less than 10% of water head

The upstream slope of all three types of earthfill dams is protected by the rip-rap which is placed over a layer of spalls to protect the slope from wave action. The rip-rap design is based on the wave height which is a function of reservoir fetch and the wind velocity. The downstream slope is protected from wind and rain erosion either by grassing or pitching.

(b) **Rockfill dams**: These are dams that are basically made of graded rock fragments or a well-graded mix of boulder, gravel and sand. It is generally constructed on

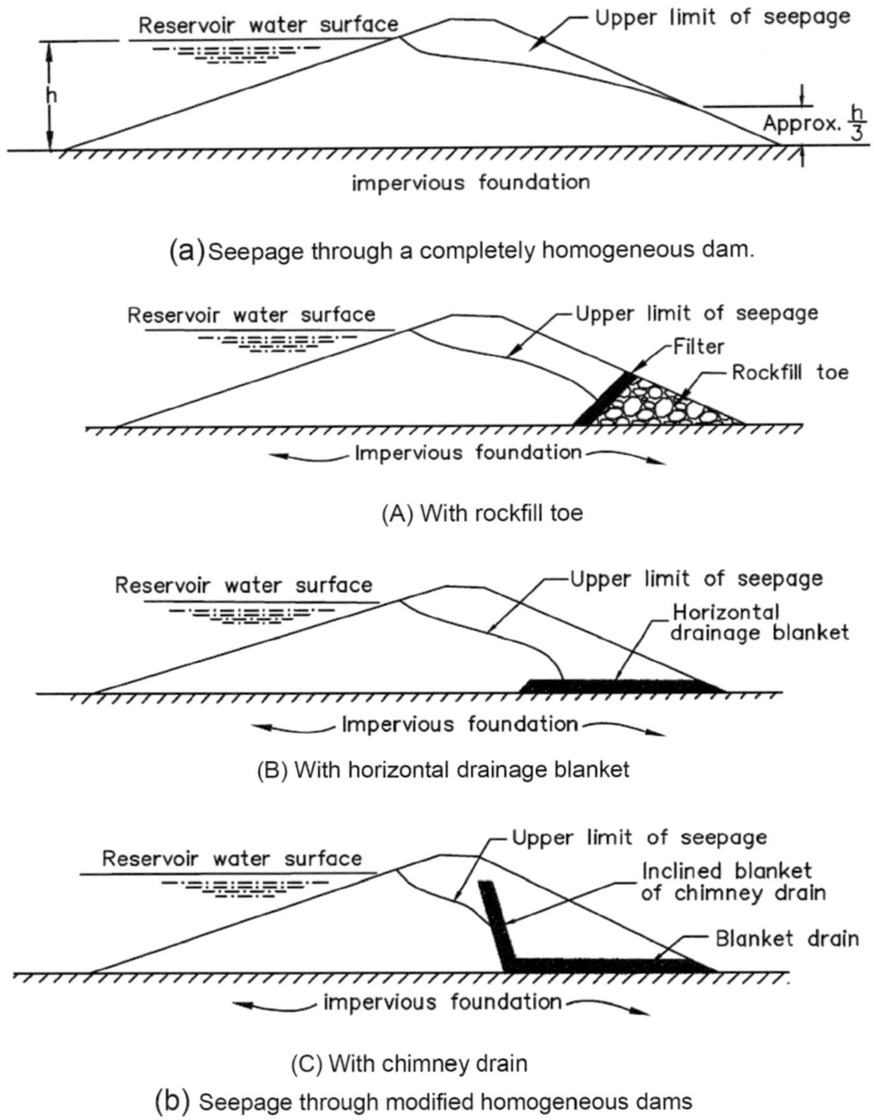

Fig. 12.20 Seepage through embankment dams

rocky foundations. Shell zones may be placed on compacted RBM. These are of two types based on the type of seepage barrier provided in the dam section.

(i) *Concrete-faced rockfill dam (CFRD)*:
This is a rockfill dam with an impervious membrane on its upstream slope to provide water tightness. It is shown in Fig. 12.22. Membranes commonly used are made of either reinforced concrete slab or asphalt concrete. In

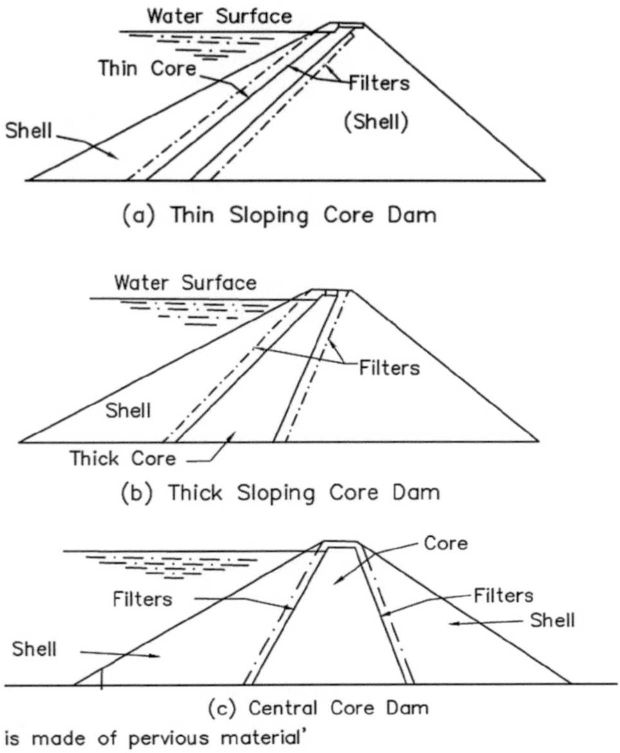

Fig. 12.21 Zoned earth dam—typical sections

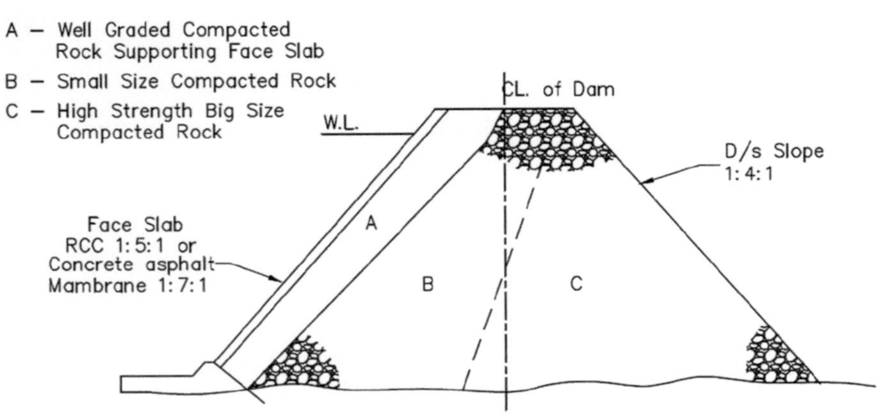

Fig. 12.22 Typical x-section of CFRD

earlier dams, wooden planks and steel plates were used on the *u/s* slope. The upstream membrane has the following advantages:

- The rock mass is free of seepage.
- It also serves as protection against wave action.
- It can be constructed after completing the dam section.
- It can be easily inspected and repaired.

This type of rockfill dam has steep slopes which may be near the natural slope of dumped rockfill. However, generally, a slope of 1.4 (H) to 1.0 (V) is provided which is found safe even in a moderate seismic condition. The upstream slope is a little flatter to facilitate the construction of the upstream membrane.

The main components of this type of rockfill dam are as follows:

- Toe slab (plinth).
- Face slab.
- Perimetric joint between toe and face slab.
- Zoned rockfill.

The design of these components are briefly given below, for details refer ICOLD bull. 141.

Toe slab: It acts as the foundation support for the face slab. It's width is generally 3–10 m but in poor rock it may be 0.3 H (H is water head). Its thickness is normally 0.3–0.4 m but it should rest on good rock strata. It is provided with temperature reinforcement of 0.3% both ways on top. It is normally anchored with rock and is used as the pad for foundation grouting. It is shown in Fig. 12.23.

Face slab: It is generally a M_{20} grade concrete RCC slab placed on the upstream slope of the dam to act as a seepage barrier. It is laid on well-compacted graded rock base to avoid excessive settlement as shown in Fig. 12.22. Its thickness is normally 0.3–0.4 m. It is reinforced at the top in both directions equal to 0.4% of concrete area. It is laid in panels with vertical and horizontal joints as shown in Fig. 12.24. Vertical joints are contraction joints and provided with a water-stop. Horizontal joints are construction joints so reinforcement is continued across these joints.

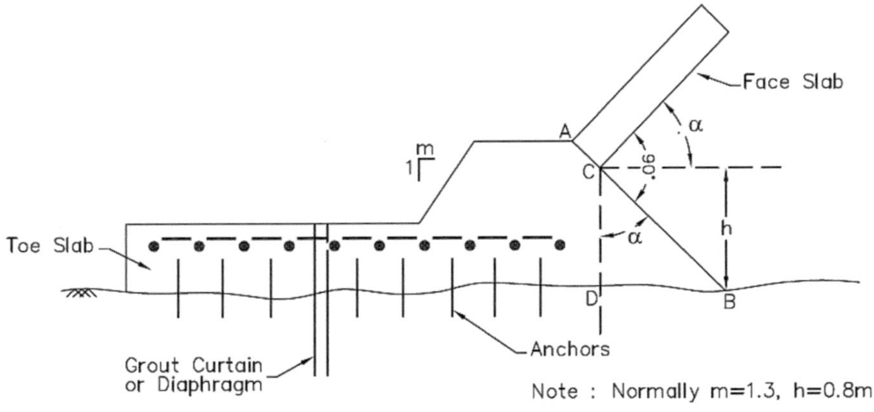

Fig. 12.23 General layout of toe slab

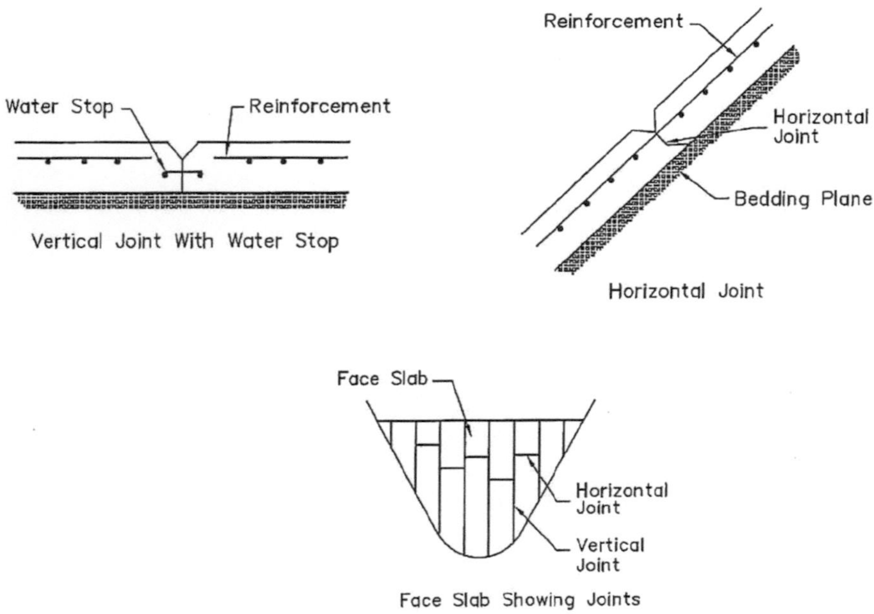

Fig. 12.24 Showing joint details is face slab

In some cases, u/s membrane is made of asphaltic concrete. The recommended upstream slope is 1.7:1 (H:V) to meet the construction requirement. The bedding support is similar to the RCC slab. The asphaltic concrete is a mix of well-graded rock pieces of maximum size of 38 mm down to rock dust and asphalt. The asphalt used is about 8–10% of weight of aggregates. It is laid is several layers. Normally the top surface of asphaltic concrete membrane is protected.

Perimetral joint: It is a joint between the face slab and the toe slab. It allows the movement of the slab due to the settlement of the bedding plane during and after filling of the reservoir. The joint has to be made watertight against the reservoir head and has to permit movement without any damage. So it needs careful designing. A joint provided at a CFRD which is about 50 m high and has been recently constructed in India is shown in Fig. 12.25 with its various components. The joint gap is filled with 15-mm-thick joint filler and is provided with two water-stops, one of PVC at the centre and the other of copper at the bottom.

Zoned rockfill: As shown in Fig. 12.22, the rockfill is divided into three zones A, B and C. Zone A is the bedding plane for the face slab. It normally consists of well-graded rock material with a maximum size of 7.5 cm down to 5% of sand size up to 0.074 mm. It's permeability should be in a range of 10^{-3}–10^{-4}cm/s. The width of this zone is 4 m placed in layers of 0.5 m. Zone B is well-graded rock and sand mix with maximum^{m} size of rock of 0.8 m, Permeability should be more than zone A and less than zone C. Zone C is made of well-graded rock of bigger sizes. The bigger size should be on or near the D/s face. It is placed in layers of 1.5–2 m.

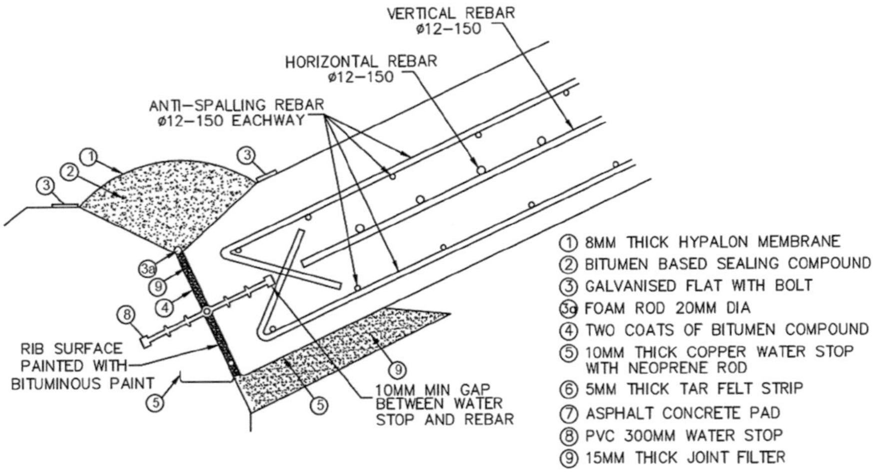

Fig. 12.25 The perimetral joint of a CFRD

(ii) *Earth Core Rockfill Dam (ECRD)*

It is a rockfill dam with a clay core as a seepage barrier. Generally, the clay core is vertical and placed centrally below the top of dam. Shell zones of rockfill are provided on both sides of the core. A properly designed filter is provided between the core and the shell. The upstream slope is protected by rip-rap against wave action. The downstream shell is self-draining but arrangements are made to collect and drain out the seepage flow. The slopes of an ECRD are generally 2–2.5:1 (H:V) and are steeper than earthfill dams and flatter than CFRD. These dams are made either on a rock foundation or on a compact RBM. The section of an ECRD is similar to a zoned earthfill dam. For the guidelines of design of an ECRD refer I.S 8826.

12.4.3 Layout of Embankment Dams

An embankment dam can be constructed on any type of foundation and in any valley shape which may be either narrow or wide. In a narrow valley, the river section is plugged by an embankment dam and the spillway is generally constructed on one side of the dam in the abutment. Such a spillway is called a chute spillway. The other alternative is to provide a tunnel spillway away from the dam. The tunnel is made in the abutment to carry the river flow downstream. Diversion tunnels can also be used for this purpose. These tunnels are connected with the FRL/MWL through vertical or inclined shafts. The provision of an inclined shaft with a diversion tunnel as a spillway in a dam is shown in Fig. 12.26. The shaft inlet may be either a horizontal crest or a tunnel-type structure (morning glory). This can be gated or ungated. The flow characteristics in the shaft vary with the head over crest and the horizontal tunnel

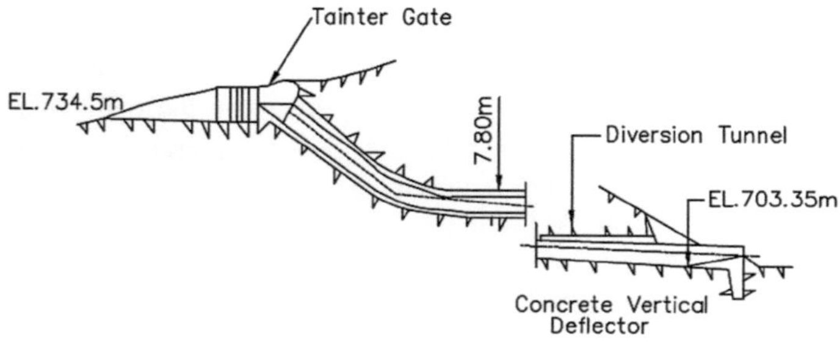

Fig. 12.26 Conversion of diversion tunnel into morning-glory spillway. Paradela Dam (Portugal)

is generally designed to run full. For details, readers may refer to Design of Small Dams—USBR.

In a chute spillway, which is commonly provided, the river discharge is conveyed from the reservoir to the river through an open channel called a chute which is constructed along the dam abutment. Flow in the chute is maintained as a supercritical flow, either at a constant or accelerating rate until it reaches the terminal structure where flow is sub-critical. Hence the terminal structure is a jump basin. The flow in the chute is regulated through a gated over flow crest provided in line with the dam axis and made perpendicular to the chute axis. The length and crest level are decided by the weir discharge formula. The sketch of a chute spillway showing the plan and section is given in Fig. 12.27. For detailed design I.S. 5186 may be referred.

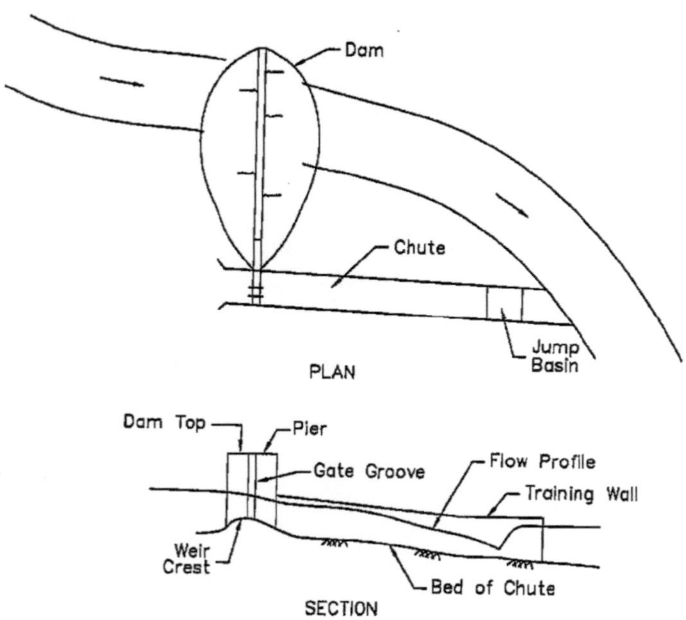

Fig. 12.27 Chute spillway

In the case of a composite dam, which is constructed in a wide river section comprising the main river channel and the flood plain, the embankment dams are provided on both sides of a centrally placed spillway on the main river channel. The spillways in such dams are generally provided as ogee-crested type concrete/masonry gravity dams. The design requirements of such spillways are the same as described above in Sect. 12.2.3. Sometimes to meet the construction requirements the spillway is provided in a location away from the main channel requiring the provision of an approach channel up to the main channel. One of the advantages in this case is that no river diversion arrangement is required for the construction of the dam. Later the river channel is plugged by the dam section. It often needs special measures to safeguard piping through the foundation of the plugged section. The interface of the abutment of spillway with earth section needs special care to check piping. Normally a 15–20-m-long concrete diaphragm from abutment wall of the spillway is made inside the core of the earthen section to break the straight seepage path.

Outlets are required to use water from the reservoir. The outlets are normally not provided in the body of embankment dams because improper interface bond between the concrete surface and soil of the dam becomes the cause of piping through the dam. It sometimes results in the failure of the dam. So normally intake structures for outlets are provided separate from the dam. These are described in a separate chapter on Intake structures.

If an outlet is provided in the earthfill dam, it should be provided with enough measures to avoid any possibility of piping. These measures shall include adequate compaction of soil at the interface of the outlet with the dam section. Collars 1.0 m high and 0.5 m wide at a spacing of 3 m as shown in Fig. 12.28 should be provided along the barrel. Filter should also be provided on downstream dam slope on both sides of the outlet. In some cases, the outlet is provided in a non-overflow block which is constructed in continuation of the concrete spillway.

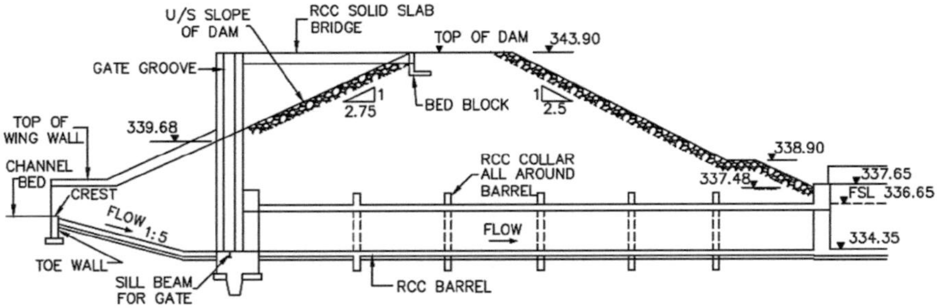

Fig. 12.28 Outlet barrel with collars in an earthfill dam

12.4.4 Design Requirements

The following requirements should be satisfied in order to ensure a satisfactory and safe design of an embankment dam:

- The slopes of the embankment should be stable under all conditions of construction and operation. The operational conditions are steady-state condition during normal operation and sudden drawdown condition which may occur during emergency emptying of the reservoir and this is critical for u/s slope.
- The embankment should not impose excessive stress on the foundation which is generally alluvium.
- Seepage through embankment, foundation and abutment must be controlled so that piping, sloughing of slopes or removal of material by solution may not occur.
- Freeboard should be sufficient to prevent overtopping of the dam
- Spillway and outlet capacity should be sufficient to prevent the overtopping of dam.
- The u/s slope should be adequately protected against wave action.

12.4.5 Design Procedure

The above requirements are further elaborated with the design procedure to achieve them in the following paragraphs.

(a) **Slopes**

The upstream and downstream slopes basically depend on the material which is used in the construction of the dam. There are several guidelines that have been suggested by individual researchers and the organisations such as Terzaghi and USBR which may be used to select the slopes and then these slopes should be analysed for safety. Generally, slopes recommended for small-height modified homogenous dams made of clayey soil are 3 to 4:1 (H:V) for u/s slope and 2.5:1 (H:V) for the downstream slope.

For zoned sections with thick clay core and shell of well-graded fragments or gravel and sand, the recommended slopes are 2.25 to 2.5:1 (H:V) for both U/s and D/s slopes. These slopes are normally analysed for various operational conditions by the method of slices in which the slip circle is divided into slices as shown in Fig. 12.29.

In this method, a circular arc with centre at O is drawn which is treated as the potential surface for the slide. If the moment of resisting force about O exceeds the moment of actuating forces, then sliding will not take place along the circular surface. For computation, the soil mass between the dam slope and the slide surface is divided into a number of slices through vertical lines as shown in Fig. 12.29. Generally, slices are made of equal width.

The forces acting on a slice are shown in Fig. 12.29b. In the Fellenius method (commonly used), the inter slice horizontal and vertical reactions are neglected and

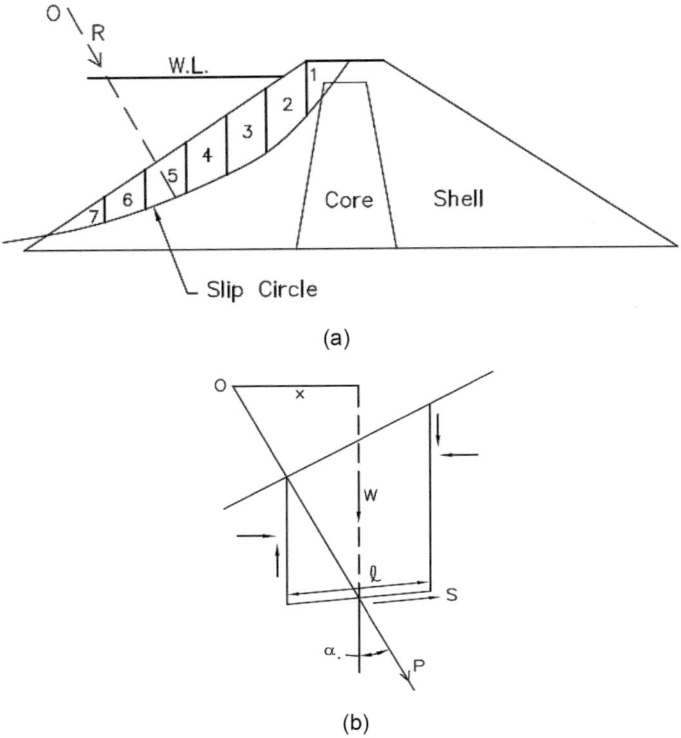

Fig. 12.29 **a** Slip circle and slices. **b** Showing forces on a slice

only weight of slice and the shear resistance S at the base of the slice are considered. So taking a moment of these two forces about O, we get

$\Sigma W.x = \Sigma S.R$, where Σ shows summation of all slices,

$S = \tau.1$, τ is shear stress and l is width of slice.

Shear strength of soil at failure τ_f by Coulombs equation is

$\tau_f = C + \sigma \tan \emptyset$ where, C and $\emptyset$ are shear parameters of soil.

The factor of safety $F = \frac{\tau_f}{\tau}$.

$$\therefore \quad \tau = \frac{1}{F}\left(C + \frac{P}{1} \tan \emptyset\right).$$

$X = R \sin\alpha$.

Hence $\Sigma W.R \sin \alpha = \Sigma S.R$

or $\Sigma W \sin \alpha = \frac{1}{F} \Sigma (Cl + P \tan \emptyset)$

$$F = \frac{1}{\Sigma W \sin \alpha} \Sigma (Cl + P \tan \emptyset).$$

or

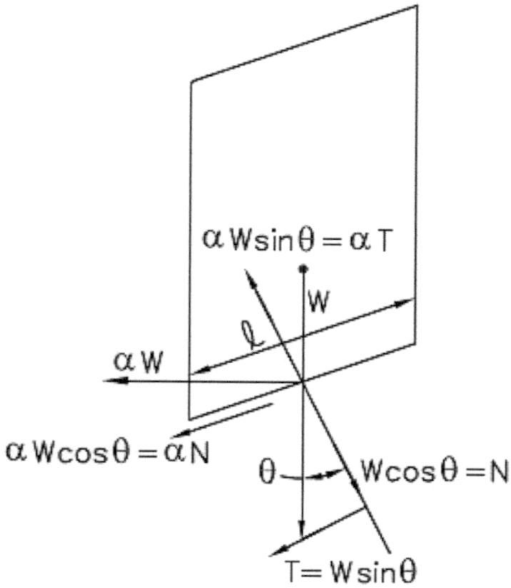

Fig. 12.30 Seismic forces on a slice

Putting $P = \mathrm{W}\cos\alpha$

$$F = \frac{1}{\Sigma \mathrm{W}\sin\alpha}\Sigma\cdot(\mathrm{Cl} + \mathrm{w}\cdot\cos\alpha\cdot\tan\emptyset).$$

F is worked out for different trial slip circles including those passing through the alluvial foundation till a circle with minimum *F* is obtained. This is the critical slip circle.

For computation, *C* and Ø are commonly used for undrained tests. This is called the total stress approach and is preferred because it is simple and eliminates the determination of pore pressures and gives reasonably reliable results for small- and medium-height dams.

The forces, considering earthquake force αW, acting on a slice are shown in Fig. 12.30. The factor of safety is given by

$$F = \frac{\Sigma(\mathrm{N} - \alpha\mathrm{T})\tan\emptyset + \Sigma\mathrm{C.l}}{\Sigma(\mathrm{T} + \alpha\mathrm{N})}.$$

The adopted slopes should ensure the following factor of safety:

Sl. no.	Test condition	Factor of safety	
		For normal loads	With earthquake
1.	End of construction	1.25	1.0
2.	Steady-state condition (D/S slope)	1.5	1.25
3.	Sudden drawdown condition (U/S slope)	1.25	1.0

A solved example illustrating the slip circle method procedure given in I.S. 7894 may be referred. Many computer programs are available for slope analysis.

(b) **Freeboard**

It is the vertical distance between the crest of the dam and the still water level in the reservoir. It should be adequate enough to avoid overtopping in any case. It depends on wave characteristics which depend on fetch (reservoir length) and the wind velocity. The wave height and the wind setup can be worked out for a reservoir by the procedure given in IS 10635.

The freeboard is taken as the sum of wave height plus wind set up plus expected settlement plus an additional margin of safety. Settlement allowance is normally taken as 1% of dam height. It is provided in the form of a camber which is maximum at the location of maximum height and varies to zero at the ends near the abutment. The freeboard is worked out both for FRL and MWL if they are different. A minimum freeboard of 2.0 m above FRL and 1.5 m above MWL are recommended.

(c) **Spillway Capacity**

Spillway capacity should be fixed on a conservative side to avoid overtopping of the embankment dam. It may be fixed according to guidelines given in I.S. 11223, for fixing spillway capacity. The length of the spillway should be decided in such a way that design discharge may pass with 10% of the bays being inoperative.

(d) **Slope Protection**

The u/s slope is protected against the damage due to wave action. The protection depends on wave height. Normally dumped rip-rap of strong rock is placed on the upstream slope. It is placed on a filter whose D_{85} should not be less than 5 cm and the ratio of D_{15} of rip-rap and/D_{85} of filter should not be more than 10. Both rip-rap and filter should be well graded. The recommended size of rip-rap for different wave heights is given below:

Max wave height (m)	D_{50} of rock size (cm)	Layer thickness (cm)	Filter thickness (cm)
0–0.6	25	30	15
0.6–1.2	30	45	15
1.2–1.8	38	60	22.5
1.8–2.4	45	75	25
2.4–4.3	60	90	30

In the rip-rap half of the rock should be larger than D_{50} but max^{m} size should not be more than 1.5 times of the D_{50} and minimum size should be 2.5 cm. Voids should be filled with spalls.

Sometimes hand-placed stone pitching or concrete blocks/slabs are also used for u/s slope protection.

The downstream slopes are protected from wind and rain erosion by favourable vegetal cover depending on soil and climate, on small- and medium-height dams. Below the turfing a 10-cm-thick suitable soil is placed above the dam slope material for growing grass. Pitching of gravel, boulder, tunnel muck, etc. is also used for slope protection. The provision of berms also protects the slope from rain erosion. It may be provided at an elevation difference of about 10–15 m. The minimum width of a berm is 3 m. A berm is desirable at the top of the rock toe. The part of the slope below the high tail water level shall be protected by rip-rap as for the u/s slope. For slope protection, I.S. 8237 may be referred to for details.

(e) **Seepage control**

In an embankment dam, seepage may take place both through the embankment and the foundation. Measures are taken to control this seepage. The seepage through soil is defined by Darcy's law.

$$V = -k \cdot i$$

where V is velocity of seepage flow through a cross section A of soil, i is the gradient denoted by h/L (h is head and L is length of seepage path) and K is permeability of soil. The seepage is given by

$$\mathrm{Q} = \mathrm{K\ i\ A}$$

K for clay is around 1×10^{-5} to $1 \times 10^{-7}\mathrm{cm/s}$ and in pervious soil it is 10^{-3} to $10^{-1}\mathrm{cm/s}$.

(i) *Seepage through dam*

The seepage through the body of dam takes place through the soil mass below the phreatic surface or line which is the upper surface of seepage established after filling the reservoir and a steady-state condition is achieved. The phreatic line in a dam section as suggested by Casagrande is a parabola. Casagrande has given a method to draw the phreatic line in different types of embankment dams and to compute the quantity of seepage. Any standard text on embankment dams may be referred to for Casagrande's method. The phreatic line in a homogeneous dam section and the modified homogeneous dam with different drainage arrangements are shown in Fig. 12.20. The phreatic line in a zoned section is drawn considering only the core portion which is like a homogeneous section.

(a) Showing rock toe.

(b) Showing horizontal drainage blanket with rock toe.

Fig. 12.31 **a** Showing rock toe. **b** Showing horizontal drainage blanket with rock toe

The seepage through a homogeneous section (see phreatic line in Fig. 12.20) causes wetting of a part of the D/s slope as the phreatic line cuts the slope. This may cause sloughing of the slope and result in the collapse of the dam. Hence, to control this condition drainage arrangements like rock toe, horizontal blanket, chimney drain, etc. are provided in the earthfill dam section which is shown in Fig. 12.20.

The rock toe is made of rock/boulder fragments at the toe of the dam section in a height of 1/3 to 1/4 of the dam height. The sloping face of the rock toe inside the dam section is made perpendicular to the downstream slope and a transition filter is provided between the soil of the dam and the rock toe as shown in Fig. 12.31a. A perforated PVC pipe of 10–15 cm diameter is placed inside the rock toe along the dam to collect and drain out the seepage. The pipe is placed open-jointed.

The horizontal drainage blanket made of filter material is placed at the dam base above the foundation and below the d/s slope of dam. It lowers the phreatic line sufficiently away from the downstream slope. The lowering of the phreatic line depends on the length of the blanket. A drain is provided at the end to collect and drain out the seepage. The blanket is generally combined with rock toe as shown in Fig. 12.31b. Generally, the length of the drainage blanket is provided up to the mid-point of the D/s slope. The blanket is made of layered filter material. The central layer is a coarse filter. The gradation of fine and coarse filters should satisfy filter design criteria. Each layer should be about 200 mm thick. The depth of the drain is about 1.5 m and its bottom width is 1.0 m and the side slopes are 1:1. The perforated PVC pipe is 15 cm in diameter and laid open-jointed at the sufficient longitudinal slope to drain out seepage in natural drainage.

A chimney filter is an inclined drainage blanket and is normally provided in the zoned embankment along the slope of the clay core which acts as a seepage barrier in the dam. The chimney filter prevents the movement of clay particles of the core in the shell and provides free drainage to seepage flow so that the D/s shell does not get wet. The chimney filter is combined with the horizontal blanket and the drain to escape seepage out of dam body. It is shown in Fig. 12.32. The chimney filter comprises fine and coarse filter layers. The fine filter layer is in contact with the core.

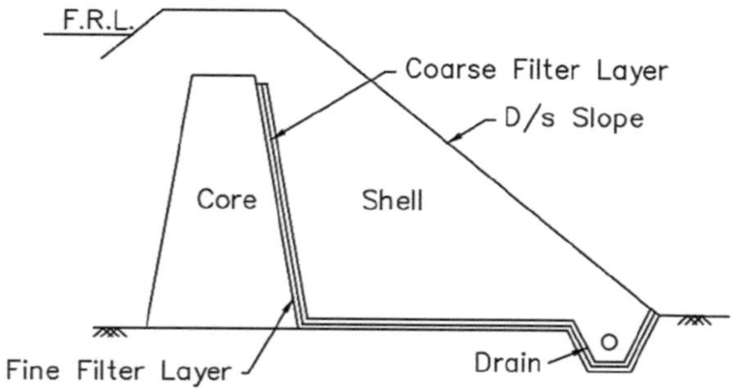

Fig. 12.32 Chimney filter in a zoned dam section

Each of the two filter layers is made 2–3 m thick for proper compaction. The filter layers should also satisfy filter design criteria which are given below.

The principal requirements for a satisfactory filter are given below:

- It must be more pervious than the protected soil.
- It must be fine enough to prevent particles of protected soil from washing.

To satisfy the above requirements, the filters are designed to satisfy the following criterion (refer I.S. 9429):

- $\frac{D_{15} \text{ of filter}}{D_{15} \text{ of soil}} > 5 - 20$ (Sherard recommends this value as 4–5).
- $\frac{D_{15} \text{ of filter}}{D_{85} \text{ of soil}} < 5 - 4$.
- Gradation curve of the filter should practically have the same shape as that of the protected material.
- Filter should not contain more than 5% fines (0.074 mm) and maximum size should be limited to 75 mm. The filter material should be cohesionless.
- When protected soil contains large sizes such as gravel, the filter should be designed on the basis of the gradation curve of the material which is less than 25 mm in size.

In the case of a rockfill dam, the following measures check seepage through dam:

- In ECRD, the measures are made similar to those of a zoned embankment which are described above.
- In CFRD, the seepage barrier is generally an RCC slab on the upstream slope of dam. It rests on a free draining bedding plane and so no extra measures are generally provided.

(ii) *Seepage through Foundation*

The measures to control seepage through foundation depend on the type of foundation which may be either rock type or alluvial. The rockfill dams are generally made on rock foundations. In this case, the seepage through the rock foundation is controlled by grouting of the foundation. The foundation grouting adopted for the two types of rockfill dams is shown in Fig. 12.33.

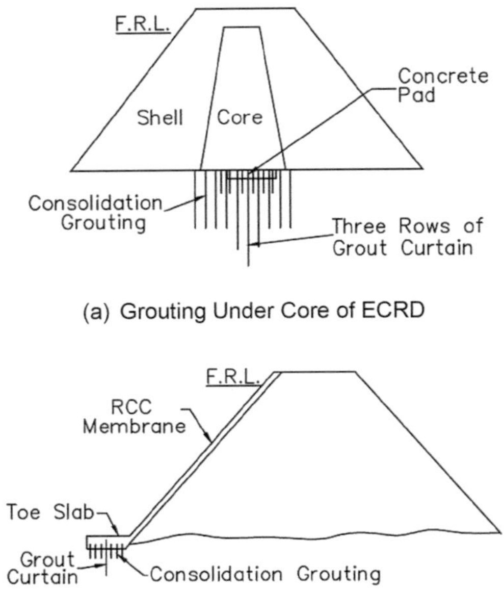

(b) Grouting for CFRD

Fig. 12.33 Grouting of foundation of rockfill dams

In ECRD, the entire area below the core is provided with consolidation grouting, and generally three rows of curtain grouting are provided in the centre of the core base. Curtain grouting is generally done through a concrete pad in the core base. In CFRD, the foundation grouting is done through the toe slab as shown in Fig. 12.33b. The grouting details are the same as described for a concrete gravity dam. In case the foundation below the rockfill is the RBM then normally a concrete diaphragm cut-off is provided.

The earthfill dams, both modified homogeneous and zoned section types, are generally constructed on an alluvial foundation. In these cases, measures are taken to reduce the seepage flow through the foundation and drainage arrangements are provided for safe exit of seepage flow. The seepage control measures are described below. The drainage arrangements are normally the same as for a homogeneous dam section, such as a horizontal drainage blanket with a drain. Relief wells are provided additionally when required.

The positive or full cut-off is provided by taking the core through the entire depth of alluvium over impervious strata or rock. It is provided when the depth of pervious foundation layer is 10–15 m. It is shown in Fig. 12.34. In case the impervious strata or rock foundation is highly jointed and fissured with high permeability then grouting is done in the rock foundation through the base of the clay core which has been made in the alluvial layer.

Fig. 12.34 Showing cut-off

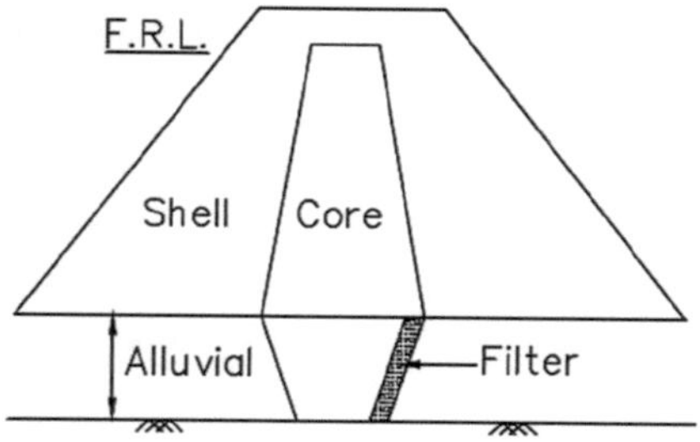

Partial cut-off (one which does not reach up to the impervious strata) is not found effective in reducing seepage significantly and so this is not preferred.

In cases where the alluvial foundation depth is large for a positive cut-off of clay, the concrete diaphragm is provided as a positive cut-off. These are 60–90 cm wide and constructed in panels 2–5 m long. A trench is excavated and filled with bentonite slurry to hold the walls of the trench. Later the slurry mud is removed, the reinforcement cage is lowered for the panel and thereafter tremie concrete is poured starting from the bottom. Alternate panels are cast. This is a specialised technique and an experienced company should be engaged. These are made both of conventional concrete as well as plastic concrete. These are made generally at the centre of the clay core.

Grouted cut-offs are also made by injecting grout material such as cement, bentonite, chemicals (such as sodium silicate) or a combination of these in the foundation below the core of the dam. These do not reduce permeability very significantly and hence generally used for the construction of coffer dams, etc.

As an alternative to the cut-offs an impervious blanket in the upstream of dam is constructed. In the downstream of the dam, relief wells are made if required. A sketch of the dam section showing the blanket and the relief well is shown in Fig. 12.35.

The length of the blanket is governed by the extent of reduction required in seepage. It can be worked out as below using Darcy's law and assuming the blanket is impervious (refer Fig. 12.35).

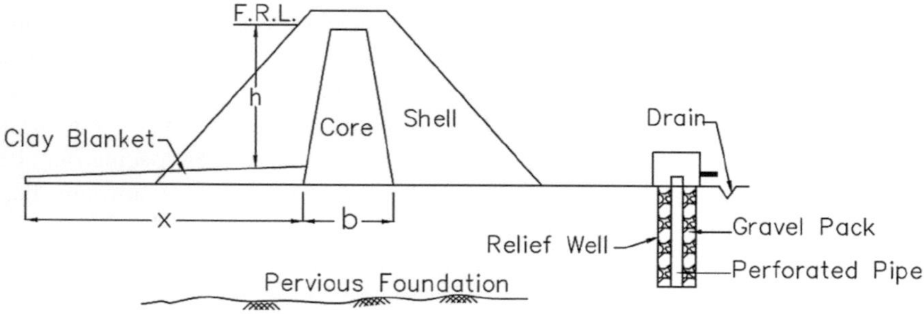

Fig. 12.35 Clay blanket and relief well

Seepage discharge without blanket:

$$q = K \cdot \frac{h}{b} \cdot d,$$

where d is thickness of pervious foundation.

Seepage discharge with blanket:

$$q_b = K \cdot \frac{h}{b+x} \cdot d$$

$$\frac{q_b}{q} = \frac{b}{b+x}$$

$$x = b\left[1 - \frac{q_b}{q}\right] / \frac{q_b}{q}.$$

Hence x is the length of blanket which depends on the value of $\frac{q_b}{q}$. The optimal length of the blanket beyond which its effectiveness does not increase appreciably is given by

$$x = \frac{\sqrt{2}}{a},$$

where

$$\frac{1}{a} = \sqrt{\frac{k_f}{k_b} z_b z_f},$$

where

k_f Permeability of foundation.
k_b Permeability of blanket.
z_b Thickness of blanket.
z_f Thickness of pervious foundation.

Normally thickness of the blanket is 10% of the water head above it with a minimum of 1 m. A length of 8–10 times of water head is generally found adequate.

A relief well, which is generally provided for the safe exit of seepage with a blanket, is a vertical hole that has a slotted pipe (generally of PVC) normally 15 cm in diameter centrally placed surrounded by the suitable graded filter of a thickness of 15 cm. The relief well spacing is normally 15–30 m but closer spacing may be provided if required to intercept seepage. At the top of the well, a non-return flap valve is provided with a suitable cover and an overflow arrangement.

(f) **Foundation stress**

The average stress being transferred to the foundation by an embankment dam is of low magnitude because of the large base width and is not of concern in case the foundation is rocky or compact RBM. It should be checked in the case of an alluvial foundation. The embankment dam on the alluvial foundation should be constructed after removing loose material and compacting the foundation if required so that excessive settlement may not take place.

References

1. Asthana, B.N., Khare, D.: Dam Engineering. Capital Pub. Co. New Delhi, India (2019)
2. Singh, B., Varshney, R.S.: Engineering for Embankment Dams. Oxford and IBH Publishing Co. New Delhi (1995)
3. Davis, C.V.: Hand Book of Applied Hydraulics. McGraw Hill, New York (1976)
4. Design of Small Dams. A water Resources Tech. Pub. (India Edition) USBR (2011)
5. Guidelines for Small Hydro Development, Standard No. 2.2.: Civil Works Hydraulic and Structural Design. AHEC, IIT, Roorkee (2013)
6. Indian Standards.: BIS, Manak Bhawan. New Delhi which are referred in the text are listed below

   (i) IS 12200.: Provision of Water Stops at Transverse Construction Joints in Masonry and Concrete Dams (2001)
   (ii) IS 6512.: Criteria for Design of Solid Gravity Dams (1998)
   (iii) IS 11223.: Guidelines for Fixing Spillway Capacity (1985)
   (iv) IS 5497.: Guide to Topographical Surveys for River Valley Projects (1998)
   (v) IS 15662.: Investigations for Gravity Dams
   (vi) IS 15686.: Preparation of Geological Maps
   (vii) IS 4453.: Exploration of Pits, Trenches, Drifts and Shafts
   (viii) IS 6934.: Hydraulic Design of High Ogee Overflow Spillways
   (ix) IS 13551.: Structural Design of Spillway Pier and Crest
   (x) IS 7365.: Criteria for Hydraulic Design of Bucket Type Energy Dissipators
   (xi) IS 12966.: Code of Practice for Galleries and other Openings in Dams
   (xii) IS 1893.: Criteria for Earthquake Resistant Design of Structure
   (xiii) IS 10135.: Code of Practice for Drainage System for Gravity Dams, their Foundation and Abutments
   (xiv) IS 7436.: Guides for Type of Measurements for Structures in River Valley Projects and Criteria for Choice and Location of Measuring Instruments (1) For Earth and Rock Fill Dams (2) For Concrete and Masonry Dams
   (xv) IS 8826.: Guidelines for Design of Large Earth and Rockfill Dams
   (xvi) IS 5186.: Design of Chute and Side Channel Spillways—Criteria (2004)
   (xvii) IS 10635.: Free Board Requirements in Embankment Dams
   (xviii) IS 7894.: Code of Practice for Stability Analysis of Earth Dams
   (xix) IS 8237.: Code of Practice for Protection of Slope for Reservoir Embankment
   (xx) IS 9429.: Drainage System for Earth and Rockfill Dams

7. RCC-Design and Construction for Hydraulic Structures.: USBR Denever, Colorado (2005)

8. RCC for Gravity Dams—State of Art. ICOLD Bulletin No. 75 (1989)
9. Sharma, H.D.: Concrete Dams. Pub. No. 266, CBI&P, New Delhi (1998)
10. Sharma, H.D.: Embankment Dams. Oxford and IBH Publishing Co. New Delhi (1991)
11. Varshney, R.S.: Hydropower Structures. Nem Chand and Sons, Roorkee (2010)

Chapter 13
Intake Structure

13.1 General

Intake is a structure to draw water from the source and to let it pass into the water conductor system in the required quantity and quality. The source of water can be a diversion structure or a storage dam on a river or a canal. The water conductor system can be either an open channel or a conduit or tunnel depending on the topography and geology of the area. The type of intake structure depends on both the source of water and the type of water conductor. The different types of intake structures that are normally adopted are briefly described in this chapter.

13.2 Functions of an Intake

The intake structure has to perform the following functions:

(i) To ensure regulated flow into water conductor
(ii) To minimise sediment entry
(iii) To prevent entry of trash and floating debris into the water conductor
(iv) To minimise head loss through intake
(v) To prevent air entry if water conductor is either a conduit or tunnel running under pressure.

13.3 Components of Intake Structure

The intake structure generally consists of the following components for satisfactory performance:

B. N. Asthana, *Planning and Design of Small Hydro Power Projects*,
https://doi.org/10.1007/978-3-031-96271-4_13

(i) The intake opening is fitted with a regulating gate to regulate the flow into the water conductor. Upstream of the regulating gate an emergency gate or stoplog is provided for repair and maintenance of regulating gate.
(ii) The intake opening is also fitted with a trash rack and its cleaning arrangement. This prevents the entry of trash and floating material into the water conductor.
(iii) The intake opening is provided with a bell mouth entry for minimizing head loss.
(iv) To minimise sediment entry through the intake, sediment exclusion devices are provided upstream of the intake structure. Generally, an excluder is provided when the source of water is a diversion structure (weir or barrage) and the water conductor is an open channel. In other cases, generally desilting basins downstream of the intake structure are provided. These are discussed in more detail in the chapter on Sediment Management.
(v) An air vent is provided downstream of the intake gate when a tunnel running under pressure is provided as a water conductor.

13.4 Planning and Design Requirements

(i) Operational requirement: The intake structure and its components and gates etc. should be planned in a manner so that they can be easily approached for quick operation when required. All the components of the intake structure should be easily approachable for inspection and maintenance. There should be enough space for storage and assembly of gates.
(ii) Hydraulic requirement: The velocities through the trash rack, gates and other passages should be limited to a value which may ensure economic design, streamlined flow and minimum head loss.
(iii) Structural requirement: The intake structure should be stable under the worst combination of loads and forces including seismic forces while operating under fully open, partially open and fully closed conditions.

13.5 Factors Affecting the Layout

The following factors affect the selection and layout of an intake structure.

(i) The type of scheme: The SHPs are generally of two types (a) run–of-river type and (b) storage type.

    (a) The run of river type is further classified as (1) diversion through a weir/ barrage into a canal and (2) diversion through a small dam or high head barrage into a tunnel.
    In the case of a weir/barrage and canal-type project a canal head regulator by the side of diversion structure and in its upstream is provided.

In the case of a small-height diversion structure and the tunnel, a tunnel-type intake by the side of the diversion structure and in its upstream is provided.

(b) A storage dam may be either a concrete dam or an embankment dam. With the storage dam, the powerhouse will be at the toe of a dam or away from the dam.

In the case of a concrete dam with a powerhouse at its toe, the intake is made in the blocks of the non-overflow dam portion. It may be a straight type for low and medium heads and a semi-circular type for high heads. If the powerhouse is away from the dam, the intake may be separate from the dam in the abutment. It may be either tower-type or tunnel-type intake.
In the case of an embankment dam, the intake is separate from the dam. It is generally either a tower-type or a tunnel-type intake in the abutment.

(ii) Topography and geology of site: The topographical and geological features of the site govern the selection of type and size of the intake structure if it is to be constructed with a diversion or storage dam.
(iii) Quality of water: The sediment content and its characteristics, which are permitted through the turbine, affect the site selection and other design parameters of the intake structure such as crest level and the provision of sediment exclusion arrangement in the upstream and downstream of the intake structure. The downstream arrangement of sediment exclusion needs extra discharge from the intake for flushing.

13.6 Design Aspects of Intake Structures

13.6.1 Intake Structure with a Weir/Barrage

(a) To divert water into a canal, a canal head regulator is provided as an intake structure. A typical sketch (plan and section) of a canal head regulator is shown in Fig. 13.1.

Generally, in a weir or barrage an undersluice with a low crest (normally at riverbed level) is provided on the canal side and the intake is provided just upstream of the undersluice. A divide wall extending up to the upstream end of intake is provided between the undersluice and the weir/barrage bays. The crest of intake is fixed as much above the crest of the undersluice as possible but normally not less than 1.5 m. The axis of the intake crest is aligned at 100° to 110° to the weir/barrage axis. The crest of intake is normally a broad-crested type. The length of the crest is worked out by the weir formula ($Q = C_d.\mathrm{L}.\mathrm{H}^{3/2}$) for the required discharge. The value of C_d is normally 1.7 for a free flow. In the case of an ogee-type crest, the value of C_d is 2.1 for a free flow. The reduction in the value of C_d shall be determined for the submerged flow (when TWL is above the crest) from standard texts.

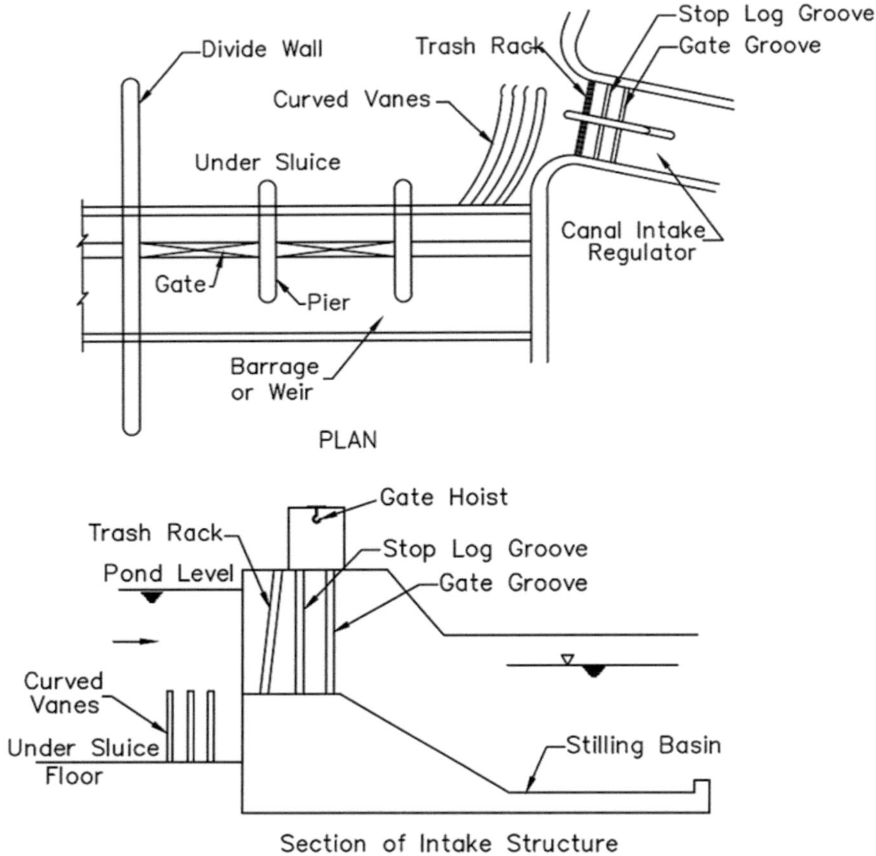

Fig. 13.1 Canal head regulator

The crest length L is divided into bays which are separated by piers. Between the piers the regulation gates, which are normally fixed wheel vertical lift gates, are provided and in the upstream of gates the provision for stop logs and trash rack is made as indicated in Fig. 13.1. The velocity through trash rack is also limited to about 1 m/s. The faces of the upstream nose of piers and abutment are made circular for a smooth entry of flow. For the energy dissipation in the downstream of intake crest, generally, a jump basin (refer to spillways in the chapter on Dam Design) is provided.

The head loss through the intake entrance is due to the following two factors:

(i) Sudden change in velocity: The velocity V_f upstream of the trash rack which is generally zero in the ponded condition is changed to velocity 'V' just after the trash rack.

The drop in water level due to a change in velocity when there is no direction change is given by

$$\frac{v^2}{2g} - \frac{v_f^2}{2g}, \quad \text{if } v_f = 0, \quad \text{it is } \frac{v^2}{2g}$$

(ii) Due to contraction at the crest of the intake.

Due to contraction at the inlet section, the head loss depends on the extent of contraction and the flow condition. It is expressed as a fraction of the velocity head i.e. $K\frac{v^2}{2g}$. For a smooth sill and streamlined shape of side walls and piers the value of K is approximately 0.3.

Hence total head loss $= K\frac{v^2}{2g} + \frac{v^2}{2g} = 1.3\frac{v^2}{2g}$.

If $V = 1$ m/s, total head loss is 6.5 cm.

Therefore, normally a heading up of 15 cm is considered adequate to force the flow into the canal through the canal regulator.

Normally the floor of the crest and the jump basin are designed for both the surface and sub-surface flows. The total length of the floor is a function of downstream cut off (to be fixed from scour consideration) and permissible exit gradient which depends on the type of strata. The surface flow studies (jump basin design) decide the length of the downstream floor and the rest of the floor-length is normally provided in upstream of the crest. The downstream floor is designed either as a gravity floor or an RCC raft against the uplift pressure caused due to sub-surface flow. The structure should also be checked for sliding and floatation. In case the river flow carries a large proportion of bed load, the curved vanes (Fig. 13.1) or the tunnel type excluder is provided in the upstream of intake crest.

The impact of the topography of the banks and the river bed sometimes requires some modification in the intake structure. If the river banks are high, contour canals are sometimes made as water conductors. The intake arrangement in such a case is typically shown in Fig. 13.2.

In this type, a high crest is provided at the diversion point to prevent the entry of bed load in the intake. The length of the crest can be worked out with the weir discharge formula described above for the diverted flow and the mean head over the weir. The value of coefficient discharge will be 0.95 times the values given above. The sediment which is deposited in the pool between the crest and the regulator is regularly flushed out through the undersluice bays. The power channel below the regulator runs almost parallel to the river for some distance and may need protection. The regulator has all the components of the intake structure.

(b) To divert water into a tunnel: In case the topography and geology of the river bank necessitate the provision of a tunnel then the intake structure may be as shown in Fig. 13.3. There should be adequate water seal below MDDL over the tunnel entry. The trash rack is provided at the raised crest and the regulation gate and stoplogs are provided at the tunnel entrance as shown in Fig. 13.3.

(c) Intake for the powerhouse on the canal: When a powerhouse is located on a canal to utilise head available at a fall in the canal, the intake is generally made as a part of the powerhouse structure. Since the head is small in such a powerhouse, either a Kaplan turbine with concrete casing or a tubular turbine or bulb turbine

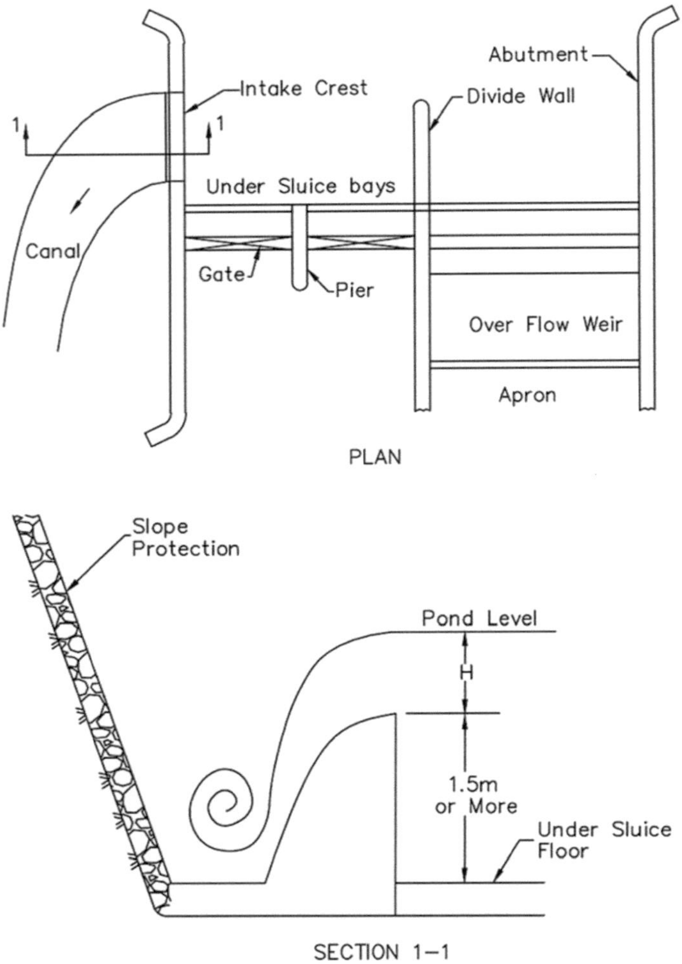

Fig. 13.2 Showing intake for a contour canal

is used. The intake as part of the powerhouse structure in the case of the Kaplan turbine is shown in Fig. 13.4. An intake with a bulb turbine is shown in Fig. 13.5. In these intakes the trash rack is provided at the entry and the arrangement for gates is also made near the trash rack. Suitable bell mouthing is also done at the entrance for minimising hydraulic losses.

(d) Forebay Intakes on Canals (Penstock intakes): In an open channel type SHP the canal is terminated into a long, wide and deep basin known as a forebay which stores the discharge for a short period (say 2 min duration) to facilitate the start of a turbine. A typical layout of forebay intake is shown in Fig. 2.14. Penstocks convey water from the forebay to the turbine. An intake structure is provided at the end of the forebay for the penstocks. This intake structure has a rash rack,

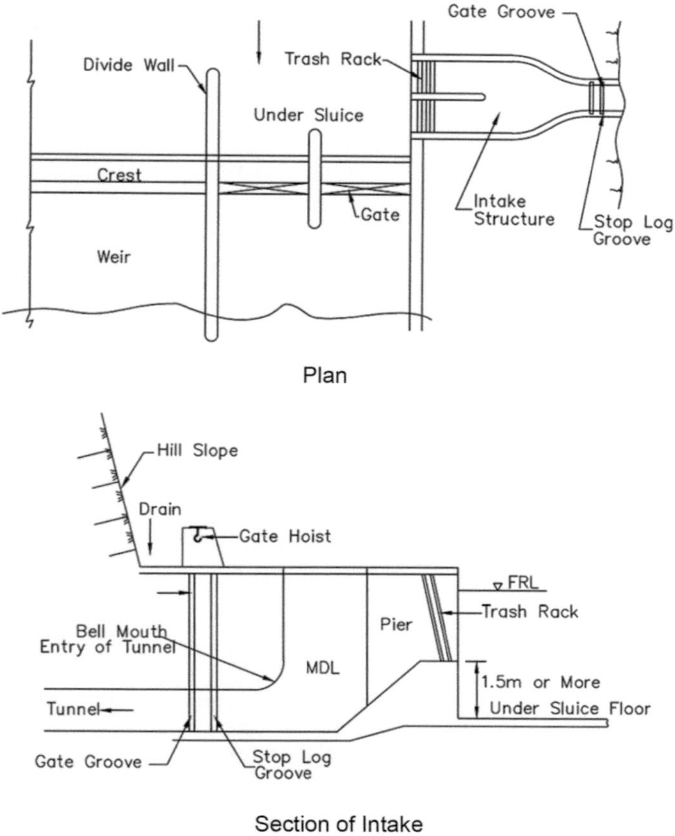

Fig. 13.3 Typical details of tunnel intake

regulation gate, and stop log arrangement at the entrance of the penstock. An air vent is provided downstream of the gate. The stream-lining of the intake entry and the transition from rectangular gate opening to circular penstock pipe in the downstream of gate are also provided. The bell mouth entry details for small diameter penstocks as suggested by CBI&P (Pub. No. 175) are shown in Fig. 13.6. The composite intake structure is a gravity structure and shall be tested for stability.

S.no.	Diameter of penstock (D) in mm	Height of bell mouth opening in mm H = 2D	Length of bell mouth opening in mm = D	Min distance of gate slot from trash rack in mm 0.75d	Width of bell mouth opening in mm B = 1.42857D
1	300	600	300	225	429
2	1000	2000	1000	750	1429
3	1400	2800	1400	1050	2000
4	1600	3200	1600	1200	2286

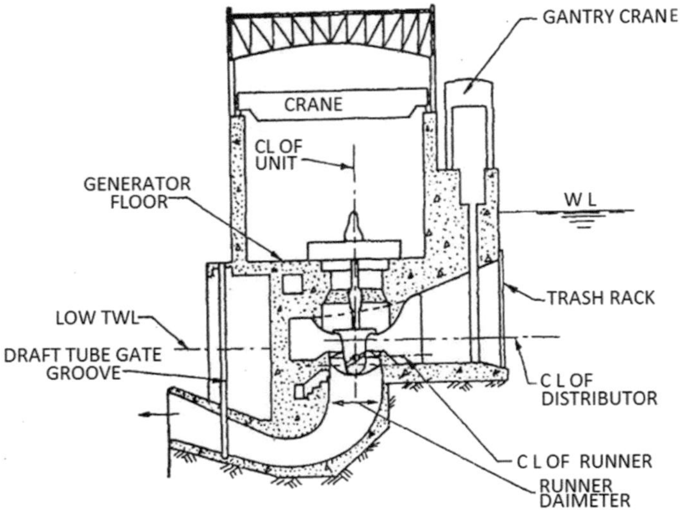

Fig. 13.4 Canal power house intakes (Kalpan Turbine) (AHEC-2.2)

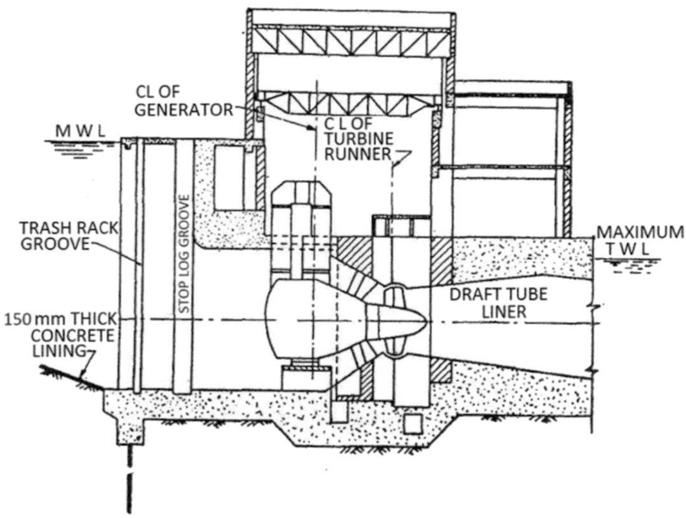

Fig. 13.5 Canal power house intakes (bulb turbine) (AHEC-2.2)

13.6.2 Intakes with Dam Projects

(a) **Diversion Dams**: For the intake with a low height diversion dam or a barrage with high headgate there may be two cases one with a canal as a water conductor and the other with a tunnel.

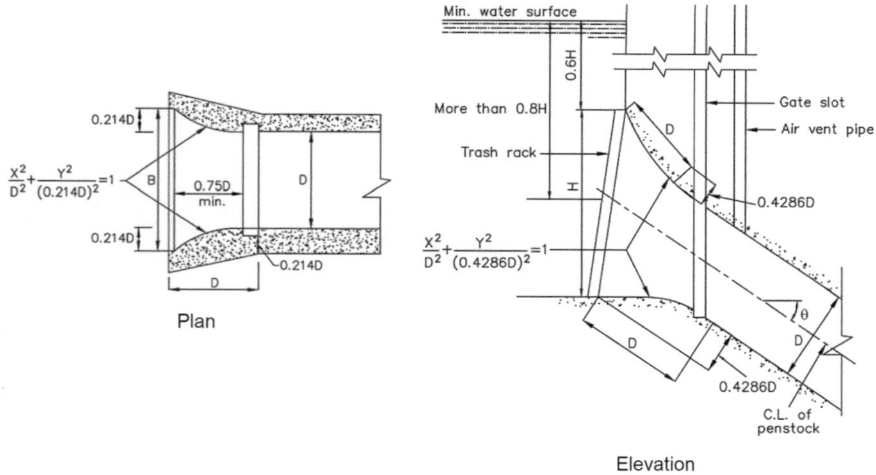

Fig. 13.6 Details of bell mouth entry (CBI&P. Pub. 175)

(i) In the case of canal, the intake structure will be similar to a canal regulator shown in Fig. 13.1. In this type the crest of the intake structure will be below the MDDL and will be sufficiently above the river bed preventing the entry of bed load in the intake. Hence, no sediment exclusion devices are placed upstream of the intake crest. For ejecting the harmful suspended load entering the intake, either an ejector or desilting basin is provided downstream of the intake.

(ii) In case the water conductor is a tunnel then a tunnel-type intake as shown in Fig. 13.3 may be adopted. It can be seen that the intake crest is kept as much above the u/s floor of the barrage as possible. In this case, for the removal of the deposited bed load in front of the intake crest some openings in the wall of the intake crest can be provided. These are connected to a pipe in the crest which will discharge the sediment-laden flow in the river downstream of the barrage.

(b) **Storage projects**: The storage dams are generally either concrete gravity dams or embankment dams. These are described in Chap. 12. The intake structures are different for the two types of dams.

(i) *Intake with Concrete Gravity Dams*

In the case of a concrete gravity dam with power house at the toe of the dam, the intake is made in one or more of the non-overflow blocks of the dam and water is conveyed to the powerhouse through penstocks from the intake to the turbine. Such an intake in low and medium head projects (head upto 30 m) is generally a straight type structure as shown in Fig. 2.7. It can be seen that in this type of intake, the trash rack is placed vertically or inclined in line with the u/s face of the dam. The opening at the entry is rectangular to place the trash rack and regulation gate which is operated

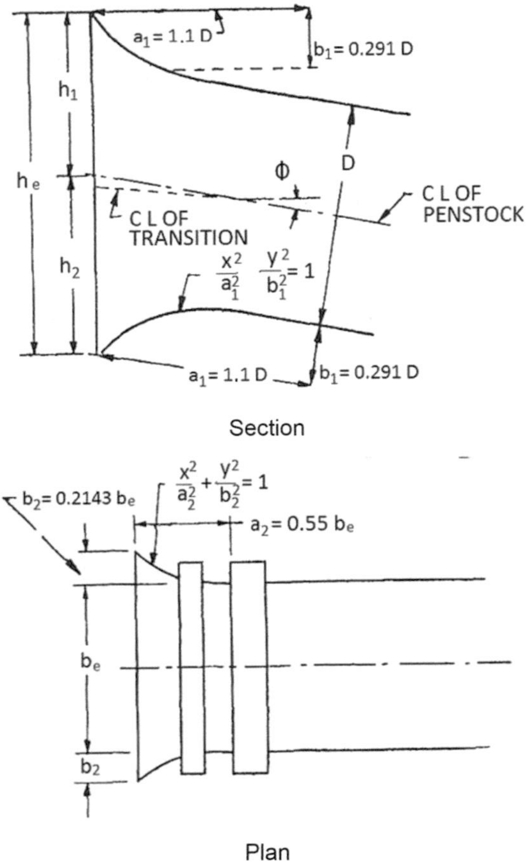

Fig. 13.7 Bellmouth details for lower and upper nappe (IS-9761)

from the top of dam through a gate shaft. The top surface and side faces of the opening are smooth curves (ellipse). The rectangular opening is gradually changed into the circular shape of the conduit or penstock. Downstream of the regulation gate, an air vent is provided. The equations of the elliptical curves at the entry are shown in Fig. 13.7.

In case of high-head concrete dams, a semi-circular intake structure is made in the reservoir joining the upstream face of the non-overflow blocks of the dam. It is shown in Fig. 13.8. The trash rack is placed in the panels on the periphery of the semi-circular cage-type structure which is made of columns and beams. The radius of the semi-circular trash rack from the face of the opening is 1.1428 times the width of the opening. The faces of columns and beams of the cage supporting the trash rack are stream-lined. The gate is placed at the rectangular opening of the conduit/penstock. The rectangular section is gradually changed into a circular section. The intake entrance has smooth elliptical surfaces on the sides, top and bottom. The equations of curves of rectangular opening are shown in Fig. 13.8.

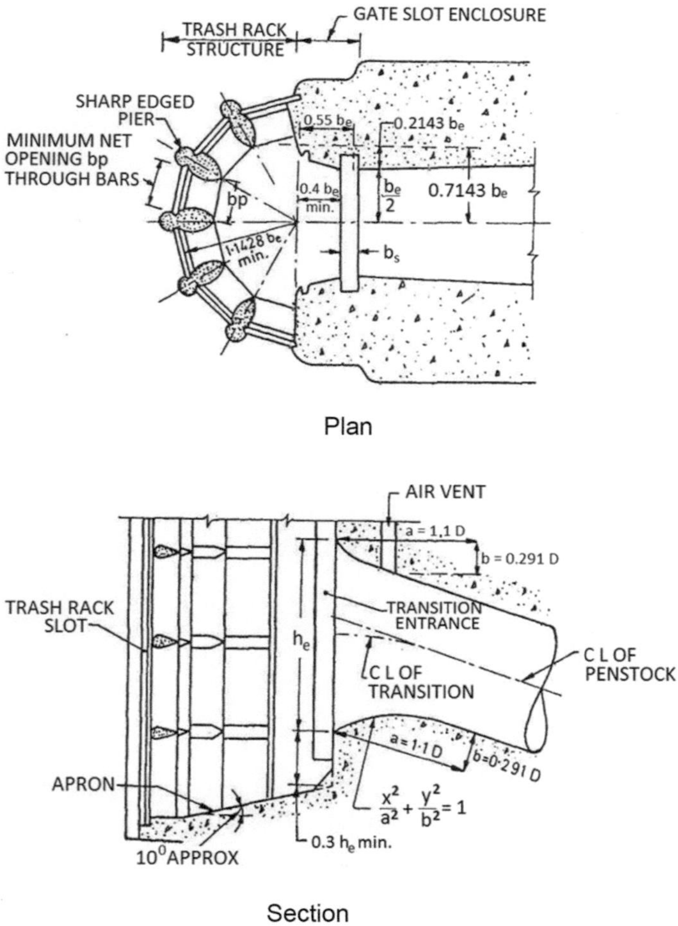

Fig. 13.8 Semi circular type intake structure (AHEC-2.2)

The bell mouth opening and transition in both the above types of intakes in dams are worked out as below.

- Size of opening: The opening area is given by

$$\text{Opening area} = \frac{\text{Penstock area}}{C_c \cos \emptyset}$$

where, C_c is coefficient of contraction ($C_c = 0.6$ for medium and high heads and $C_c = 0.7$ for low heads) and Ø is the angle of inclination of the centre line of penstock from the horizontal, it will be zero for a horizontally laid penstock.

The height of opening (h_e) is worked out as below (Ref. Fig. 13.7).

$$h_1 = D\left[\left(1.21\tan^2 \emptyset + .0847\right)^{1/2} + \frac{1}{2\cos\emptyset} - 1.1\tan\emptyset\right]$$

$$h_2 = D\left[\left(\frac{0.791}{\cos\emptyset} + 0.077\tan\emptyset\right)\right]$$

$$h_e = h_1 + h_2$$

$$\text{Width of opening } b_e = \frac{\text{Area of opening}}{h_e}$$

- Shape of opening: The inlet is streamlined to minimize losses. The profile of roof and bottom (refer Fig. 13.7) is generally given by following the equation of ellipse.

$$\frac{x^2}{(1.1\text{D})^2} + \frac{y^2}{(0.291\text{D})^2} = 1$$

The profile of the sides follows the following equation

$$\frac{x^2}{a_2^2} + \frac{y^2}{b_2^2} = 1 \text{ where, } a_2 = 0.55b_e \text{ and } b_2 = 0.2143b_e$$

- Transition: The rectangular section of the entrance is provided up to a certain distance downstream of the gate groove. Thereafter the rectangular section is gradually changed into the circular section of penstock/conduit. The transition should be designed with the following requirements:
  - The transition should be made about the centre line of mass flow and should be gradual.
  - Side walls should expand at a maximum of 5° from the centre line of mass flow.

The centre line of intake should be sufficiently below the minimum draw down level (MDDL) to avoid vortex formation at the entry of the intake. It depends on the approach flow conditions, velocity and Froude no. $\left(\frac{V}{\sqrt{\text{gD}}}\right)$.

For large intakes with $F_r \leq 1/3$, the submergence 'h' above the centre line of penstock should be 1–1.5 times 'h_e' (height of intake) and for small size intakes ($F_r \geq 1/3$) the submergence 'h' should be $(0.5 + 2F_r)$ D. In any case, the submergence above the centre line of the penstock should not be <0.8 h_e.

Sometimes the flow approach conditions and location of intake produce vortex formations. In that situation, anti-vortex devices are provided. These may be evolved on a hydraulic model.

In these two types of intakes, the gate shaft and conduit/penstock opening in the dam are structurally designed as cavities in mass concrete. The trash rack cage in the semi-circular intake is designed as a multi-storey structure made of columns and beams. The cage is normally closed with a slab at the top above the MDDL.

(ii) *Intakes with Embankment Dams*

In the case of embankment dams generally, the intake is located away from the dam. It is normally located in the abutment depending on the topography and geology and on the location of the water conductor which is generally a tunnel. The two types of intakes generally adopted are as below.

(a) Tower type intake: The tower type intake as shown in Fig. 13.9 has a circular tower type trash rack supporting structure generally resting on a sound rock bench made at a suitable location in the abutment rock. There is a vertical shaft below the tower-type rack structure. The shaft has a circular bell mouth entry. The shaft is connected with the tunnel through a bend/elbow. The vertical shaft should preferably be located in good rock. The flow in the tower intake is generally controlled either by a single cylindrical gate or by a number of gates in the tower.
(b) Shaft type intake: This type of intake is provided independent of the dam. In this type of intake, as shown in Fig. 13.10, the trash rack is provided at the entrance of the tunnel at the slope of the abutment. The trash rack may be straight type, along the slope or semi-circular type depending on the head. The gate is provided much downstream of the entrance and is operated from a vertical shaft approachable from hilltop. The bell mouth entry, size of opening and the transitions in this type of intake are similar as described above. The transition is

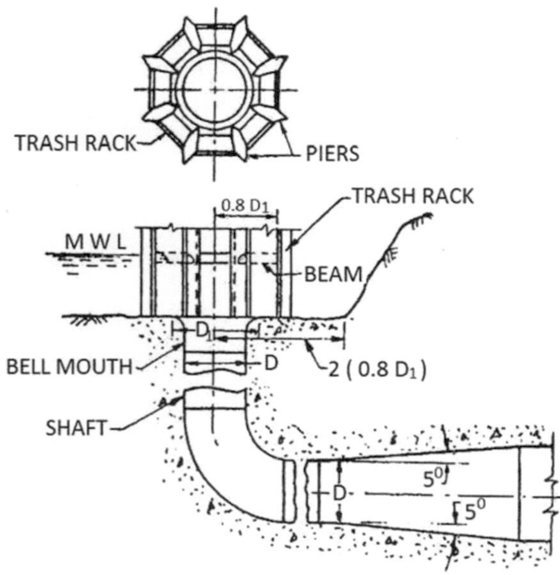

Fig. 13.9 Tower type intake (I.S. 9761)

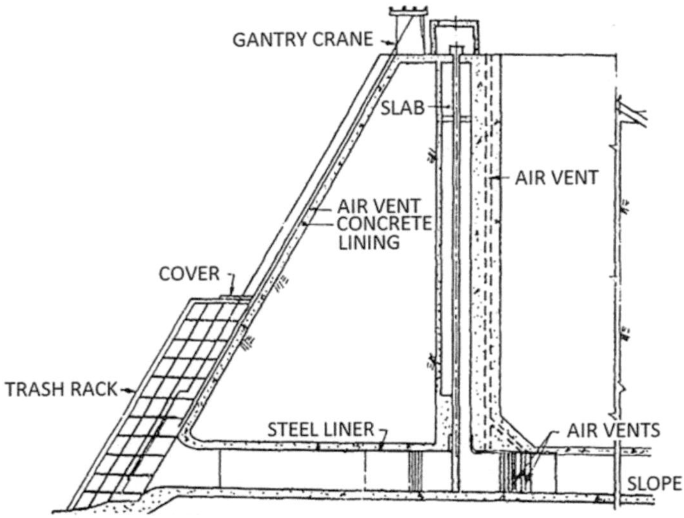

Fig. 13.10 Shaft type intake (P.S. Nigam)

provided from the downstream of the gate shaft. This type of intake needs good abutment rock for the construction of the gate shaft. For detailed hydraulic and structural design of such an intake, a reference may be made to The Book of Hydroelectric Engineering by P.S. Nigam.

An independent shaft separate from the abutment rock can also be provided for the tunnel intake if the abutment rock is not strong enough for a shaft. A typical layout of this type of intake is shown in Fig. 13.11.

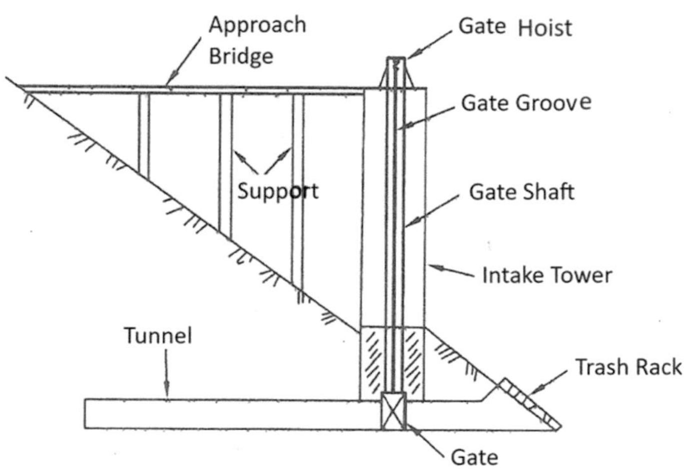

Fig. 13.11 Tower type intake with independent shaft

13.6.3 Trash Rack

In all types of intake structures, a trash rack is provided at the entrance of the water conductor to prevent the entry of the trash floating with the river flow. Where floating trash is made of big trees and logs, additional protection is provided in the form of log booms or skimmer walls in the upstream of trash rack. The shape and location of trash racks differ with the type of intake and these have been shown with the intakes in the preceding paras. The design aspects of a trash rack are briefly discussed below.

(i) *General*

The trash rack is a closely spaced mesh of bars. The bars are supported rigidly on the steel frames which slide in grooves provided in the piers. The trash racks are made in small segments of 1.2–1.5 m in width and of a length not too long for handling. The trash racks are coarse or fine depending on the spacing of bars. The spacing of bars in coarse trash rack ranges from 5 to 30 cm depending on the type of trash to be prevented. The fine trash rack is provided to protect the turbines from the trash and so the bars are spaced with a net opening of at least 5 mm less than the minimum opening between turbine runner blades.

The trash racks are placed either in an inclined or a vertical position depending on the type of intake in which these are installed and on the method of cleaning. Normally the trash racks in semi-circular and tower-type intakes are placed vertically. For manual cleaning, the inclined trash racks are preferred. The inclined trash racks are also generally provided in low-head intakes. The inclination of such trash racks is normally 10–15° from the vertical.

(ii) *Hydraulic Design Considerations*

The velocity of flow through the trash rack is invariably calculated on the basis of gross area and is kept small to keep the head loss at the minimum and also for the ease of cleaning the rack. In many projects, the velocity through the trash rack is found to vary from 0.4 to 1.5 m/s. As per USBR, the port area should be such that net velocity through racks does not exceed 0.61 m/s. Mosonyi suggests that to avoid eddies and vortices formation the entrance velocity should be kept significantly low. It may be worked out by

$$V \leq 0.075\sqrt{2\mathrm{gH}}$$

In high head intakes, higher velocities through the trash rack (up to 1.5 m/s) are permitted to limit the total area requirement of the trash rack. Adequate weightage should also be given to operational and economical considerations while limiting the area of trash rack. The I.S. 9761 recommends normal velocity of flow through the racks as below.

$$V = 0.75 \text{ m/s (For hand racking)}$$

$$V = 1.5 \text{ m/s (For mechanical cleaning)}$$

The head loss through the trash rack is given as $K\frac{V^2}{2g}$,, where K is a coefficient that depends on a large number of factors such as thickness, spacing, inclination, shape of bars, type of cleaning, etc. and V is velocity through racks. Many investigators have developed expressions to evaluate K (refer P.S. Nigam). One simple formula is as below.

$$K = 1.45 - 0.45A - A^2$$

where,

$$A = \frac{\text{net area of rack}}{\text{gross area}}$$

To illustrate the use of this formula.
Assume approach velocity to rack = 0.6 m/s, and
ratio of bar area to total (gross) area of trash rack = 0.3
Then $A = 0.7$

$$\text{Velocity through rack} = \frac{0.6}{1 - 0.3} = 0.87\text{m/s}$$

$$\text{K} = 1.45 - 0.045 \times 0.7 - 0.7^2 = 0.645$$

$$\text{(head loss)} = 0.645 \times \frac{0.87^2}{2\text{g}} = 0.025\,\text{m}$$

The USBR practice to work out head loss is to assume it equal to $V^2/8$, where V is velocity through rack.

The head loss is suggested as below.

V (m/s)	Head loss (cm)
0.3	3.0
0.45	9.0
0.60	15.0

(iii) *Some Aspects of Structural Design*

The trash rack bars of rectangular sections with depth '*d*' and thickness '*t*' are generally preferred. The depth to thickness d/t ratio which is normally used is between 5 and 12. The unsupported length of bar should not be more than 70 times the thickness. For higher flow-through velocity in the trash rack, the unsupported length of bars should be less because higher velocity may cause vibrations. The spacing of bars as mentioned above should be adjusted so that forcing frequency to natural frequency of bars is less than 0.6. Normally the bar section and unsupported length suggested above satisfy the frequency criterion to avoid resonance.

The trash rack and the supporting structures must be designed to withstand the water pressure imposed upon them when the racks become clogged with trash. The loads on trash racks are normally small but these are subjected to high differential heads created due to clogging of the bars. In the case of racks that are regularly cleaned a 1.5 m differential head is often used for design. Otherwise higher differential head should be considered for design. I.S. 9761 recommends a minimum differential design head of 3–6 m depending on the efficiency of the trash rack cleaning arrangement. Normally arrangements are made for manually cleaning the trash at regular intervals and mechanical cleaning arrangements are made for continuous cleaning when floating debris is abundant. Racks in cold regions are clogged by ice. To keep the racks free for flow in such a situation, a heating arrangement of racks should be provided.

13.6.4 Air Vent

Air vents are provided downstream of the control gates for the following requirements:

(i) To admit air into the conduit when the gate is closed and the conduit is emptied.
(ii) To escape air when the conduit is being filled.
(iii) To reduce sub-atmospheric pressure downstream of the gate during partial opening.

There are a large number of methods to evaluate the air vent size and these methods give a vent size varying from 30 to 200 cm^2/m^2 of the gate opening. Some methods are given below.

USBR: The capacity of the air vent may be taken as 25% of conduit discharge and the area of the vent can be based on air velocity which may be assumed to vary from 45 to 90 m/s.

Another simple criterion is to provide the area of the air vent equal to 10% of the control gate area.

I.S. 9761 suggests the following formula for the area of the air vent

$$F = \frac{Q\sqrt{S}(D/t)^{3/2}}{750000\ \mathrm{C}}$$

where,

- F Area of air vent in m^2
- Q Maximum discharge through penstock
- D Diameter of penstock in m
- t Thickness of penstock shell in m
- S Factor of safety against collapse of pipe (normally assumed 3–4)
- C Coefficient of discharge through inlet (0.5 for intake valves and 0.7 for short inlet pipe).

References

CBI&P.: Small hydro stations standardization. Pub No. 175, CBI&P, New Delhi (1985)

Guidelines for Small Hydro Development, standard No. 2.2 and 2.3.: Civil Works—Hydraulic and Structural Design. AHEC, IIT Roorkee (2013)

IS 9761.: Hydropower Intakes-Criteria for Hydraulic Design. BIS, Manak Bhawan, New Delhi (1995)

Nigam, P.S.: Hand Book of Hydro-Electric Engineering. New Chand and Sons, Roorkee (1985)

Chapter 14
Design of Water Conductor System

14.1 General

A water conductor is a structure that carries water from the source (river or any other water body) to the powerhouse. It is generally either an open channel or a tunnel/conduit or a combination of the two. The choice depends on topography, geology, economy and the resources available for the construction. The open channel ends in a forebay and the long tunnel running under pressure ends in a surge tank. Some layout details of the two types of water conductors are described in Chap. 2. The details of the layout and design aspects of the open power channel and the tunnel as water conductor are given in the following paras.

14.2 Power Channel

14.2.1 Considerations for Layout

The power channel in a run-of-river scheme is generally aligned along the hill slope following the contours and so its length is normally more than the possible alternative alignment of a tunnel. The power channel also has to cross several drainages so proper drainage crossings have to be provided. These structures may be either an aqueduct or syphon depending on the levels of the bed of the channel and the drainage. While aligning the power channel, reaches of the unstable hill slopes should be avoided, the bed of the channel should be in cutting and sharp bends should be minimal. The curves should be provided with a minimum radius say 150–200 m for lined power channels of small or medium discharge say up to 40–50 cumec (for details refer I.S. 5968) and for unlined channels it may be more say 500–800 m. In case topography requires a sharp bend then superelevation in the channel bed should be provided. The

B. N. Asthana, *Planning and Design of Small Hydro Power Projects*,
https://doi.org/10.1007/978-3-031-96271-4_14

channels should be provided with the minimum required slope in the bed in order to utilise the maximum head for power generation.

14.2.2 Channel Section

The channel sections are either rectangular or trapezoidal in shape. The power channels are normally lined. Lining reduces seepage losses and requires a section smaller than that required for an unlined channel because of velocity limitations. Higher velocities are permitted in lined channels. Sometimes unlined alluvial channels are made for the canal-based SHP utilising a canal fall for power generation.

The lined power channel section is dimensioned by satisfying the following two equations.

$$V = \frac{1}{n} R^{2/3} S^{1/2}$$

and

$$Q = A.V.$$

where V is velocity (m/sec) 'n' is Manning's coefficient of rugosity,

R is hydraulic mean depth $\left(\frac{\text{Area}}{\text{Perimeter}}\right)$, S is slope of channel, A is area in m and Q is design discharge in cumec.

Depending on the type of lining, the limiting value of V is decided. The value of 'n' also depends on the type of lining. The value of 'n' as per I.S. 10,430 for various types of lining is given.

S. no	Surface characteristic of lining	Value of n
1	2	3
(i)	Concrete with surface Formed, no finish/PCC tiles or slabs Trowel float finish Gunite finish	0.018–0.020 0.015–0.018 0.018–0.022
(ii)	Concrete bed trowel float finish and slopes as below Hammer dressed stone masonry Coarse rubble masonry Random rubble masonry Masonry plastered Dry boulder lining	0.019–0.021 0.018–0.020 0.020–0.025 0.015–0.017 0.020–0.030
(iii)	Brick tile lining	0.018–0.020

Higher values of n should be adopted in curved channels and lower values are recommended for higher discharges.

For unlined channels in alluvial soils a value of n = 0.02 for Q > 15 cumec and 'n' = 0.0225 for channels with Q < 15 cumes are recommended in I.S. 7112. The limiting safe velocities for the type of lining as a guide are as below.

Boulder lining	1.5 m/sec
Burnt clay tile lining	1.8 m/sec
Cement concrete lining	2.7 m/sec

Normally a velocity of 2 m/sec is considered safe for concrete lining.

In an unlined channel depending on the type of soil, the permissible velocity is in a range of 0.6 to 1.0 m/sec.

The side slopes in a trapezoidal lined channel are generally 1:1 (H:V) but in a channel made by cutting a hard strata the slopes are 1:2 (H:V). In an unlined channel in alluvium, the side slopes are 1.5:1 (H:V).

After deciding the type of lining, value of 'n', the permissible velocity and the side slopes, the above two equations for dimensioning the lined section of a power channel can be used as below.

In a rectangular section (Fig. 14.1a).

D = depth of flow, B is bed width

$$A = BD, \quad P = B + 2D, \quad R = \frac{BD}{B + 2D}.$$

Fig. 14.1 **a** Rectangular channel section. **b** Trapezoidal channel section

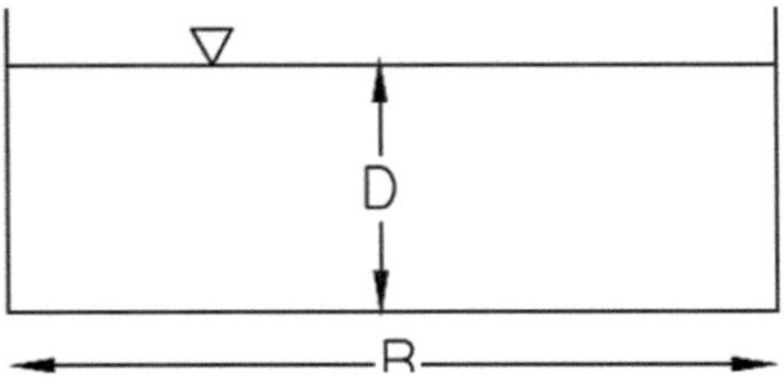

(a) Rectangular Channel Section

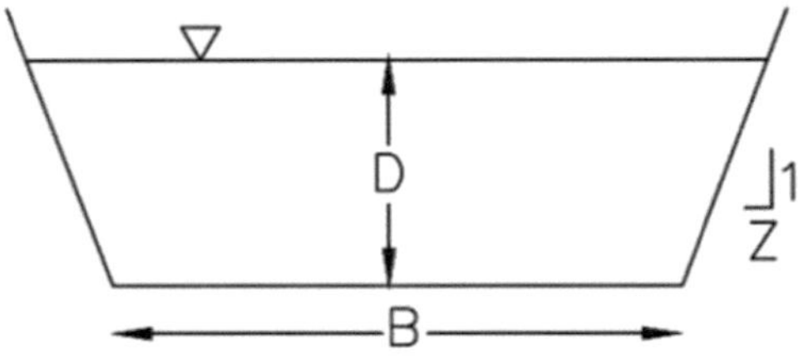

(b) Trapezoidal Channel Section

In a trapezoidal section (Fig. 14.1 b)

$$A = D(B + DZ)$$
$$P = B + 2\left(1 + Z^2\right)^{1/2} D$$

If $Z = 1$ $A = D(B + D)$ and $P = B + 2\sqrt{2D}$
Knowing Q and V, A is worked out as $A = \frac{Q}{V}$.

In Manning's equation of velocity, V and 'n' are known and R is in terms of B & D. If S is fixed on the basis of topography then a combination of B and D can be worked out by trial to satisfy the value of velocity.

In a lined channel normally B is equal to D, hence in that case no trial is required for B and D.

If B is decided from site consideration, then the required bed slope can be worked out from the use of Manning's equation of velocity.

The CBI Pub. 175 has given typical sections for discharges ranging from 1 to 12.5 cumec. These are assumed lined with LDPE covered by tiles and are cut along the hill slope with side slopes 1:2 (H:V). A 3 m wide service road is provided. In channels of higher discharges, the inspection road should be wider say 5 m or more. A catch water drain is provided to drain the hill slope. For these channel sections refer to Fig. 2.11.

Rectangular sections are excavated out of good rock slopes for the power channels of run-of-river SHPs. The section depends on the topography and the geology. Two typical sections one in complete cutting and the other in partial cutting are shown in Fig. 14.2. The service road should preferably be 3.0 m in width but a minimum width of 1.5 m should be provided. The valley side natural slope should be checked for stability because a stable slope is necessary for the safety of the canal. Drainage of hillside slopes is also very essential for safety.

In the channels, a freeboard is provided which is defined as the difference between the FSL and the top of the bank. A minimum freeboard of 0.5 m is provided for discharges less than 10 cumec and 0.3 m for discharges of 1 cumec or less. For a discharge of more than 10 cumec a freeboard of 0.6 m is recommended for lined channels (Refer IS. 10,430).

While closing the machine in the powerhouse a surge wave is generated which travels in the channel upstream of the forebay resulting in the rise of the water level of the channel (refer to Fig. 14.3). Hence, in the upstream of forebay an allowance for this rise in water level should be made in the freeboard of the channel.

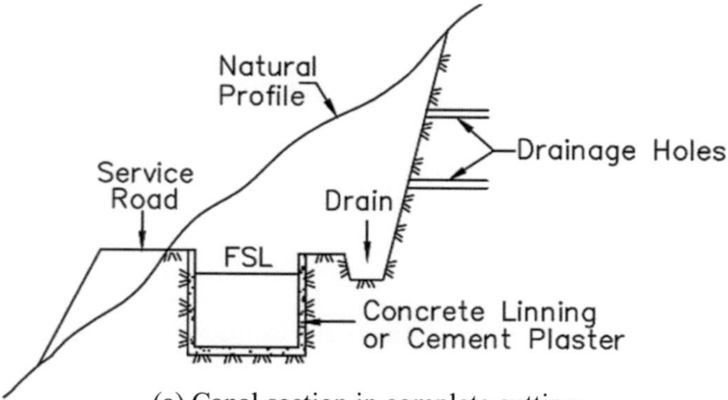

(a) Canal section in complete cutting

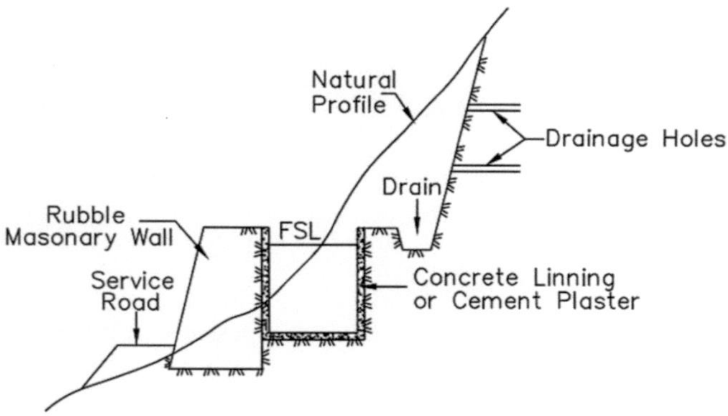

(b) Canal section in partial cutting

Fig. 14.2 Showing rectangular section on hill slopes

Fig. 14.3 Sketch showing surge

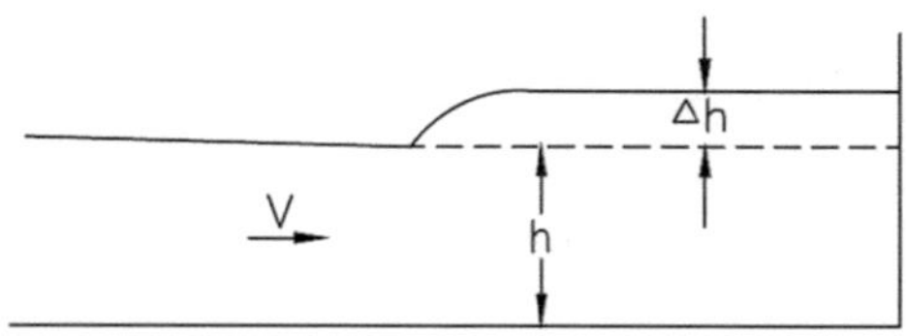

The maximum rise in water level for sudden closure is given by

$$\Delta h_{max} = K + \sqrt{K^2 + 2kh}$$

where $K = \frac{V^2}{2g}$, h is $\frac{\text{area of flow}}{\text{top width of water surface}}$ and V is mean velocity of flow.

For gradual closure.

$$\Delta h_{max} = \frac{K}{2} + V\sqrt{\frac{h}{g}}.$$

The increasing volume of water may over top the canal bank. To avoid such a situation an automatic escape is provided in the forebay whose capacity is made equal to canal discharge. The height of the surge wave shall be added to the head above the escape crest.

14.2.3 Canal Lining

As stated in 14.2.2 power channel sections are normally lined. The advantages of lining are as below.

- (i) Reduction in seepage losses.
- (ii) Since higher velocities are permitted in lined sections, the canal discharge requires a smaller section. This saves in the cost of land, reduces the quantity of earthwork and size of canal structures, etc.
- (iii) The smooth surface of the lining with less value of rugosity coefficient reduces hydraulic losses.
- (iv) Higher velocities prevent silting of the channel section.
- (v) Operation and maintenance costs are less.

(a) **Types of Lining**

Depending on the type of material used, the types of lining are categorised as below.

(1) Hard surface type

- (i) In situ concrete lining.
- (ii) Precast concrete slab lining.
- (iii) Brick or tile lining.
- (iv) Shotcrete lining.
- (v) Asphalt concrete lining.
- (vi) Stone masonry lining.

(2) Membrane lining

- (i) Prefabricated asphaltic membrane.
- (ii) Plastic film lining.

(3) Earth lining

- (i) Compacted earth lining.
- (ii) Soil cement lining.

(4) Porous type lining

- (i) Boulder or brick pitching.

(b) Selection of Type of Lining

Hard surface type linings such as in situ concrete, brick tile and precast concrete slab type are widely used because of impermeability, better hydraulic efficiency due to less value of rugosity coefficient and resistance to erosion due to sediment in the flow. While selecting a hard surface type of lining it should be kept under consideration that brick tile and precast slab-type linings are more easily replaceable and repairable than in situ concrete lining.

The in situ concrete lining is best done both mechanically and manually when the thickness is 8–10 cm. So it is preferred when the bed width is 8 m or more. In smaller bed width channels brick, brick tile or precast slab type lining is preferred. The flexible membrane (LDPE) lining is often adopted for small channels with a bed width of 1–2 m. (Refer IS 10430 for the selection of type of lining).

(c) Design of Lining

Canal linings as a rule are not designed to resist any load or force. It is primarily provided to reduce rugosity, to allow higher velocities and to prevent seepage.

No general rule can be stated for working out the thickness of concrete lining. Normally concrete linings are unreinforced. For small channels a thickness as small as 4 cm has been adopted. For large canals with higher velocities, a thickness of about 10 cm has been provided. It also depends on the mode of construction. Generally M_{20} grade concrete is used for lining. Reinforcement is sometimes provided where excessive cracking of concrete is a possibility due to exposure to the atmosphere or unexpected settlement of subgrade or hydrostatic pressure. The amount of reinforcement generally provided is 0.2% of the area in both longitudinal and transverse directions. Contraction joints are provided both in longitudinal and transverse directions at a spacing varying from 5 to 10 m to control cracking due to shrinkage caused by a drop in temperature or moisture loss. If reinforcement is provided it is discontinued at the joints. The joints are provided with a groove at the top to be filled with bitumen to make them watertight (Refer IS. 13,143).

Brick or brick tile lining is often done in two layers with a 15 mm thick sandwich layer of plaster of 1:3 cement mortar to make it watertight. The brick or tile joints are filled with mortar. If the bed is on a good subgrade sometimes only one layer of lining is done with plaster at the top. No contraction joints are provided. Refer IS. 3872 for details.

Precast concrete slabs of size 500 × 250 × 50 mm are generally used for lining (IS. 10,646). Joints are filled with 1:3 cement mortar. In hill channels stone slabs are also used for lining. For details of slab lining refer to IS. 3873.

Shotcrete lining is done in channels that are cut in the rock. A suitably proportioned mix of cement, sand and gravel is applied through a pneumatic pump on the channel surface. The thickness is normally 15 cm which is applied in layers. As an alternative to shotcreting the channel sides are made of random rubble masonry plastered with 20 mm thick 1:3 cement mortar and the bed is made of 10 cm thick M_{20} grade concrete.

The membrane lining consists of a thin waterproofing compound. The membrane after being placed in the channel is covered with a protective layer of brick or brick tiles. Sometimes a thick layer of good earth is also used as a protective layer. The protective layer prevents the membrane from damage. This type of lining can prevent seepage but the permissible channel velocities are the same as for an unlined channel. For details refer IS. 9698.

(d) **Under Drainage**

When a lined canal passes through an area where the seasonal ground water table is higher than or likely to be higher than the water level inside the canal or where subgrade is sufficiently impermeable to allow free drainage to the seepage from the canal, suitable drainage arrangement should be provided to protect the lining. Depending on the water table position and nature of the fill the following cases arise for drainage arrangement.

- If the water table is below the canal bed and the fill is free draining no drainage arrangement is required. If fill is of low permeability a well-designed drainage arrangement is required to arrest seepage.
- If the water table is between bed level and the FSL, drainage arrangement is required irrespective of the nature of the fill.
- If the water table is above the canal FSL, an extensive, drainage arrangement comprising longitudinal drains in bed, drain or inverted filter behind lining on slopes with pressure release valves is required.

The drains are made of perforated PVC pipe and laid open joined surrounded by filter material. The pressure release valves have a hinged flap at the top and are placed inside a filter pocket. The drainage arrangements should be planned and designed in accordance with IS. 4558.

14.2.4 Small Rectangular Channels

In many SHPs, which utilise small discharges, and are located in reaches of steep and relatively unstable hill slopes, the water conductor is provided in the form of covered rectangular channels of RCC or of masonry for safety and economy. These are shown in Fig. 14.4. These are designed as open channels for a maximum velocity of 2 m/sec for a reasonably small section and head loss. The required slope should be provided. A freeboard of 300–500 mm is provided. The RCC channels are designed as box structure and adequate reinforcement is provided. In RCC channels contraction joints are normally provided at a spacing of about 20–30 m and the joints are provided with waterstops. Such joints are not required in masonry channels which normally have stone masonry sides walls and bottom. The periphery is provided with a 20 mm thick 1:3 cement sand plaster. Sometimes instead of plaster 75 mm thick M_{20} grade concrete is placed with nominal reinforcement. At the top a RCC slab is provided. The slab is simply supported on the side walls. The side walls are designed for thickness as cantilever walls. The burried channels are provided with manholes of size 800 × 800 mm in the top slab at a spacing of about 50–100 m for inspection and maintenance.

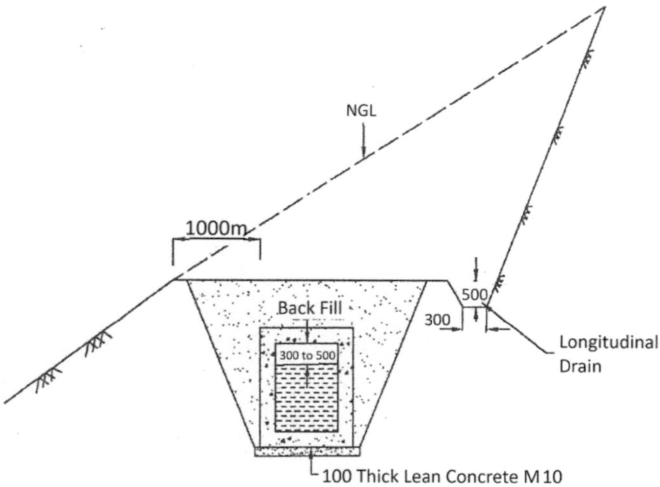

(a) Typical section of RCC channel in cut & cover.

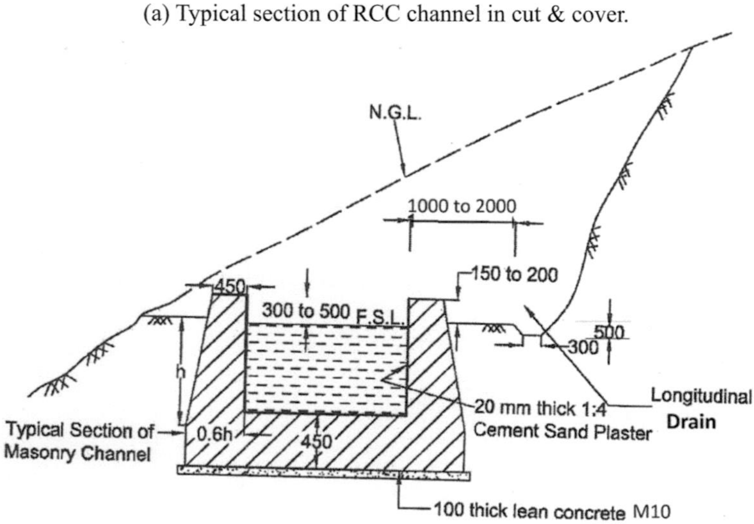

(b) Typical Section of Masonry Channel.

Fig. 14.4 Typical canal sections of RCC and stone masonry in hill slopes (AHEC—2.2)

14.2.5 *Head Loss in Channels*

Head loss in the power channel comprises the following:

(1) Friction Loss in the Channel

$$h_f = L.S$$

where S is slope and L is length of channel.

(2) Loss at Bends

$$h_b = S\,L_B + 2b\frac{V^2}{2g.r}$$

where S is slope, L_B length of bend, b is bed width, V is average velocity and r is radius of curve.

(3) Losses in Expansions

$$h_{ex} = \left(1 - \frac{A_1}{A_2}\right)^2 \cdot \frac{{V_1}^2}{2g}$$

where A_1 is u/s area of flow and A_2 is d/s area, V_1 is velocity at upstream end of the expansion (Fig. 14.5a).

(4) Losses in Contraction

$$h_c = 0.2\frac{{V_2}^2}{2g}$$

where V_2 is velocity at downstream end of contraction (Fig. 14.5b).

Fig. 14.5 **a** Loss in expansion. **b** Loss in contraction. Trash Rack Loss

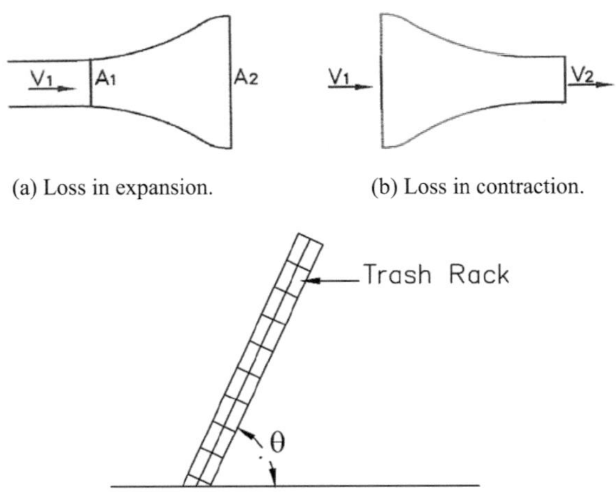

(a) Loss in expansion.

(b) Loss in contraction.

(c) Trash Rack Loss.

(5) Losses at Canal Head

$$h_e = 0.05\frac{V_2{}^2}{2g} + \frac{\left(V_2{}^2 - V_1{}^2\right)}{2g}$$

where V_1 is velocity upstream of entrance and V_2 is velocity at its downstream in channel.

(6) Trash Rack Losses (Refer Fig. 14.5c)

$$h_{tr} = 2.3\frac{t}{b}\sin\theta \cdot \left(\frac{V^2}{2g}\right)$$

where

t = Thickness of bars.
b = Spacing of bars.
V = Velocity of flow approaching bars.
θ = Inclination of rack with horizontal.

The major head loss is friction loss, i.e. L.S. and all other losses are of small magnitude. At the planning stage of the project, all other losses may be assumed approximately equal to 10% of friction loss.

14.2.6 Conduits and Pipes

At many SHP projects, conduits of either hume pipes or steel pipes are found best suited in the region of steep hill slopes. Pipes of high-density polyethylene (HDPE) and glass fibre reinforced plastic (GRP) are also used which are available in different sizes (300–2000 mm). These are smooth with a rugosity coefficient of 0.01. These are buried in a cut-and-cover trench (refer Fig. 2.12). The advantage is that the hill slope is disturbed to the minimum. These conduits can also cross the small drainages and gullies without disturbing the regime of drainage.

The conduit or pipe is placed in a trench and is covered with the compacted backfill. At the drainage crossings, the conduit is covered with backfill concrete as shown in Fig. 2.12 to avoid damage from the drainage flow. At the bends of the pipe, anchors are provided. In case of exposed long reaches of steel pipes proper expansion joints are also provided.

The conduits are designed to either run fully under pressure or partially full. A partially running conduit has a maximum water depth equal to 3/4th of the diameter. It is designed as a free flow channel and the required slope is provided. When the conduit is running full the diameter depends on the velocity permitted in the conduit which is generally kept low say 3–4 m/sec because high velocity will result in large friction loss which is worked out by

$$h_f = \frac{fLV^2}{2g\,D}$$

where f is coefficient of friction in pipe, V is velocity, L is length and D is diameter of conduit.

The slope of the conduit running full is not the hydraulic requirement. Hence, the slope is provided as required by construction considerations.

14.3 Tunnels

Tunnels are also provided as water conductors depending on topographical and geological conditions. Generally, a small diameter tunnel is required for SHP projects as the discharges are not very large. These are limited to 4–5 m in diameter. The minimum diameter for the mechanical construction of a tunnel is about 3.0 m. The tunnels may be designed as running full under pressure or as partial running, i.e. free flow tunnel which is hydraulically similar to an open channel.

The tunnels are normally preferred when the side hill slopes are unstable/weak and steep. The sharp bends in the river also make it difficult to construct a contour channel and in such a situation tunnel is favoured because of its much shorter length. But for a tunnel good quality rock is required. A tunnel besides being shorter in length eliminates the cross drainage works and requires practically no maintenance. So the selection between the canal and a tunnel should be based on careful consideration of these aspects along with the technical feasibility, available resources and economy.

14.3.1 Layout of Tunnel

Geological mapping and investigations through geophysical surveys, borings, test pits, shafts, drifts, etc. are necessary before deciding the location and layout of a tunnel. The investigations should also identify the stratification of rock and the location and nature of geological structures such as faults and shear zones. These also need careful consideration while deciding the tunnel layout.

The tunnel should be aligned through sound rock with favourable stratification as far as possible. The alignment should ensure the shortest route and adequate top and lateral rock cover over the tunnel. As a thumb rule, the minimum rock cover should

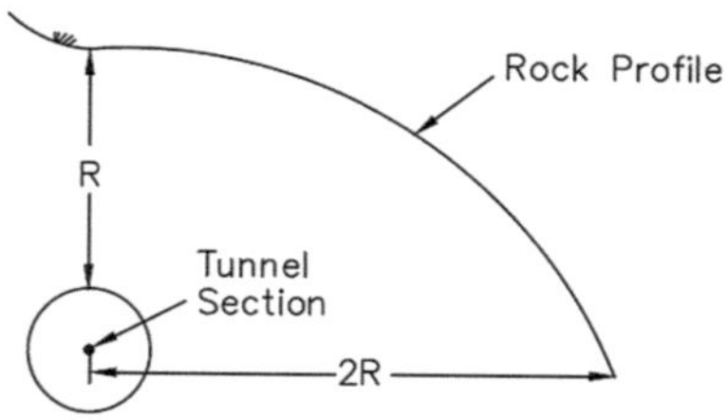

Fig. 14.6 Showing minimum rock cover

be three times the diameter of the tunnel in reasonably sound rock. Another criterion for rock cover is as shown in Fig. 14.6 where R is equal to max. internal pressure head in metre inside the tunnel.

The tunnel may be deviated from the shortest route to ensure adequate cover and to avoid faults, shear zones, water-bearing strata, etc. It may also be sometimes economical to deviate from the straight route to provide approach adits of short length which are normally required from construction considerations in a long tunnel.

The tunnel portal location should be selected with great care. The location should have minimum overburden and the hill slope should be steep with favourable geology. If geologically an open cut to locate the portal is unavoidable, its length should be as small as possible.

14.3.2 Tunnel Slope

In free-flowing tunnels, the gradient should be according to the hydraulic requirement to pass the discharge at the allowable velocity which in concrete-lined tunnels is 2–2.5 m/sec. The depth of flow in free-flowing tunnel should be limited to 3/4th of the diameter of the tunnel.

In pressure tunnels, the gradient should be such that the entire tunnel length remains under positive pressure at all discharges. The long tunnels running fully under pressure are provided with a surge tank at the end. The location of the surge tank and the surge heights also control the tunnel gradient. The tunnel end must get an adequate water seal at the maximum down surge level in the surge tank. The gradient should also be as minimum as possible because there is no hydraulic gradient requirement. Generally, from construction and maintenance requirements a positive gradient of 1/500 in the direction of flow is provided.

The inlet of the pressure tunnel should have an adequate water seal below MDDL to avoid entry of air with the flow.

14.3.3 Shape of Tunnel

The desirable shape of a tunnel from the considerations of the stresses which are induced in the tunnel section due to rock load only is elliptical and a circular section is structurally better for hydrostatic loading. Internal water pressure makes conditions

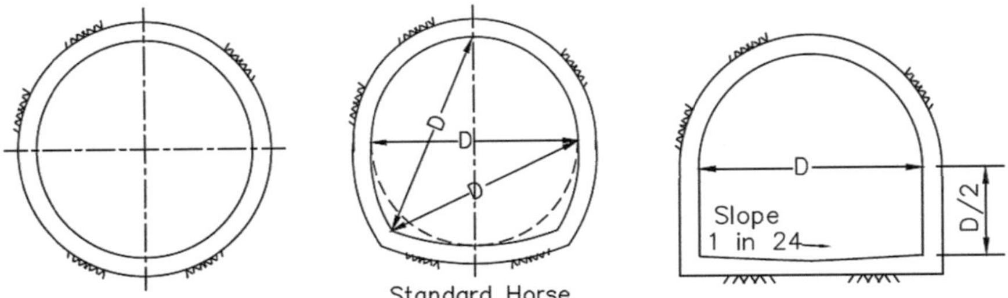

Fig. 14.7 Tunnel sections

favourable for adopting a circular section. Hence for a hydraulic tunnel, the ideal section is circular. This shape is suitable from the consideration of the rock strains which are caused by internal water pressure and due to natural rock pre-stressing. If rock tends to creep or squeeze a circular section will be self-stabilising.

Construction of a circular section is difficult especially when the size is small which is generally the case in SHP projects. A wider base width makes excavation and concreting easy. Hence a D-shaped or horseshoe shape tunnel section is preferred for small size tunnels. These tunnel shapes are shown in Fig. 14.7.

Horseshoe section is commonly used for both free flow and low-pressure tunnels in good rock. It has a semi-circular roof with arched sides and invert. It provides a wider base for construction ease. If a structural requirement needs a circular section, it can be achieved by lining in a circular shape at a minimum extra cost in a horseshoe section.

For free-flow tunnels in good rock and where the risk of failure due to external pressure from water or rock upon the lining is less, it is convenient to provide a D-shape section with a semi-circular roof, vertical sides and relatively flat invert. It is advisable to provide fillets at right-angle corners in a D-section. For free-flowing tunnels in good rock, a partly lined D-section with lining up to FSL may be adopted.

14.3.4 Spacing Between Adjacent Tunnels

In cases where two parallel tunnels are provided, the clear spacing between the two tunnels should normally be three times the diameter of the tunnels. In good rock, the spacing may be reduced to 2 times the diameter.

14.3.5 Curves

Normally curves should be avoided in the layout of a tunnel. But in case these are to be provided then the minimum radius of the curve should preferably be 5 times the diameter of the tunnel. For the ease of movement of the construction equipment in the tunnel, a radius more than the minimum should be provided.

14.3.6 Hydraulic Design Considerations

1 Size of Tunnel

In the case of free-flowing tunnels, the hydraulic design is the same as for an open channel by using Manning's equation. The coefficient of rugosity (n) is assumed same as for a concrete-lined channel. The required slope should be provided in the tunnel.

The size of a pressure tunnel depends on the permissible velocity. The average permissible velocity in a concrete-lined tunnel is 4–5 m/sec which may be reduced to 3 m/sec or so if water carries abrasive sediment significantly. The diameter of tunnel D is worked out by.

$$Q = \frac{\pi}{4} D^2 V$$

where Q is maximum discharge and V is permissible velocity.

The head loss in tunnels, which is of concern in power tunnels, is a function of $\frac{V^2}{D}$. The higher velocity requires a small diameter but results in greater head loss which in turn results in loss to power generation whereas smaller velocity requires a bigger diameter tunnel resulting in increasing the cost of the tunnel. Hence, an economic diameter should be worked out. It is a diameter at which an increase in cost balances energy loss. This can be expressed as

$$\frac{dc_1}{dD} \geq \frac{dc_2}{dD}$$

where c_1 = annual cost of tunnel, and c_2 is annual energy loss due to friction.

One expression for determining economic diameter which is derived by equating revenue loss per year due to energy loss and the capital annual cost of the tunnel including O&M cost is given below (refer P.S. Nigam)

$$D^{7.33} = \frac{19.35 \cdot Q^3 n^2 \eta C_0 10^5}{(C_e + 0.36 \cdot C_1) \cdot a}$$

where D is economic diameter of tunnel (m), Q is discharge through tunnel (m^3/sec), 'n' is coefficient of rugosity (assumed 0.014 for concrete lining), η is efficiency of generating equipment (taken as 0.85), C_0 is the selling rate of power (Rs/unit), C_e is cost of tunnel excavation (Rs/cum), C_1 is cost of lining (Rs/cum) and 'a' is annual cost of tunnel (as % of total cost)

It can be seen that the determination of the economic diameter of the tunnel is dependent on a number of factors and rates of work items whose realistic evaluation at the design stage is difficult. Hence, for an SHP generally, the tunnel diameter is decided on average velocity which is generally taken as 3–3.5 m/sec.

2 Transitions

Transitions from one shape of the tunnel to the other are generally required at the inlet and downstream of the gate shaft. Transitions are required at the contraction and expansion of the tunnel section which are provided from geological or structural considerations. The smooth transitions are designed to minimise head loss and avoid cavitation.

(a) At the Entrance

A bell mouth entrance normally gives the best entrance shape. It is defined by an ellipse. The criteria given in the chapter on Intake may be followed to design the bell mouth entrance. The design of elliptical curves may also conform to the criteria given in I.S. 4880 (Part III).

(b) At the Contraction and Expansion

To minimise head loss and avoid cavitation along tunnel surfaces, contraction and expansion transitions to and from the gate control section in the pressure tunnel should be gradual. The length of transition is determined by the angle of convergence.

In contraction, the maximum convergent angle is given by the relation

$$\tan \propto = 1/\mathrm{U}$$

where $\propto$ is the angle of the tunnel wall surface with respect to its centre line and 'U' is defined as below.

$\mathrm{U} = \mathrm{V}/\sqrt{\mathrm{gD}}$ where V and D are average values of depth and velocity at the beginning and end of transition.

Expansion should be more gradual than contraction and so the angle in this case is defined as

$$\tan \propto = \frac{1}{2.\mathrm{U}}.$$

However, '$\propto$' in expansion should not be more than 10^0 to keep the losses low.

(c) At Exit End

When a circular tunnel flowing partly full empties into a chute, the transition from the circular section to one with a flat bottom may be made in the open channel downstream from the tunnel portal or it may be made within the tunnel so that the bottom will be flat at the portal section. The transition should be made by gradually decreasing the circular quadrants from full radius at the upstream end of the transition to zero at the downstream end. Empirical rule which may be used for velocities up to 6 m/sec is as follows.

$$L = 2.187\,V.D$$

where L is length of transition (m), V is exit velocity (m/sec) and D is tunnel diameter (m).

3 Losses in Tunnels

Hydraulic losses in tunnels are due to friction, trash racks, intake, bends, gates and exit. Out of these friction losses are more significant and need careful estimation. These losses can be worked out as below.

(i) **Friction Losses**: The following two approaches are generally used for estimating friction losses.

(a) Manning's equation—It can be used for both free flow and pressure flow tunnels.

$$V = \frac{1}{n} R^{2/3} S^{1/2} \text{ and } h_f = S.L = \frac{V^2 n^2 L}{R^{4/3}}.$$

(b) Darcy-Weisback formula – This is used for pressure flow tunnels only

$$h_f = \frac{f.L.V^2}{2\,gD}.$$

In these expressions

- V Velocity in tunnel (m/sec)
- D Diameter of tunnel (m)
- L Length of tunnel (m)
- R Hydraulic radius (m) $= \frac{\text{flow area}}{\text{wetted perimeter}}$
- g Acceleration due to gravity (m^2/sec)
- S Slope of energy line both in pressure tunnel and free flow tunnel
- n Manning's coefficient of rugosity
- f Friction factor.

The value of 'n' for concrete-lined tunnels is normally taken as 0.014. The value of 'n' for unlined tunnels is recommended as below for various rock surface conditions.

Very rough surface	0.04–0.06
Surface trimmed	0.025–0.035
Surface trimmed and invert concrete lined	0.02 to 0.03

The friction factor 'f' is a function of Reynold's number and relative roughness defined by K_s/D, where K_s represents the average non-uniform roughness of the tunnel. For the tunnels which are concrete lined or steel lined and the flow is in the range of Reynold's No. between 3000 and 10,000 which is generally found in power tunnels the value of 'f' can be estimated by using the following formula.

$$\frac{1}{\sqrt{f}} = 2.0 \log\left(\frac{1}{2E}\right) + 1.74$$

where $E = K_s/D$

The value of K_s may be taken from the following table depending on the type of concrete lining.

Surface	K_s(mm)
Unusually rough due to poor form work or erosion or poor alignment of joints	0.6–6.0
Rough due to erosion, marks due to form work or spalling of laitances	0.4–0.6
Granular, surface in good condition with good joints	0.18–0.4
New or fairly new smooth surface with steel forms and smooth joints	0.06–0.18

I.S. 2951 (Part-I) recommends the use of a friction factor formula and provides working charts.

For free flow tunnels, Manning's formula is generally used. The maximum depth of flow should be 0.75 D so that ample freeboard for pulsations in the flow is available. Design of small Dams (USBR) gives hydraulic properties for different flow depths in circular and horse shoe tunnels and can be used to compute hydraulic radius (R).

(ii) **Trash Rack Losses**

There are several relations for trash rack losses. One is given below.

$$h_t = k_g \frac{V^2}{2g}$$

and

$$k_g = 1.450 - 0.45\frac{a_n}{a_g} - \left(\frac{a_n}{a_g}\right)^2$$

where

V Velocity through net area of trash rack.
a_n Net area through trash rack.

and

a_g Gross area of the trash rack and supports.

(iii) **Entrance Losses**

Bell mouth entrances are normally provided to reduce losses. The entrances are either circular or square to provide a trash rack or gate at the entry. The losses are given by

$$h_e = k_e \frac{V^2}{2g}$$

where V is velocity, k_e is loss coefficient and its values are assumed as below.

Type of entrance	Max^m	Min^m	Average
Circular bell mouth	0.10	0.04	0.05
Square bell mouth	0.20	0.07	0.16

For other types of entrances the value of k_e is much higher.

(iv) **Transition Losses**

At the entrance generally gradual contractions are provided. The loss is given by

$$h_{et} = K_{et}\left(\frac{V_2^2}{2g} - \frac{V_1^2}{2g}\right)$$

where K_{et} is the loss coefficient, V_1 is velocity at entry and V_2 is velocity at exit end of transition. K_{et} is assumed as 0.1 for gradual contraction where the flare angle does not exceed from those defined above for transitions. It may be 0.5 if the contraction is abrupt.

(v) **Bend Losses**

Curves in tunnel are provided with a certain radius R_b and deflection angle $\propto$. The bend losses are estimated by the following relation.

$$h_b = \frac{2 \propto}{\left[\pi^2\left\{\log\left(\frac{R_b}{D}\right) + \propto\right\}\right]}$$

If the bend is 90° ie $\propto$ is $\frac{\pi}{2}$ then

$$h_b = \frac{1}{\left[\pi\left\{\log\left(\frac{R_b}{D}\right) + \frac{\pi}{2}\right\}\right]}$$

The bend loss will be practically insignificant if R_b is large as compared to the diameter of tunnel (D).

(vi) **Gate Loss**

No gate loss is assumed if the gate is installed at the entrance. It is estimated as below if the gate is placed inside the tunnel

$$h_g = K_g \frac{V^2}{2g}$$

where K_g is loss coefficient and V is velocity.

In case, the gate is mounted on either side of the thin headwall, the sides and bottom of the flow jet are suppressed and the top is contracted, the value of K_g may be obtained as below.

Conditions	Value of K_g		
	Max^m	Min^m	Average
Gate in thin wall with unsuppressed contraction	1.8	1.0	1.5
Gate in thin wall-bottom and sides suppressed	1.2	0.5	1.0
Gate in thin wall-corners rounded	1.0	0.1	0.5

In case the gate is so mounted in the tunnel that the floor, sides and roof upstream and downstream of the gate are continuous with the gate opening the value of K_g is assumed as 0.1.

(vii) **Exit Losses**

In case where the pressure tunnel discharges freely, or is submerged or supported on a downstream floor, the velocity head coefficient is taken as 1.0 and head loss is equal to $V^2/2g$.

14.4 Air Locking and Remedial Measures

Air locking: Air may enter the tunnel during the filling and emptying of the tunnel or due to inadequate water seal or unusual approach conditions at the entrance or due to jump formation with partial gate opening. The trapped air causes damage to the lining, reduces discharge, increases head loss and affects adversely the efficiency of turbines. The remedial measures are as follows.

- Avoid vortex formation and provide adequate water seal at the entrance.
- Partial gate opening and jump formation should be avoided.
- The entire length of the tunnel should be below the energy gradient line.
- No negative pressures should develop due to surges.

If entry of air is unavoidable a deaeration arrangement should be provided. Normally an air vent is provided just downstream of the gate (refer to Intake design chapter).

14.4.1 Structural Design Aspects

The structural design of a tunnel is normally considered in the following two parts.

(i) Design of supports required during construction.
(ii) Design of lining.

These are described briefly below.

(a) **Design of Supports**

The support system has to withstand the rock loads during the period of construction. The rock load is the height of the rock mass which tends to drop out of the roof. If no supports are installed the mass of rock will drop into the tunnel during construction forming a cavity.

The following geological features and other parameters should be considered in evaluating the rock load on a tunnel:

(i) Characteristics and strength of rock formation and its elastic properties.
(ii) Dip and strike of rock formation in relation to tunnel alignment.
(iii) Prestressing of rock.
(iv) Tunnel shape and size.
(v) Time gap between excavation and support placement.

Terzaghi has expressed rock loads in terms of tunnel dimensions and rock conditions. These are given in Table 14.1.

The rock load is independent of rock cover if it is more than 1.5 times the combined width and height of the tunnel. The below sketch (Fig. 14.8) explains the rock load.

There are other methods to evaluate rock loads and type of supports based on rock quality. These are briefly described below. However, Terzaghi's method is simple and is generally used.

Table 14.1 Rock Load

S. no	Rock condition	Rock load H_p (m)	Remarks
1	Hard and intact	Zero	Light lining is required only if spalling or popping occurs
2	Hard stratified or schistose	0–0.50 B	Light support
3	Massive, moderately jointed	0–0.25 B	Load may change erratically from point to point
4	Moderately blocky and seamy	0.25–0.35 (B + H_t)	No side pressure
5	Very blocky and seamy	(0.35–1.10) (B + H_t)	Little or no side pressure
6	Completely crushed but chemically intact	1.10 (B + H_t)	Considerable side pressure, softening effect of seepage towards bottom of tunnel, Requires either continuous support for lower ends of ribs or circular ribs
7	Squeezing rock, moderate depth	(1.10–2.10) (B + H_t)	Heavy side pressure. Invert struts required. Circular ribs are recommended
8	Squeezing rock, great depth	(2.10–4.50) (B + H_t)	-do-
9	Swelling rock	Up to 80 m irrespective of value of (B + H_t)	Circular ribs required. In extreme cases use yielding support

*Note 1*The above Table has been arrived on the basis of observation and behaviour of supports in Alpine tunnels where the load was derived mainly on a loosening type of rock
Note 2 The roof of the tunnel is assumed to be located below the water table. If it is located permanently above the water table, the values given for types 4 to 6 may be reduced by fifty percent
Note 3 Some of the most common rock formations contain layers of shale. In an unweathered state, real shales are no worse than other stratified rocks. However, the term shale is often applied to firmly compacted clay sediments that have not yet acquired the properties of rock. Such so-called shales may behave in the tunnel like squeezing or even swelling rock
Note 4 If rock formation consists of a sequence of horizontal layers of sand stone or lime stone and of immature shale, the excavation of the tunnel is commonly associated with a gradual compression of the rock on both sides of the tunnel, involving a downward movement of the roof. Furthermore, the relatively low resistance against slippage at the boundaries between the so called shale and rock is likely to reduce very considerably the capacity of rock located above the roof to bridge. Hence, in such rock formations, the roof pressure may be as heavy as in a very blocky and seamy rock

Fig. 14.8 Sketch showing rock load

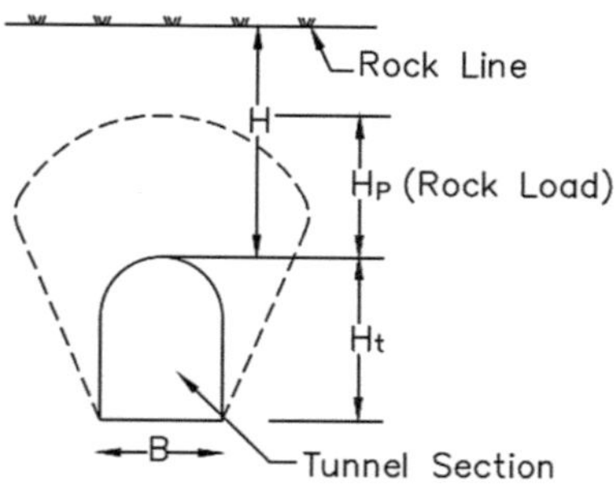

- **Rock Quality System (RQD)**

The rock quality system defines the relationship between the engineering quality of rock mass and the RQD index as given below. The RQD is defined as the ratio of the sum of sound rock core pieces that are greater or equal to 100 mm and the length of the core run.

RQD (%)	Rock quality
≤ 15	Very poor
25–50	Poor
50–75	Fair
75–90	Good
90–100	Excellent

- RMR-System (see Ref. AHEC,-2.2)

This classification of rock mass utilises the following six parameters of rock mass, all of which are measurable in the field and can also be obtained from bore hole data.

1. Uniaxial compressive strength of intact rock mass.
2. Rock quality designation (RQD).
3. Spacing of discontinuities.
4. Condition of discontinuities.
5. Orientation of discontinuities.
6. Ground water conditions.

Once the above parameters are determined, the importance ratings under the worst condition are assigned to each parameter. The importance ratings of these parameters are summed to obtain the total RMR of rock mass. The rock mass classification according to the RMR of rock is given in Tables 14.2.

Table 14.2 Engineering properties of rock mass (Beniawski (1973) and IS codes)

S. no	Parameters	Rock Mass Rating (RMR)				
	Properties of rock mass	100–81 (I)	80–61 (II)	60–41 (III)	40–21 (IV)	< 20 (V)
1	Classification of rock mass	Very good	Good	Fair	Poor	Very poor
2	Average stand up time	10 years for 15 m span	6 months for 8 m span	1 week for 5 m span	10 h for 2.5 m span	30 min for 1 m span
3	Angle of internal friction	> 45°	35° – 45°	25° – 35°	15° – 25°	15°
4	Cohesion of rock mass (Mpa)	> 0.4	0.3–0.4	0.2–0.3	0.1–0.2	< 0.1

Support load (P) can be determined from RMR as

$$P = \frac{100 - \text{RMR}}{100} \text{r.B}$$

where B = tunnel bore diameter (m), r = density of rock mass (kg/m^3)

Figure 14.9 gives the range of roof span, RMR and stand up time in which supports are required. The supports as per RMR system are given in Table 14.3.

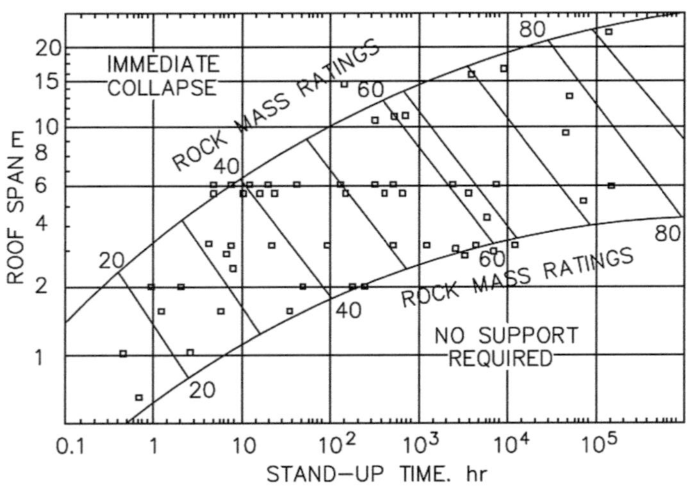

Fig. 14.9 Classification of rock masses and support requirement

- **Bartons Quality Index (Q-System)**

This is a qualitative direct method of designing a support system. Borton et al. (1974) observed that Bieniawski ignored three important parameters such as roughness of joints, frictional strength of joint filling material and cover pressure. They therefore included these parameters and proposed an elaborate and extensive classification after studying 200 case records. The classification takes into consideration, the RQD, joint set number (J_n), joint roughness number (J_r), joint alternation number (J_a), joint water

Table 14.3 Guidelines for excavation and support of rock tunnels.as per the RMR system

Rock mass class	Excavation	Support		
		Rock bolts (20 mm Dia. Fully Grouted)	Shotcrete	Steel sets
Very good rock I RMR 81–100	Full face 3 m advance	Generally, no support required except for occasional spot bolting		
Good rock II RMR 61–80	Full face 1.0–1.5 m advance Complete support 20 m from face	Locally, bolts in crown 3 m long, spaced 2.5 m, with occasional wire mesh	50 mm in crown where required	None
Fair rock III RMR 41–60	Top heading and bench. 1.5–3 m advance in top heading, Commence support after each blast, Complete support 10 m from face	Systematic bolts 4 m long, spaced 1.5–2 m in crown and walls with wire mesh in crown	50–100 mm in crown and 30 mm in sides	None
Poor rock IV RMR 21–40	Top heading and bench, 1–1.5 m advance in top heading, Instal support concurrently with excavation 10 m from face	Systematic bolts 4–5 m long, spaced 1–1.5 m in crown and walls with wire mesh	100–150 mm in crown and 100 mm in sides	Light to medium ribs spaced 1.5 m where required
Very poor Rock V RMR < 20	Multiple drifts, 0.5–1.5 m advance in top heading, Instal support concurrently with excavation, Shotcrete as soon as possible after blasting	Systematic bolts 5–6 m long, spaced 1–1.5 m in crown and walls with wire mesh, Bolt invert	150–200 mm in crown, 150 mm in sides and 50 mm on face	Medium to heavy ribs spaced 0.75 m with steel lagging & fore polling if required. Close invert

reduction factor (J_w) and stress reduction factor (SRF). Based on it the rock mass quality (Q) is evaluated from the relationship.

$$Q = \frac{RQD}{J_n} \cdot \frac{J_r}{J_a} \cdot \frac{J_w}{SRF}$$

Classification of rock based on Q values is as below.

Q	Group	Classification
10–40	1	Good
40–100		Very good
100–400		Extremely good
400–1600		Exceptionally good
0.10–1.0	2	Very poor
1.0–4.0		Poor
4.0–10		Fair
0.001-0.01	3	Exceptionally poor
0.01–1.0		Extremely poor

Relation Used for Support

$$\text{Equivalent dimension } D_e = \frac{\text{Excavation size}}{\text{ESR}}$$

where ESR (excavation supporting ratio) relates to the importance of underground opening

The ESR values for various categories of excavation are

1.6 for water conveying tunnels.

1.3 for storage room, road tunnels, etc. and

1.0 for power house, major railways and road tunnels.

Depending on the value of Q and D_e, the suitable support requirement can be obtained from Fig. 14.10.

Empirical relations given by Borton are also given below.

(i) Maximum^m unsupported space = 2 (ESR) $Q^{0.4}$.
(ii) Roof support pressure

$$W_r = \frac{2.Q^{-0.22}}{\mathcal{J}_r} \text{ for 3 or more sets of joints}$$

and

$$= \frac{2.\mathcal{J}_n^{0.5} Q^{-0.33}}{J_r} \text{ for less than 3 joint sets}$$

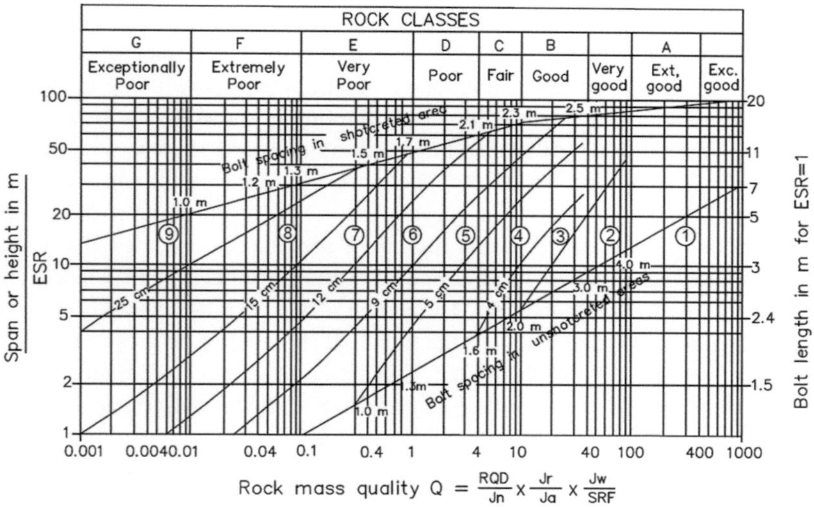

REINFORCEMENT CATEGORIES:

1) UNSUPPORTED
2) SPOT BOLTING
3) SYSTEMATIC BOLTING
4) SYSTEMATIC BOLTING, (AND UNREINFORCED SHOTCRETE, 4–10 CM)
5) FIBRE REINFORCED SHOTCRETE AND BOLTING, 5–9 CM
6) FIBRE REINFORCED SHOTCRETE AND BOLTING, 9–12 CM
7) FIBRE REINFORCED SHOTCRETE AND BOLTING, 12–15 CM
8) FIBRE REINFORCED SHOTCRETE > 15 CM, REINFORCED RIBS, SHOTCRETE AND BOLTING
9) CAST CONCRETE LINING

Fig. 14.10 Estimated support categories based on the rock mass quality index Q (AHEC–2.2)

(iii) Length of rock bolt $L = 2 + 0.15\frac{\text{Excavation width}}{\text{ESR}}$.

(iv) Anchor length $= 0.4\frac{\text{Excavation width}}{\text{ESR}}$.

(v) Wall pressure Q_{wall}

$$\begin{aligned} Q_{wall} &= 5.0\,Q \quad && \text{for}\,Q > 10 \\ &= 2.5\,Q && \text{for}\,Q\,0.1 < Q < 10 \\ &= 1\,Q && \text{for}\,Q < 0.1 \end{aligned}$$

Ribs: The above methods suggest the supports in the form of steel ribs (I-joists), rock bolts and rock bolts with shotcrete. Ribs are designed for rock load which results in thrust and bending moment. The I-sections of 100 mm size are normally used for small tunnels of diameter up to 4 m and spacing is worked out which is in the range of 50–100 cm. The ribs are placed in small segments which are rigidly connected either by bolts or welding. Ribs are supported at the ground on the base plates. The two adjacent ribs are connected by tie rods. The space between the rock surface and the rib is filled either with concrete or timber blocks which transfer rock load to the rib.

Rock bolts: The rock bolting fastens the loose rock near the surface to solid rock by means of anchor bolts. These also support the roof and act as load-carrying elements.

These are generally 25–32 mm diameter grouted rods. There are other types of rock bolts. For their details refer I.S. 5878 (Part IV). Rock bolts are sometimes used with shotcreting of the rock surface. Shotcreting may be done with wire mesh, chain link fabrics or reinforcement. Rock bolt with shotcreting is known as the New Austrian Tunnelling Method (NATM). Depending on design requirements shotcreting thickness generally ranges from 5 to 15 cm. Shotcreting is done by pumping a high slump concrete mix (maximum size of aggregate used is 20 mm) on the rock surface in layers of 5 cm thickness. Generally, rock bolts are placed soon after the tunnel is excavated. The rock supports itself after excavation for some time depending on the type of rock due to arching action. The length of the rock bolt should be equal to the excavated unsupported length of the tunnel. It should also be greater than L/3 or L/4, where L is the width of the tunnel. The rock bolts should preferably be placed across the joints in the rock. For the design of rock bolts I.S. 4880 (Part VI) may be referred.

In poor and very poor rock, rock bolting and shotcreting should be done immediately after blasting and muck removal. Thereafter the required steel supports should be placed.

(b) **Design of Lining**

Generally unreinforced concrete lining is provided in tunnels to reduce friction losses. This type of lining can withstand outside pressure from rock but not the internal water pressure. So these linings under high water pressure get slightly cracked and are not watertight. To improve the situation, the technique followed is compressing the lining by rock grouting. There are two approaches to the design of tunnel lining.

(i) To treat the lining as an independent structural member subject to internal and external pressure.
(ii) To take help from the surrounding rock mass by assuming the lining as the part of the rock mass ring which may act as a structural member to resist the loading. This requires suitable rock treatment with high pressure grouting so that the rock is mobilised to share the internal pressure.

The second approach is generally adopted as it leads to a considerable economy. The design of concrete lining may be carried out as per I.S. 4880 (Part IV).

The thumb rule for the thickness of the unreinforced lining is to provide 6 cm thickness per metre of the finished diameter of the tunnel. Thus, a 3 m diameter tunnel may be provided with 18 cm thick PCC lining. The minimum thickness of PCC lining is recommended as 150 mm. In weak and sheared rock zones generally, RCC lining is designed to take the loads. The minimum thickness of RCC lining is recommended as 300 mm.

Grouting: As stated above grouting of rock behind the lining is the design requirement so that a ring of rock may act with the lining to take internal pressure. The following two types of grouting are carried out.

(i) Contact grouting: It is a low-pressure (2–3 kg/cm^2) grouting to fill the gaps between the lining and the adjoining rock surface. The grouting is carried out through the holes drilled through the lining piercing through the rock by about 300 mm. The holes are made staggered at a spacing of 3–5 m in both directions.

(ii) Consolidation grouting: This is a high-pressure grouting to strengthen the rock so that it may share the internal water pressure. The grouting is done at a pressure of about 1.5 times the internal water pressure. The holes are made staggered at a spacing of 3–5 m in both directions up to a depth of 3–5 m in the rock.

The grout for grouting is a mix of cement and water (1:5 by weight). Thick mixes are used following the thin mixes depending on the intake. The choice of mix may depend on water tests. The thickening of the mix may be done by adding pozzolona, fine sand, bentonite, etc.

(c) **Drainage**

Drainage holes are often provided in the crown of free-flowing tunnels. These are also provided in pressure tunnels to drain the rock mass at the time of dewatering of tunnels. Normally 300 mm deep holes at the crown and the springing are provided at a longitudinal spacing of 3 m. In case fine material of the surrounding rock is expected to be washed with drainage discharge then filter material around the holes may be provided.

14.4.2 Portals and Plugs

1. **Portals**

When a tunnel construction is planned and started, the first item of work is to construct the tunnel portal. It is the entry point of the tunnel in the rock mass. It is preferable to locate the portal at a site where the rock is of good quality and the hill slope is steep. If it is located in a poor rock zone long open-cut approach would be required before the portal is located in firm rock (Fig. 14.11).

The portal is generally an RCC portal frame which is designed to carry the rock load from the top. The frame is analysed for the loads by the method of moment distribution to evaluate the fixed end moments. The top beam is designed for the bending moment and the vertical legs are designed for vertical load and bending moment at the base. Sometimes instead of a concrete portal frame a steel portal made up of R.S. Joist is provided. The two types of portals are shown in Fig. 14.12.

2. **Plugs**

The power tunnels which are long in length require adits for construction. These adits after construction are plugged at the junction with the main tunnel. Generally plugging is done with a solid concrete plug. In some adits concrete plug with an inspection gate is provided. In dam projects where diversion tunnels are made for the construction of the dam, they are later plugged with concrete plugs. These are shown in Fig. 14.13.

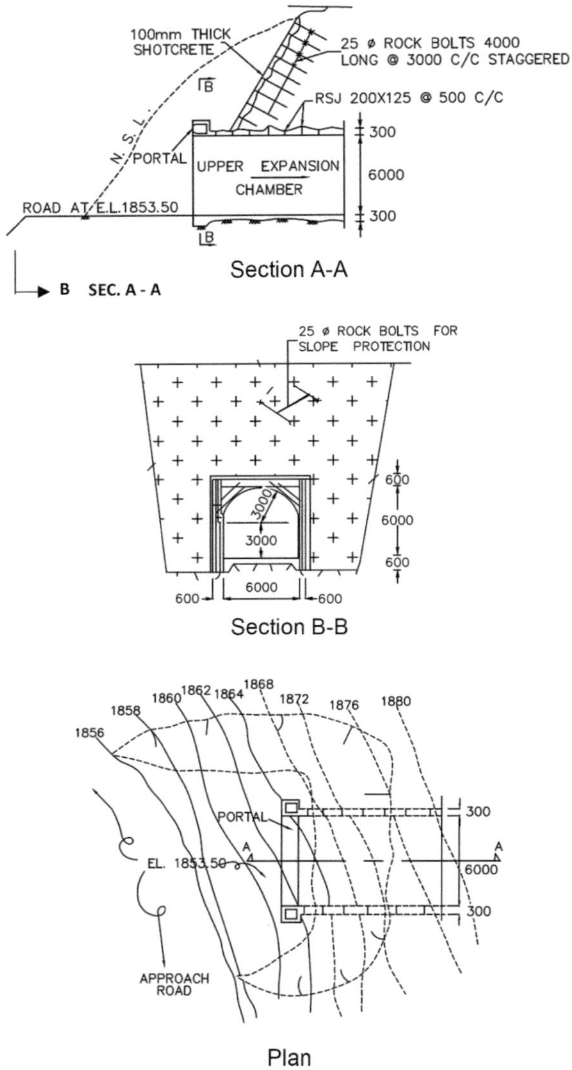

Fig. 14.11 Details of inlet portal of upper expansion chamber (Chuka project)

The plug is designed for the water load corresponding to the water head acting on the plug. For the plug on the diversion tunnel, the water load corresponds to the reservoir level and in the case of a plug in a tunnel adit, it corresponds to the pressure head in the tunnel at that point plus the water hammer effect, if any.

The length of the plug should not be less than the diameter of the tunnel. The length of plug L may be worked out by the following expressions.

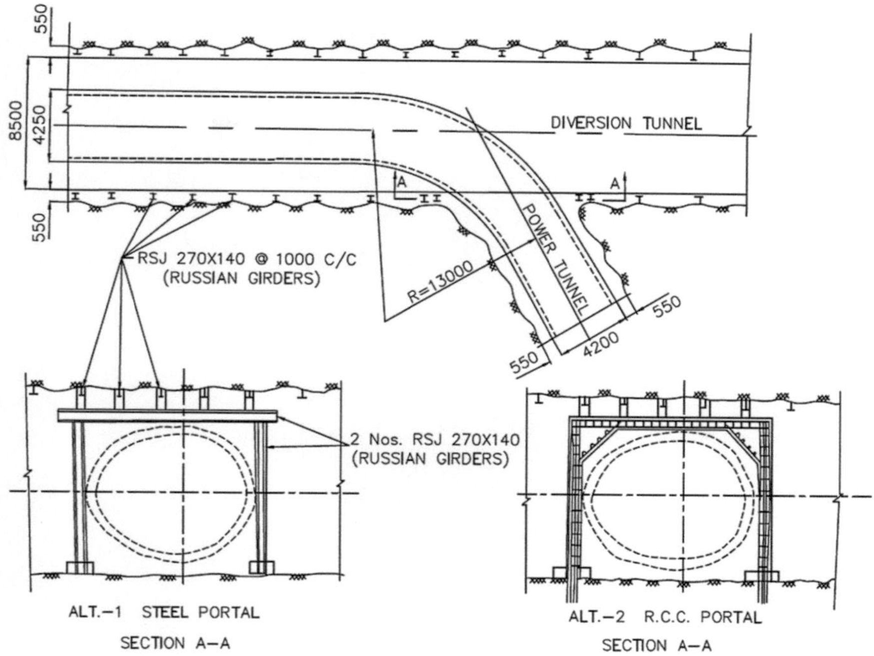

Fig. 14.12 Portal at junction of diversion tunnel & power tunnel (Salma project)

$$L = HF/(SS \times P)$$
$$L = HF/FR$$
$$L = HF/(FR + SRA)$$

where

HF Hydrostatic force on plug
SS Permissible average shear stress (10 to 15 kg/cm^2). Refer I.S. 456 for the stress value of the grade of concrete
FR Frictional resistance of plug (Depends on friction coefficient of rock)
SRA Shear resistance of anchors
P Perimeter of tunnel section.

The procedure generally adopted is indicated below.

The friction coefficient μ is taken as tan ϕ, where ϕ is the angle of internal friction of rock. Normally μ for good rock is taken as 0.75. The total frictional resistance $F_1 = \mu W$, where W is total weight of plug.

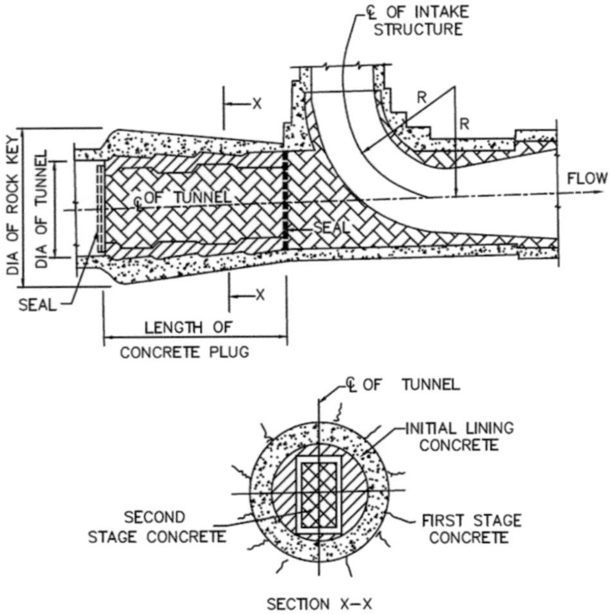

(a) Showing concrete Plugs in Stages.

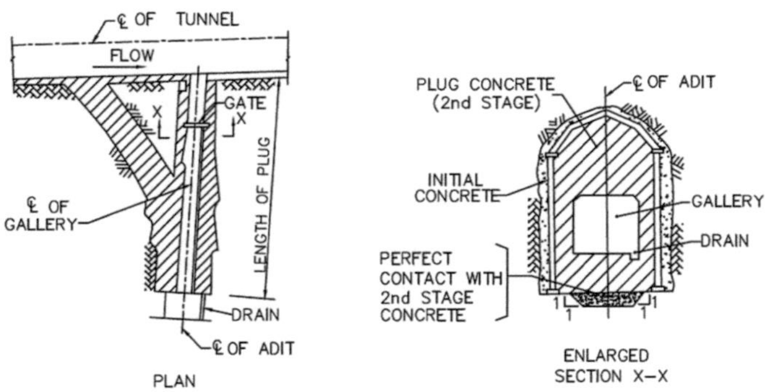

(a) Plan showing Gated Plug.

Fig. 14.13 Showing plugs

The other resisting force F_2 = average shear stress between rock and concrete multiplied by the contact area of plug which is equal to the length of plug multiplied by its perimeter.

The factor of safety is worked out by

$$FoS = \frac{F_1 + F_2}{HF}$$

It is usual to provide a high value of factor of safety because of uncertainty in the values of F_1 and F_2, The plug is considered safe if FoS is between 4 and 5. If FoS < 4 then length of plug should be increased.

Normally wedge shaped keys are provided at the interface of the rock and concrete of the plug to ensure effective plug action. The depth of keys may be calculated from the average bearing stress between concrete and rock in case the plug is in contact with the rock surface. The bearing stress in concrete normally governs as the rock is generally stronger.

$$\text{Average bearing stress} = \frac{HF}{\text{Effective projected bearing area}}$$

The effective projected area is normally assumed as 75% of the total projected area. The permissible bearing stress for concrete may be taken from I.S. 456. The depth of the key should not be less than 500 mm.

Contact grouting between the plug and concrete lining should be done to fill shrinkage gaps. To reduce shrinkage, the concrete of the plug should be precooled.

References

1. Bieniawski, Z.T., "Engineering classification of Jointed Rock Mass". Trans. South African Inst. Civil Engineers 15. (1973)
2. Borton N.R., Lien R and Lunde J, "Engineering classification of Rock Masses for the Design of Tunnel support", Journal of Rock Mechanics, 6 (4) (1974)
3. CBI&P.: Small hydro stations standardization. Pub. No. 175, New Delhi (1985)
4. Guidelines for small hydro development. 'Standard No. 2.2 and 2.3, Civil Works-Hydraulic and structural Design. AHEC, IIT, Roorkee (2013)
5. 'Indian Standard Codes' B I.S., Manak Bhawan New Delhi are listed:
   (i) I.S. 10430.: Criteria for design of lined canals and guidance for selection of type of lining, (2000)
   (ii) I.S. 10646.: Canal linings—cement concrete tiles—specification, (1991)
   (iii) I.S. 4558.: Underdrainage of lined canal—code of practice, (1995)
   (iv) I.S. 4880 (Part 3 & 4).: Structural design of concrete lining of water conveying tunnels—code of practice, (2005)
   (v) I.S. 4880 (Part 6).: Design of tunnels supports—code of practice, (2005)
   (vi) I.S. 5878 (Part 4).: Construction of tunnels—tunnel supports—code of practice, (2005)
   (vii) I.S. 2951 (Part 1).: Recommendations for estimation of head loss in straight pipes due to friction resistance, (2003)
   (viii) I.S. 7112.: Criteria for design of unlined canal in alluvial soils, (2003)

(ix) I.S. 3872.: Code of practice for lining of canals with burnt clay tiles, (2002)
(x) I.S. 3873.: Laying cement concrete/stone slab lining on canals, (2004)
(xi) I.S. 5968.: Guide for planning and layout of canal system for irrigation, (2003)
(xii) I.S. 9898.: Lining of canals with polyethylene films, (2005)

6. Mosonyi, E.: Water power development. Publishing House of the Hungarian Academy of Science, Budapest (1965)
7. Nigam, P.S.: Handbook of hydroelectric engineering. Nem Chand and Sons, Roorkee (1985)
8. USBR.: Design of small dams. Oxford and IHB Publishing company, (2001)

Chapter 15
Design of Structures on Water Conductor

15.1 General

The water conductor in a run-of-river type SHP is either a power channel or a tunnel. The tunnel may be a free flow or a pressure flow type. A power channel normally has to cross a number of drainages. Cross-drainage structures are provided over the drainages. The tunnel does not need such structures. Design considerations of cross-drainage works are given in this chapter.

In both types of water conductors normal desilting arrangements are provided. These are described in a separate chapter. In an SHP, generally desilting basins and vortex-type desilting arrangements for small discharges are provided. The design aspects of these arrangements are described in this chapter.

Sometimes if the topography permits a balancing reservoir is provided at some location on the water conductor which may act as a storage for peaking and as a settling tank to settle fine sediment. If it is at the end of the water conductor and close to the power house location, then it can act as a forebay also. The design aspects of balancing reservoir are mentioned briefly in this chapter.

15.2 Cross Drainage Works

Generally, an aqueduct is provided in the power channel which crosses the drainage. The bottom of the aqueduct is normally kept above the high flood level (HFL) of the drainage. There should be adequate headroom between the HFL of drainage and the bottom of the aqueduct, preferably it should be about 1 m or more.

B. N. Asthana, *Planning and Design of Small Hydro Power Projects*,
https://doi.org/10.1007/978-3-031-96271-4_15

15.2.1 RCC Aqueduct

The RCC aqueducts crossing major drainages are of two types:

(a) A rectangular power channel normally crosses the drainage without any fluming or reduction in its section. The rectangular RCC trough across the drain is supported on the abutments at the banks of the drainage and on the piers across the drain. The foundation of abutments and piers is designed on the basis of bearing capacity and expected scour. The pier spacing is normally between 8 and 12 m and depends on the economical design of the RCC trough carrying water. Generally, the trough is provided with a slab at the top for supporting the bridge.

The side walls of the trough are designed as a beam to carry water load as well as live and dead load of top and bottom slabs. The slabs are designed to be supported on side walls for all the loads. The design should also be checked as a beam supported on piers. The beam may be in the shape of a U bend or box. The trough should be made of M25-grade concrete. A typical section is shown in Fig. 15.1.

(b) If the power channel is trapezoidal in shape, it is normally flumed into a rectangular trough while crossing the drainage. The fluming in the bed width is limited to 60%. The velocity in the RCC rectangular section is also limited to about 4 m/sec. The typical plan and L-section of such an aqueduct are shown in Fig. 15.2. At the entry side the section is contracted. The contraction is done in 3:1 and at the exit end the expansion is done in 5:1 as shown in Fig. 15.2. The water profile is worked out by using Bernoulli's equation between 1 and 2 of contraction reach and Sects. 15.3 and 15.4 of expansion reach marked in the plan of Fig. 15.2. The bed slope in the rectangular trough is worked out by using Manning's equation. For details reader may refer to any standard textbook on Irrigation Structures. The other design considerations of the RCC trough are the same as given in para (a) above.

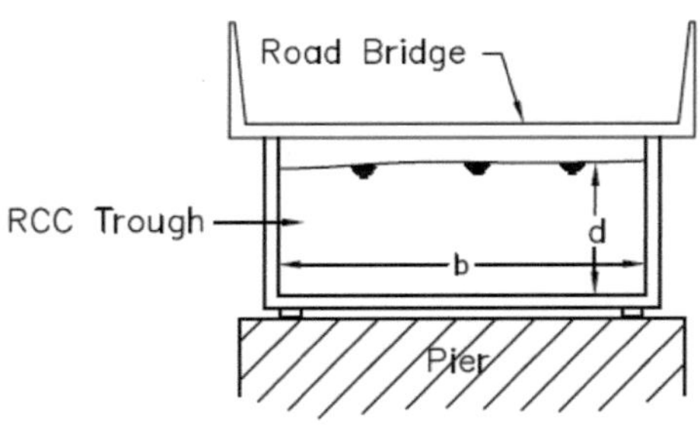

Fig. 15.1 Typical section of RCC aqueduct

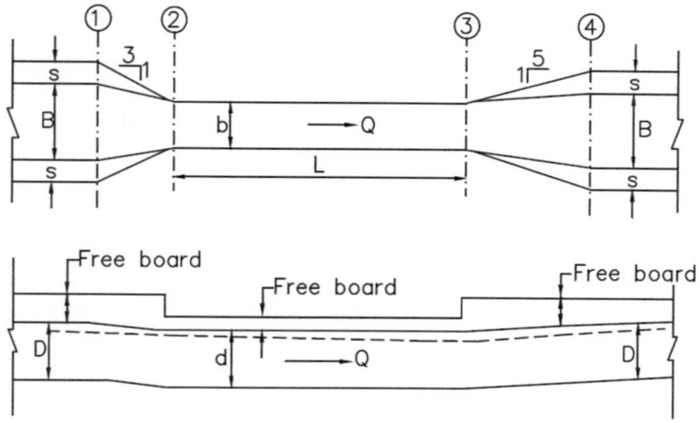

Fig. 15.2 Geometric layout of flumed aqueduct (AHEC–2.2)

15.2.2 Pipe Aqueducts

The water conductor can cross the drainage in a pipe. If the water conductor is a pipe the same can cross the drainage. In other cases, where the water conductor is a channel it can be changed into a pipe while crossing the drainage. This pipe can be both free flow or pressure flow. In the case of a pressure flow pipe, the velocity is normally limited to around 4 m/sec. The pipe will span between two piers as well as between the pier and abutment. The distance between the two piers should be around 5–6 m for an economical design of the pipe. Steel pipes are generally used. In case pier spacing is to be increased, a structure of RCC or steel like a bridge may be provided between the piers to carry the pipe over it to cross the drainage.

15.2.3 Syphon Type Drainage Crossings

In case adequate headway between the bed of the aqueduct and the HFL of the drainage is not available, a syphon drainage crossing may be adopted. In this case, the pipe or the RCC box can be laid below the drainage bed. The pipe or the RCC box should be placed sufficiently below the drainage bed. The depth between the drainage bed and top of the pipe or box should be around 1.5 m. Sufficient protection should be provided at the drainage bed in the form of CC blocks or gabions. A typical L-section and plan of a syphon drainage crossing are shown in Fig. 15.3.

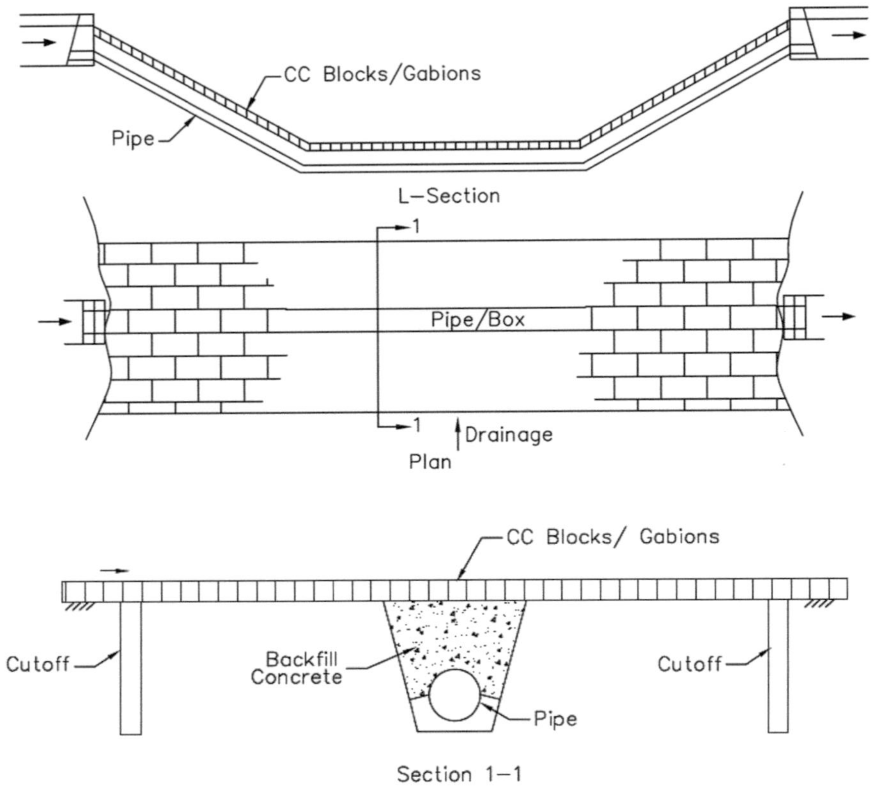

Fig. 15.3 Syphon drainage crossing

15.3 Desilting Devices

In run-of-river type SHPs, despite the measures taken to reduce the entry of sediment through the intake the suspended sediment as well as some of the bed load find entry into the water conductor. To avoid damage to the turbine parts from the abrasion due to sediment, desilting devices are provided in the water conductor close to the intake so that the water conductor may not get silted. Normally desilting basins and vortex-type devices are provided in SHPs. The design aspects of these devices are briefly given in the following paras.

15.3.1 Desilting Basin

Desilting basin is the most commonly used device to reduce sediment to the permissible level required for a turbine. The permissible size and concentration of sediment govern the design of the desilting basin. There is no well-defined criterion to fix permissible sediment size and concentration for a turbine. However, on the experience of existing power plants, some guidelines are as below.

The maximum permissible size of sediment particles depending on the type of turbine is as below.

Pelton wheel	0.1 mm
Francis turbine	0.3–0.5 mm
Kaplan turbine	0.5–0.7 mm

The limit on permissible particle size as suggested by Mosonyi is 0.5–0.2 mm for medium-head plants and 0.1 mm for high-head plants. BHEL recommends that sediment particles greater than 0.25 mm and having hardness more than 5 on Moh's scale are harmful and if the concentration is more than 200 ppm of harmful particles of size 0.25 and above, then a desilting device should be provided for all heads. The desilting device should have the extraction efficiency of the harmful particles between 85 and 95%.

(a) **Design Principle**

The extraction of sediment from the desilting basin is achieved through a reduction in flow velocity by increasing the flow area. This makes bigger size particles settle to the bottom of the basin which are removed continuously through a flushing arrangement provided at the bottom of the basin. A typical arrangement of a desilting basin with hoppers as flushing arrangement is shown in Fig. 15.4.

The length of the basin is worked out on the basis of the particle size to be extracted and its fall velocity. If the fall velocity of the particle is w_o then the time required to settle the particle to the bottom of the basin is h/w_o, where h is the depth of flow. The length of basin L is worked out as $\frac{u.h}{w_o}$ where u is flow velocity. Because of turbulence the fall velocity reduces and the length required for the basin increases. There are many approaches to assess the impact of turbulence and sediment concentration. An ad-hoc increase of 20% is recommended for the turbulence effect (Garde et al. 1985).

(b) **Design Procedure**

Many desilting basins have in the past been designed and tested on hydraulic models. The model study results have shown that the basins designed on Mosonyi's Approach generally perform satisfactorily. This approach is described.

The three basic equations in this approach are

$$Q = b \cdot h \cdot u$$

$$t = h/w$$

$$L = u \cdot t$$

where

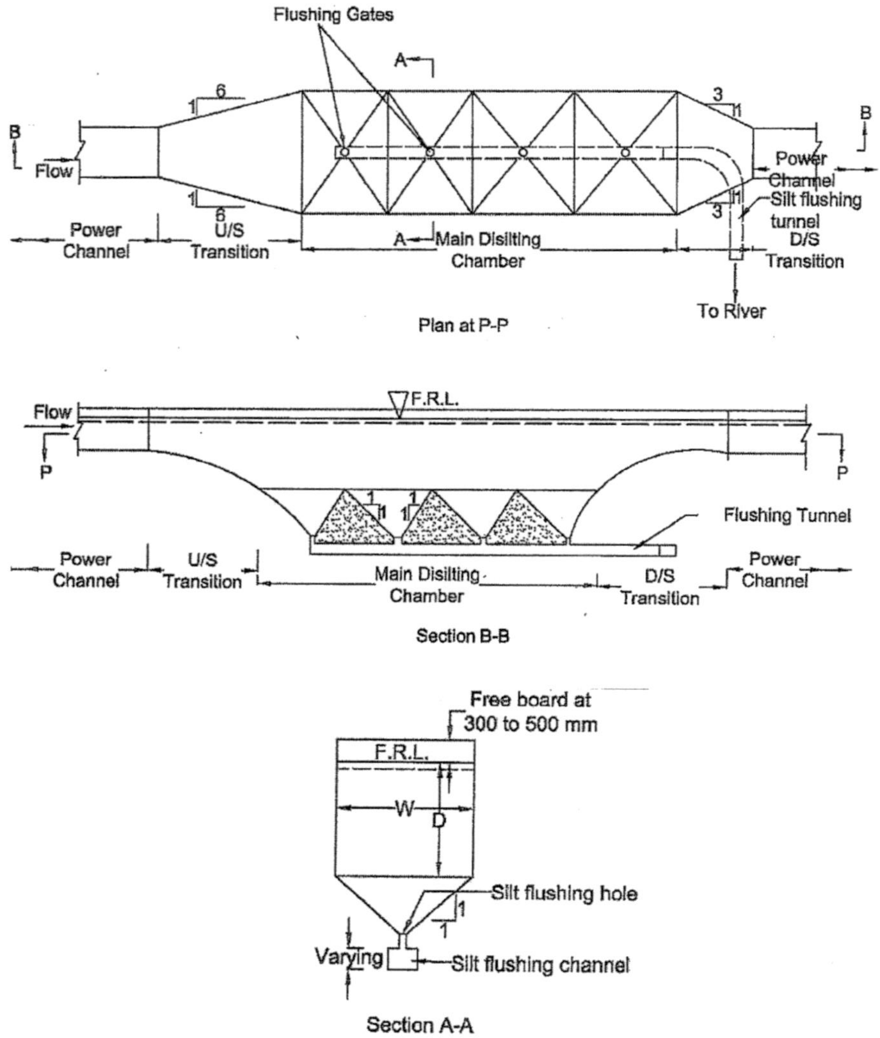

Fig. 15.4 Typical arrangement of desilting chamber (AHEC–2.2)

Q = Discharge through basin (m^3/sec).
b = Width of basin (m).
h = Depth of flow (m).
u = Flow velocity (m/sec).
L = Length of basin (m).
w = Fall velocity of particle in stagnant water (m/sec).
t = Settling time in seconds.

Eliminating ‘t’ we get the two equations

$$Q = b.h.u \text{ and } L.w = u.h$$

For the solution of these two equations, four variables should be known. Q is known and u, w and h are generally assumed on the following criterion:

Mosonyi suggested the value of 'h' between 4 and 8 m. To decide flow velocity 'u' Mosonyi recommends the adoption of critical flow through velocity which should not entrain the particles once settled at the bottom. He has recommended the use of Camp's relation.

$$u = a\sqrt{d}(\text{cm/sec})$$

where d is diameter of the particle in mm and 'a' is the coefficient which is taken as given below.

$$a = 36 \text{ for } d > 1\text{mm}$$

$$a = 44 \text{ for } 1\text{mm} > d > 0.1\text{mm}$$

$$a = 56 \text{ for } d < 0.1\text{mm}$$

If the flow-through velocity computed from Camp's criteria is very low, hydraulic short circuiting may occur. To avoid it, Mosonyi has recommended a flow-through velocity in the range of 0.4–0.6 m/sec.

The fall velocity w for the particle size in stagnant water can be obtained from standard curves. Sudry's curve for fall velocity for flows having different densities is given in Fig. 15.5. This value from the curve can be corrected for turbulence effect as below.

$$w_e = w - w'$$

where w_e is effective fall velocity.

$$w' = \alpha.u$$

$$\alpha = 0.132\sqrt{h}$$

where h is depth of flow is m.

Thus,

$$L = \frac{h \cdot u}{w_e} = \frac{h^{3/2} \cdot u}{h^{1/2} \cdot w - 0.132 \cdot u}(m)$$

and

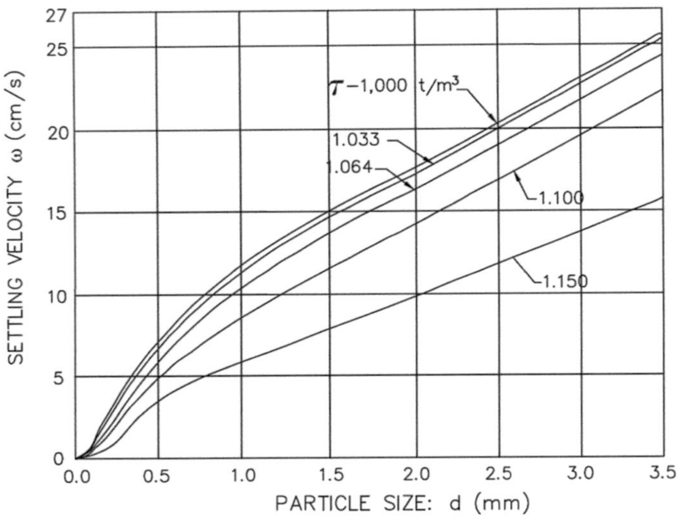

Fig. 15.5 Settling velocity in stagnant water (Sudry's Curve)

$$b = \frac{Q}{h \cdot u}.$$

The efficiency of the above-dimensioned basin can be determined by using Camp's curve (Fig. 15.6). The efficiency as shown in the curve is dependent on two parameters $\frac{w \cdot D^{1/6}}{n \cdot u\sqrt{g}}$ and $\frac{w.L}{u.D}$ where D is depth of flow and 'n' is Manning's rugosity coefficient.

The above design approach by Mosonyi is illustrated by the following example.

- A medium head SHP is designed for a discharge of 50 cumec.
- Design discharge = 50 cumec.
- Flushing discharge (20% of design discharge) = 10 cumec.
- Discharge in the desilting basin = 60 cumec.
- Particle size for removal = 0.25 mm.
- Settling velocity in stagnant water = 0.043 m/sec.
- Flow through velocity as per Camp's relation = $44\sqrt{0.25} = 0.22\text{m/sec}$.
- Since it is a low velocity; hence it is assumed as 0.4 m/sec.
- Basin depth assumed as 8 m.
- Width of basin = $\frac{60}{8 \times 0.4} = 20\text{m}$.

The length of basin with settling velocity correction is

$$L = \frac{h^{3/2}u}{h^{1/2}w - 0.132u}$$

$$= \frac{8^{3/2} \times 0.4}{8^{1/2} \times .043 - 0.132 \times 0.4} = 134\text{m}.$$

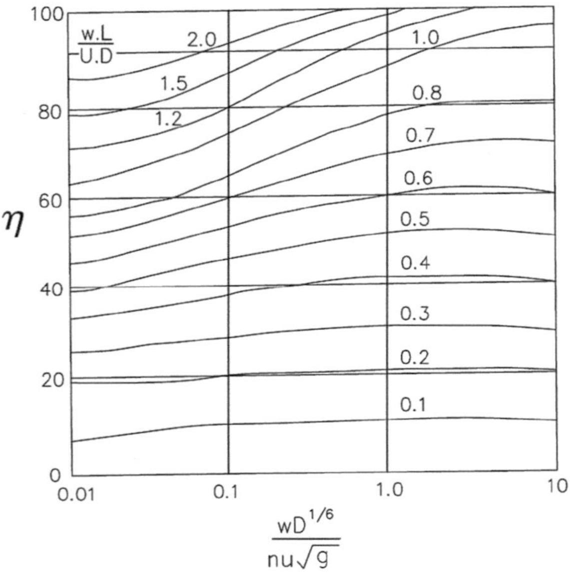

Fig. 15.6 Camp and Dobbins relation of efficiency of desilting basin

Provide length as 140 m.
Efficiency of this basin from Camp's curves worked out as
Evaluate the following two parameters.

$$\frac{\mathrm{wL}}{\mathrm{uD}} \text{ and } \frac{\mathrm{wD}^{1/6}}{\mathrm{nu}\sqrt{\mathrm{g}}}$$

$$\frac{\mathrm{wL}}{\mathrm{uD}} = \frac{.043 \times 140}{0.4 \times 8} = 1.8$$

$$\frac{\mathrm{wD}^{1/6}}{\mathrm{nu}\sqrt{\mathrm{g}}} = \frac{.043 \times 8^{1/6}}{0.014 \times 0.4 \times \sqrt{9.18}} = 3.5.$$

Camp's curve gives 100 efficiency for the above dimensioned basin.
If we do not correct the settling velocity for turbulence, then basin length works as

$$\mathrm{L} = \frac{8 \times .4}{.043} = 80\mathrm{m}.$$

If we increase length by 20% for turbulence then the length of the basin works as 100 m. The efficiency of this length of the basin from Camp's curve works out as

$$\frac{\mathrm{wL}}{\mathrm{uD}} = \frac{.043 \times 100}{0.4 \times 8} = 1.3.$$

The other parameter will be 3.5 (as worked above).

Efficiency from the Camp's curve = 98%.

Considering the economy in the cost of the basin a length of about 120 m may be adopted.

The dimensions of desilting basins for various discharges suggested by CBI&P (Pub 175) are as follows.

Discharge (Cumec)	Length (m)	Width (m)	Depth (m)
Up to 1	10	3.5	2.5
1–5	30	3.5	3.0
5–7.5	40	3.5	3.5
7.5–12.5	50	4.5	4.0

(c) **Flushing Discharge and Arrangement**

(i) Flushing Discharge

The flushing discharge for efficient removal of sediment from the desilting basin is in the range of 20–30% of the design discharge. For efficient functioning of a desilting basin, flushing should be continuous during the monsoon period when the sediment is excessive in the river flow.

(ii) Flushing Arrangement

A hopper-type flushing arrangement is shown in Fig. 15.4. The construction of hopper type arrangement for flushing is found costly and time-consuming. Another arrangement has been developed in which a continuous flushing channel is provided at the bottom of the basin. At the downstream end of the flushing channel, a regulation gate or valve is provided to control the discharge. This arrangement is recommended for SHP projects of small discharges.

The section of the flushing channel is normally 0.5 × 0.5 m which increases from the upstream to the downstream end of the basin. At the end, the depth is increased to 0.7 m. It is provided with a sufficient bed slope of the order of 1 in 50 to 1 in 100. The typical desilting basin of this type recommended by CBI&P (Pub 305) is shown in Fig. 15.7.

In many cases, the flushing channel is covered at the top with a slab in which holes of varying sizes are provided. Normally size and number of holes are decided on a hydraulic model. A typical basin of this type which has been adopted for the Larji project is shown in Fig. 15.8.

(d) **Other Aspects of Design**

(i) Inlet Arrangement

A transition from the channel to low velocity basin is provided in such a manner that no separation shall take place. The inlet divergence should be flatter than 1:4.

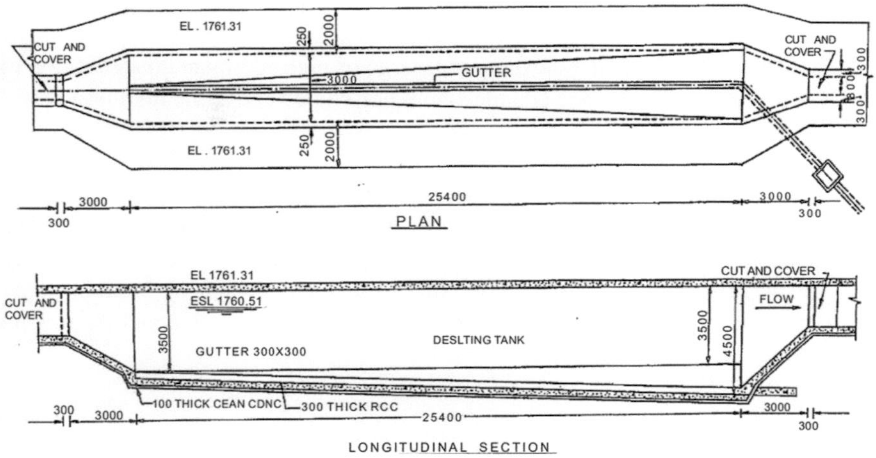

Fig. 15.7 Typical desilting tank (CBI&P–305)

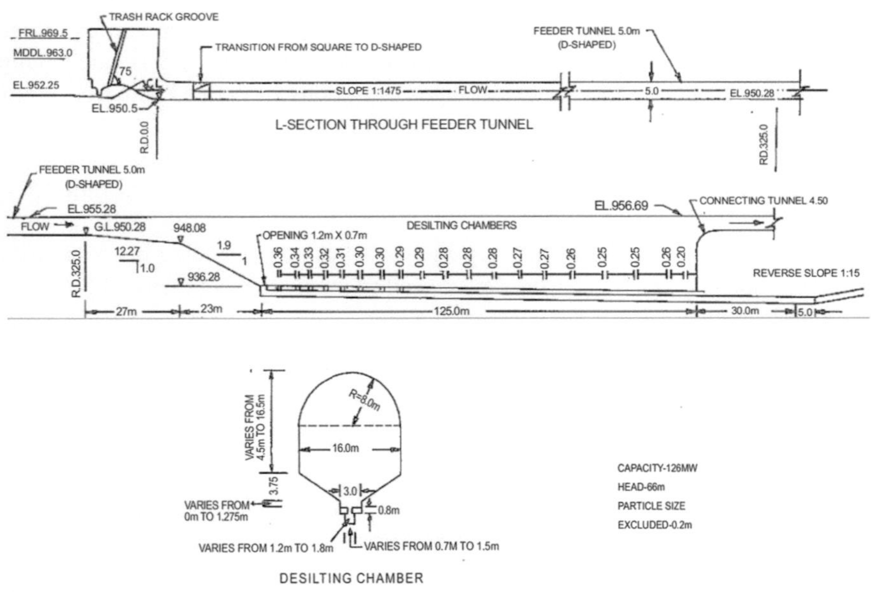

Fig. 15.8 Desilting arrangement at Larji project (HP)

A flat vertical expansion is not desirable. A bed slope of 1:2 to 1:3 has been found satisfactory.

(ii) Outlet Arrangement

The outlet crest should be placed at as high a level as possible. A proper convergence should be provided from the basin width to the outlet channel.

(iii) Escape Channel

It should be designed to carry flushing discharge of high sediment concentration to the river or a drain. The escape channel should discharge above the high flood level of the river.

(iv) Structural Considerations

The desilting basin is an RCC structure normally constructed in M20 concrete. The bottom slab should be placed on firm and strong foundation strata. The walls are designed as retaining walls for earth pressure from outside when the basin is empty and also for the condition that water pressure is acting from inside and there is no earth pressure from the outer side.

15.3.2 Vortex Settling Basin

A vortex settling basin is a desilting device that works on the principle of vortex flow to extract sediment from the flow. It consists of a cylindrical basin with an orifice at the centre of its bottom to remove sediment-laden flow. The flow is tangentially introduced in the basin. The combined vortex conditions are created with free vortex near the orifice and forced vortex conditions in the outer region.

Many investigators have carried out model tests and developed relationships for configuration, dimensions of basin, extraction ratio and sediment removal efficiency. For details refer CBI&P Pub 301.

These basins are smaller in size requiring low flushing discharge in comparison to other sediment extraction basins/devices and so these are recommended for SHPs that utilise small discharges for power generation.

The definition sketch of a vortex settling basin is shown in Fig. 15.9.

(a) **Design Parameters**

(i) Basin Diameter

Many investigators have suggested relations for basin diameter but these are found close to 5B (five times the bed width ‘B’ of the inflow channel).

(ii) Flushing Discharge

Investigators have suggested flushing discharge ranging between 5 and 10% of the inflow discharge. The sloping basin floor increases flushing discharge. Higher sediment removal efficiency has been observed with increased flushing discharge but an increase in efficiency is not found significant for flushing discharge greater than 10% of inflow discharge. A valve is provided at the outlet of the flushing pipe to control flushing discharge.

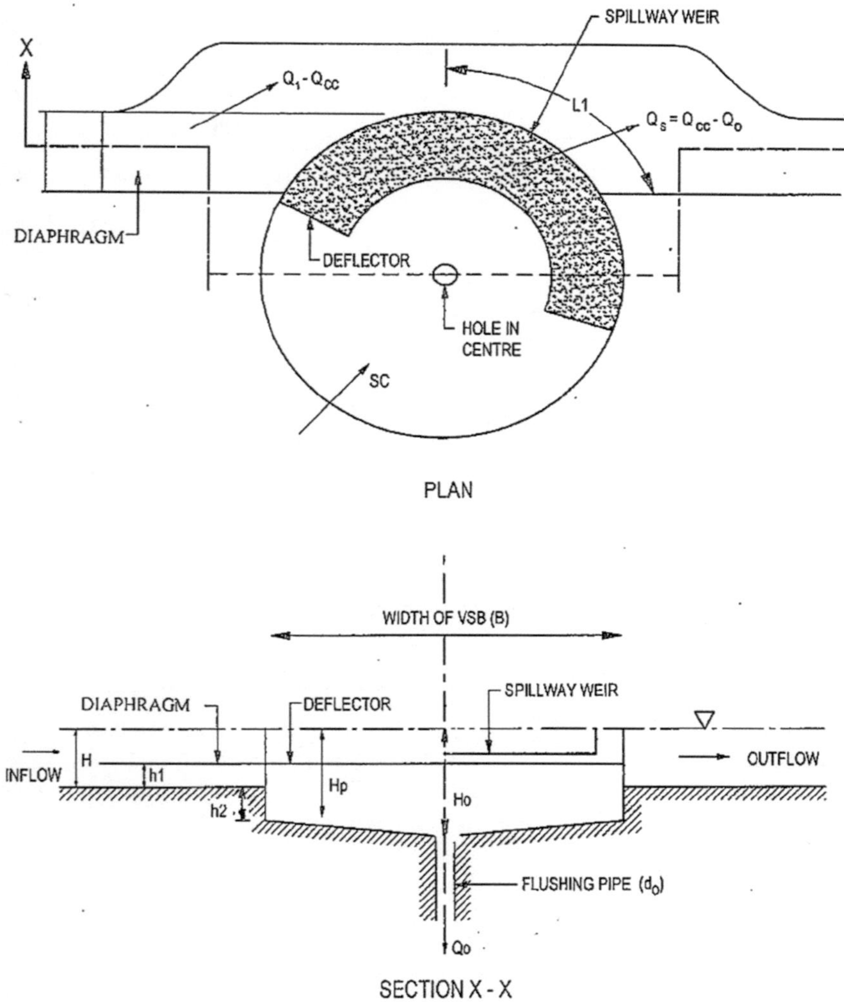

Fig. 15.9 Definition sketch (Plan and section) of a vortex settling basin showing diaphragm and deflector (CBI&P-301)

(iii) Orifice Size

It depends on the flushing discharge and depth of flow at the orifice. But the size of the orifice 'd_o' is limited to $22.5 \leq \frac{D}{d_o} \leq 40$ where, D is diameter of basin.

(iv) Basin Depth

The basin depth below the channel bed h_2 is suggested as 0.6 H (H is flow depth in the channel) and h_1 (the height of diaphragm) equal to 0.2 H for efficient sediment removal.

The depth of flow in basin H_p is $H + h_2$. The height of side wall should be equal to depth of flow $H_p + a$, where 'a' is the freeboard which is generally 0.3 m.

(v) Slope of Basin Floor

The slope of basin floor 'S_c' should not be steeper than 1:10 (V:H) because a steeper slope increases the flushing discharge without a significant gain in sediment removal.

(vi) Deflector

Many investigators have recommended for providing a deflector or diaphragm over the vortex basin (Fig. 15.9) to allow only lower portion of depth of flow carrying high sediment concentration to the basin. The upper portion of the flow is bypassed to the outlet channel without entering the basin. The diaphragm is provided at a height of 0.2–0.3 H (H is depth of flow in inlet channel). It is extended into the channel in a length of 20–50 H. Some investigators do not suggest a deflector as it causes afflux in the inlet channel.

(vii) Outlet Channel

It is suggested that the bottom of outlet channel should be higher than the bottom of the inlet channel. The difference should be about 10% of depth of flow in the basin.

(b) **Examples**

A vortex basin provided at the Gaj project in Himachal Pradesh which is shown in Fig. 9.6 has been found working satisfactorily. A diaphragm-type vortex basin has been recommended for Jal Dhaka HEP by Department of WRDM, I.I.T (Roorkee).

15.3.3 Vortex Tube

A vortex tube is a circular tube laid across the width of the channel below the bed with an opening at the bed level. It is placed either normal to the flow or at an angle of 30–45° to the flow. To control the flow out of the tube, its downstream end is fitted with a valve. As water passes over the tube a shearing action across the open portion sets up a vortex motion within the tube which carries the sediment out of the tube. The vortex tube is found to trap the lower depth flow of the channel which carries the coarse sediment as bed load. These are found efficient in extracting sediment of size 0.5 mm or more.

(a) **Design Parameters**

A lot of research work both theoretical as well as experimental has been carried out in different countries of the world since 1933 when Parshall designed the first vortex tube in the USA.

Robinson (1960) carried out a dimensional analysis of the problem and found that the trap efficiency of a vortex tube is dependent on several dimensionless parameters as given.

$$E = f\left(\frac{d}{D}, R, C, \frac{d_s}{D}, F\right)$$

where

$E =$ Trap efficiency.
$d =$ Depth of flow in channel at vortex tube.
$D =$ Width of opening.
$R =$ Extraction ratio.
$C =$ Sediment concentration $= \frac{\text{sediment discharge}}{\text{channel discharge}}$.
$d_s =$ Size of sand fraction.
$F =$ Froude No.

After carrying out series of experiments Robinson gave following criteria for satisfactory design and operation of a vortex tube shown in Fig. 15.10.

- The flow over the vortex tube should correspond to a Froude Number of about 0.8.
- The flushing discharge which depends on flow parameters and geometry of the tube ranges from 5 to 10% of the total discharge.

Fig. 15.10 Flow analysis in vortex tube sediment extractor

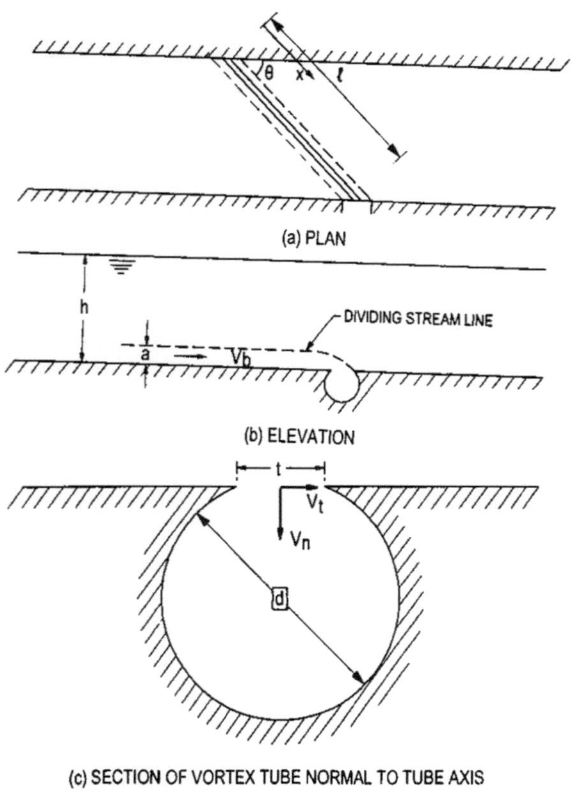

- The width of the opening should range from 0.15 to 0.3 m.
- The ratio of length to width of opening L/D should not exceed 20 with the maximum length of tube (L) being about 4 m.
- The angle of vortex tube θ should be 45° with flow direction.
- Straight tubes function equally well.
- Elevation of the upstream and downstream lips of the tube may be the same.
- The required area of tube can be approximated by 0.06DL sin θ.
- The tube is expected to remove 80% of sediment sizes equal and greater than 0.5 mm. The removal efficiency of smaller particles is considerably low.

Many studies thereafter validated the above criteria. However, several investigators proposed an analytical solution to the problem. Atkinson (1994) developed an analytical procedure to work out the trapping efficiency of a vortex tube (Refer CBI & P. Pub. 301). He has suggested the following criteria.

- It is preferable to have a straight tube.
- High trapping efficiency can be expected at Froude No. less than 0.8.
- If one tube is not effective or sufficient or is of large size or head loss is more then, more than one tube may be adopted.
- Ratio of width of the opening to tube diameter may range between 0.15 and 0.3.
- The vortex tube should be located away from canal head where turbulence generates uniform sediment distribution across the depth of flow. The minimum distance between canal head and vortex tube L_a can be determined from the curve given in Fig. 15.11 which is between dimensionless parameters and $\frac{L_a W_j}{\overline{U}.h}$ and $\frac{W_j}{U_*}$.

  where

W_j = Settling velocity of the particle to be extracted.
U_* = Shear velocity.
$\overline{U}$ = Mean velocity of flow.
H = Depth of flow.

The distance between the canal head and vortex tube should be 1.5 times the L_a obtained from the curve. This criterion may also be adopted for locating the sediment ejectors and desilting basins.

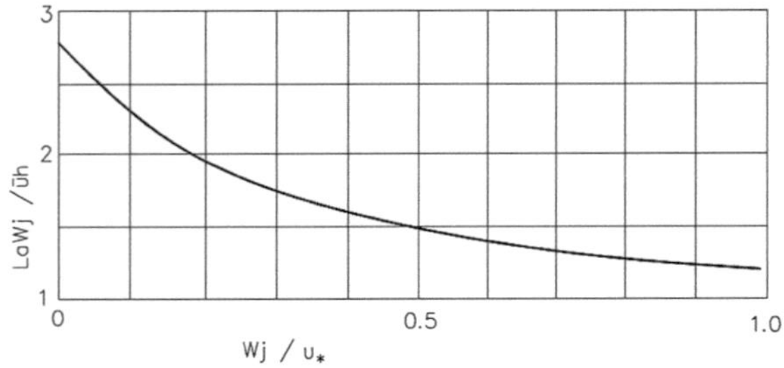

Fig. 15.11 Non-dimensional adaptation length as function of W_j/U_* function of (*Source* Atkinson 1994)

The distinct advantages of a vortex tube as the desilting measure in a small hydro project are easy construction, reasonable efficiency with a low extraction ratio and low cost.

(b) **Examples**

A vortex tube desilting measure has been adopted at the Ganvi Project (22.5 MW, 393 m head) in the Himachal Pradesh. It was developed through hydraulic model studies to extract particles bigger than 0.2 mm at IRI Roorkee. It is shown in Fig. 15.12. It was commissioned in the year 2000 and is reported to function satisfactorily. Another vortex tube system was designed and adopted at the Gumma Project (3 MW, 183 m head) in Himachal Pradesh. It is shown in Fig. 15.13.

15.4 Balancing Reservoir

It is a pond that either exists naturally or is created artificially in the alignment of the power channel. Creation of pondage is preferred in cutting. It may be created by constructing an embankment around a naturally depressed land. It should be ensured that the embankment does not get overtopped. The inner face of the embankment which is in contact with water should be lined to prevent seepage. It should be provided with a spilling arrangement for escapage.

The pond can be used to store 2–4 h of channel flow depending on the storage capacity and the power requirement. The purpose is to store water during off-peak hours and to supply the same for the peaking operation of the power plant. Such a pond is usually called a balancing reservoir. If such a natural pond is feasible, it can reduce the requirement of diurnal pondage at the diversion structure. The balancing reservoir will have an intake as well as an outlet structure.

Such a pond can also act as a settling tank where fine sediment of small size (–0.15 mm) may settle due to a long retention period. The deposited sediment in the long run will reduce the storage capacity. Hence flushing arrangement should be provided in the balancing reservoir. Alternatively, the balancing reservoir should be desilted periodically either manually or mechanically. In major ponds, continuous dredging arrangements may be made.

If the balancing reservoir is located near the powerhouse it may also be made to function as a forebay with penstocks at the outlet end. A typical balancing reservoir is shown in Fig. 2.15 (Chap. 2).

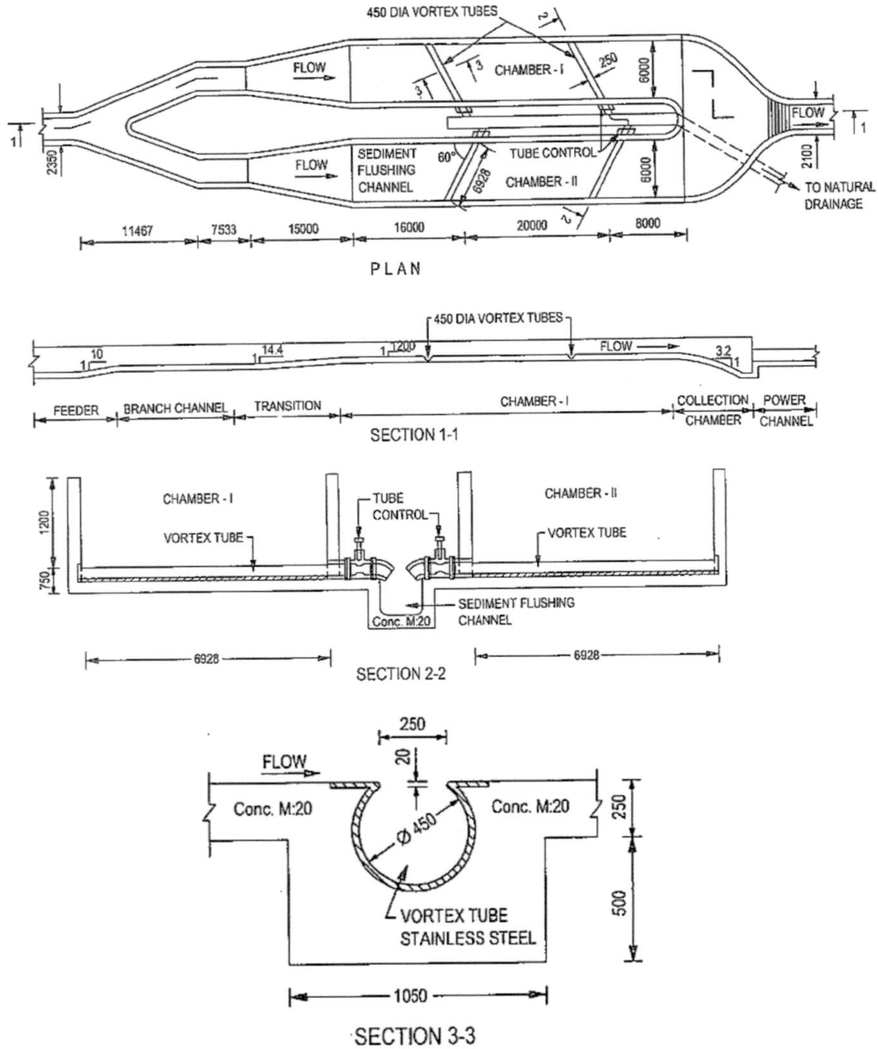

Fig. 15.12 Desanding basin-Ganwi HEP

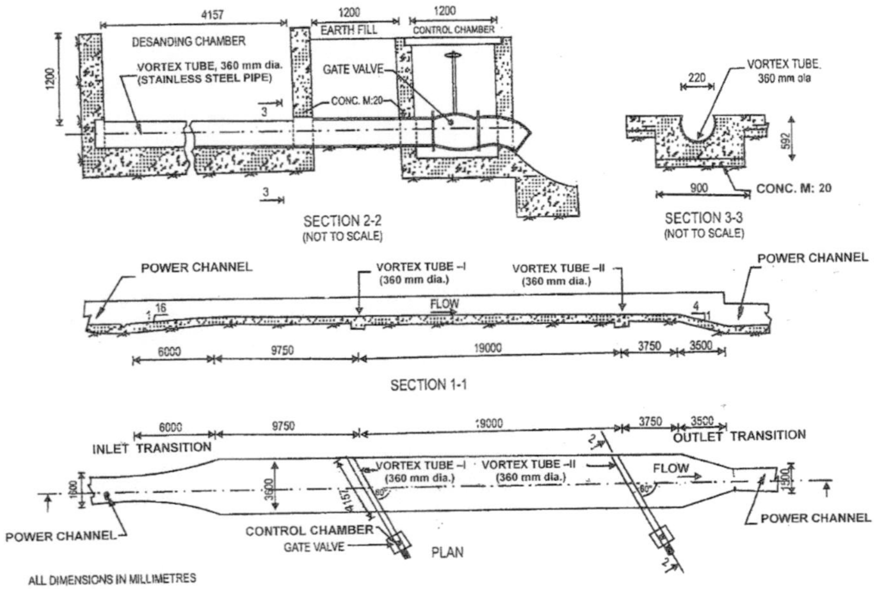

Fig. 15.13 Desanding basin-Gumma HEP

References

1. Asthana, B.N.: Sediment management in water resources projects. Pub. No. 301, CBI&P, New Delhi (2007)
2. Atkinson, E, ‘Vortex Tube Sediment Extractors-II, Design,’ Journal of Hydraulic Engineers ASCE, (1994)
3. CBI&P.: Small hydro stations standardization. Pub. No. 175, New Delhi (1985)
4. CBI&P.: Manual, on planning and design of small hydroelectric schemes. Pub-No. 305, New Delhi (2009)
5. Garde, R.J., Ranga Raju, K.G.: Mechanics of sediment transport. New Age Publishers, New Delhi (1985)
6. Guidelines for Small Hydro Development, Standard No., 2.2 and 2.3. ‘Civil Works-Hydraulic and Structural Design’, AHEC, IIT, Roorkee (2013)
7. Mosonyi, E.: Water power development-II. Publishing House of the Hungarian Academy of Science, Budapest (1965)
8. Robinson A.R., “Vortex Tube Sand Trap” Journal of Irrigation and Drainage Division, ASCE, (1960)
9. Singh, B.: Fundamental of irrigation engineering. New Chand and Bros. Roorkee, (2005)

Chapter 16
Design of Surge Tank and Forebay

16.1 General

At the end of a long water conductor in a run-of-river SHP, either a surge tank or a forebay is provided. A surge tank is provided at the end of a tunnel running under pressure. However, the requirement of a surge tank is dependent on the length of the tunnel, the head on the turbine and time taken by the wicket gates of the turbine. In case of a power channel or free flow tunnel/conduit a forebay is provided. These structures have some storage which is released at the time of starting a machine. These supply flow to the turbine for the time by which the flow is accelerated in the water conductor. Therefore, these structures are located as near to the powerhouse as possible. This reduces the length of the penstock which is subjected to high internal pressure. It is illustrated in Fig. 16.1. The layout and design aspects of these structures are briefly described in this chapter.

16.2 Surge Tank

A surge tank at the end of a long pressure tunnel is normally a large size circular shaft. It acts as a reservoir with storage space to regulate flow to the turbine during the acceptance or rejection of load. Besides flow regulation, the surge tank also serves to relieve the water hammer pressure within the penstock under conditions of load rejection and acceptance. The pendulation of water levels during small or large load variations is damped due to the damping effect of frictional resistance.

B. N. Asthana, *Planning and Design of Small Hydro Power Projects*,
https://doi.org/10.1007/978-3-031-96271-4_16

Fig. 16.1 Design pressure diagram

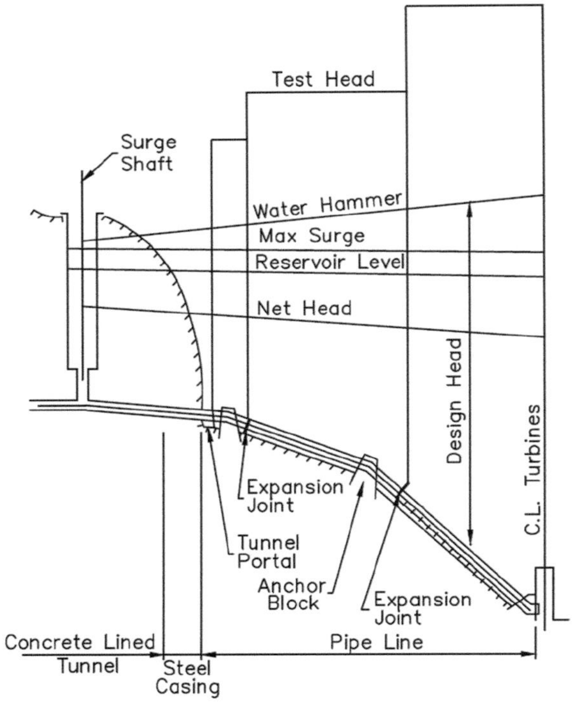

16.2.1 Functioning of a Surge Tank

As mentioned above the functioning of a surge tank during load acceptance and rejection is explained in Fig. 16.2.

When there is a sudden load rejection, the guide vanes in the turbine are closed and flow reduces. This initiates a water hammer wave. It travels to and fro between the turbine and surge tank. The entire mass of water oscillates in the penstock and the surge tank. This effect is of short duration and dies away before the surge reaches a significant proportion. Thus, while evaluating surges the water hammer effect is disregarded. The surge causes a rise in the water level in the surge tank. When the water level in the surge tank exceeds the reservoir level it produces a retarding effect which arrests and reverses the flow in the upstream conduit/tunnel. The maximum water level achieved in the surge tank is called the upsurge level. This water level in the surge tank will then drop and when it reaches below the reservoir level a reversal of flow will start in the water conductor and this cycle is repeated till surges are damped out by frictional forces. This phenomenon is shown in Fig. 16.2a.

During load acceptance, the water in the water conductor is at rest, the surge tank supplies water to the turbine and the water level in the surge tank falls rapidly below reservoir level (known as down surge level) creating sufficient head to accelerate flow in water conductor. The flow, because of head more than normal head, being in excess of turbine requirement will cause rise in water level in surge tank. This rise

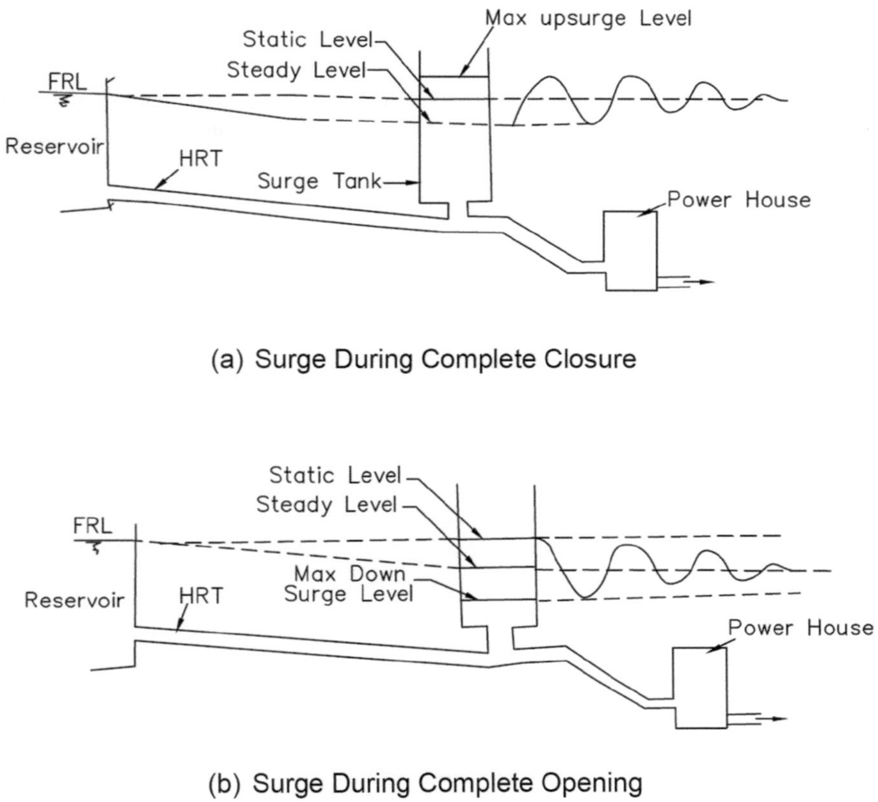

Fig. 16.2 Showing surge tank operation.

in level will retard the flow in surge tank and a swing in water level will be set-up until damped out by friction. It is shown in Fig. 16.2b.

This behaviour occurs at partial load variation also. The surge amplitude is approximately proportional to the surge tank diameter.

When there is full load rejection the surges do not affect the operation of the turbine. However, at load fluctuations, the operation of the turbine is affected by changing water levels in the surge tank. The guide vanes are opened or closed to maintain the required discharge under varying water levels in the surge tank.

If the surge tank is not properly proportioned, it is possible at small load changes that the governor imposes forced oscillation on the mass of water in the tank and conduit. In such cases, small surges instead of being damped continue indefinitely. Therefore, a tank in which surges are damped in the above conditions is called a stable surge tank. If the surge tank is not stable, there will be continuous hunting of the turbine and governor. Hence the diameter of the surge tank should be properly designed to make it a stable surge tank.

16.2.2 Diameter of a Surge Tank

Normally, Thoma's criterion is followed for determining the diameter of a stable surge tank. It is for an isolated power station. Thoma's relationship for the area of the stable surge tank is as below

$$A_{th} = \frac{L \cdot A_t \cdot V_1^2}{\beta V_0^2 \cdot H_0 \cdot 2g}$$

where A_{th} = Area of surge tank
A_t = Area of tunnel.
L = Length of the tunnel.
βV_0^2 = Minimum head loss in tunnel.
V_1 = Velocity in the tunnel upstream of surge tank.
H_0 = Head on turbine.

To work out the diameter of surge tank 'A_{th}' is doubled from practical considerations including the factor of safety for a simple surge tank. For restricted orifice type Thoma's area is multiplied by 1.6.

16.2.3 Location of Surge Tank

The surge tank, as shown in Fig. 16.1, should be near the powerhouse. It should also be located on high ground to reduce the height of the shaft.

16.2.4 Types of Surge Tank

Basically, surge tanks are divided into three types, viz.

(i) Simple surge tank,
(ii) Restricted orifice type and
(iii) Differential type.

(i) **Simple Single Tank**: A simple surge tank is just a small reservoir connected to the penstock in such a manner that allows flows into and out of the tank without any loss of head. In this case, the area of the riser pipe connecting the tunnel with the surge tank is at least equal to the area of the tunnel (Fig. 16.3).

The simple surge tank requires an area more than the other types. There is a gradual change in the water level in this type of surge tank. The action of a simple surge tank is sluggish and requires more time for damping the surges. Hence, it is

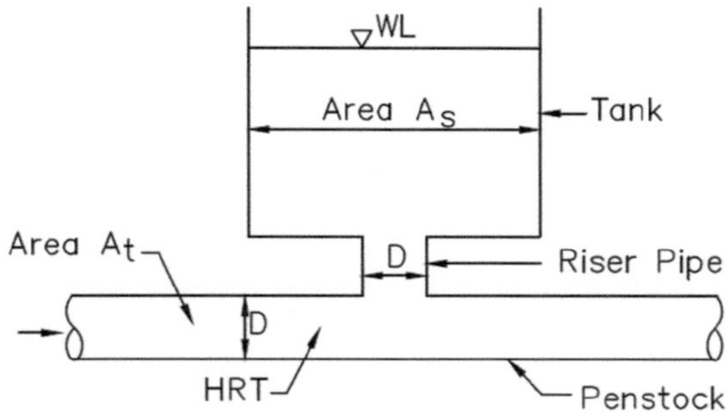

Fig. 16.3 Simple surge tank

ideal with slow-acting governors. Because of large area requirement, it is often more expensive.

It is preferred in small hydro projects being a simple large diameter pipe/shaft. If site conditions are suitable, these can be made inclined by laying large-diameter steel pipes along a hill slope. This type of surge tank which is laid on a hill slope has been used in the Lower Lagyap H.E. project (Fig. 2.20).

(ii) **Restricted Orifice Type Surge Tank**: The distinctive feature of this type of surge tank is a restricted orifice placed between the tunnel and the tank (Fig. 16.4). The purpose of the orifice is to generate significant friction loss when water flows to and from the tank.

When load is rejected, the surplus water passes through the orifice, a retarding head equal to the friction loss due to the restricted orifice is built up in the conduit. Under the condition of load acceptance, the orifice tends to develop an accelerating head in the conduit more quickly than it would develop in a simple surge tank.

The size of the orifice is proportioned in such a way that at full load rejection, the initial retarding head is approximately equal to the ultimate rise of the water surface in the tank. If the size of the orifice is too large it will tend to be a simple surge tank and if it is too small the retarding head is equal to the water hammer in the conduit with no surge tank.

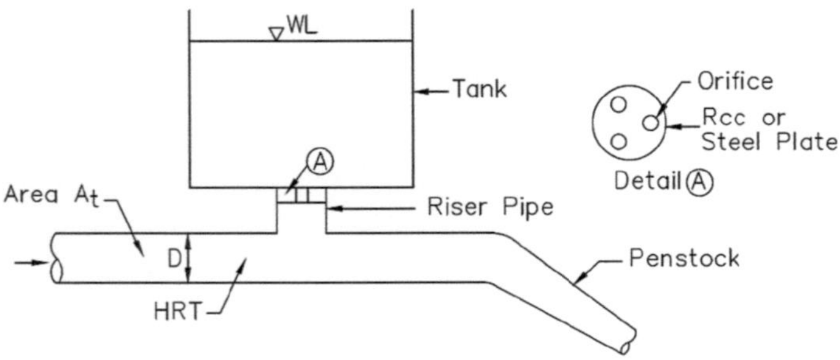

Fig. 16.4 Restricted orifice type surge tank

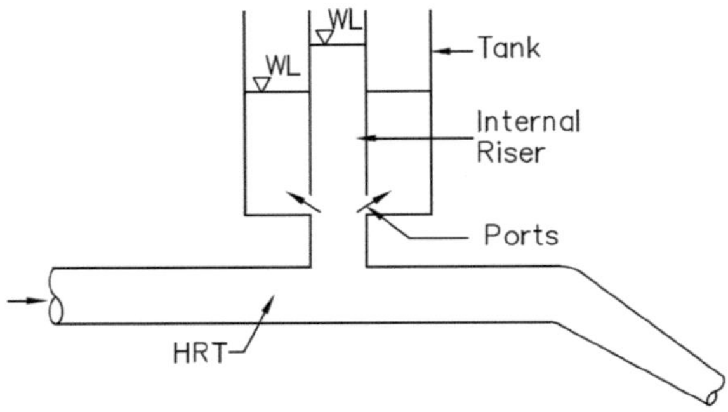

Fig. 16.5 Differential surge tank.

In this type of surge tank where the accelerating and the retarding heads are generated quickly on conduit discharge, less water is required to be stored or released from the tank. Hence, in this type of surge tank smaller storage capacity for water is required. It is therefore quick acting and economical as compared to a simple surge tank. It is used with a quick-acting governor. It is normally used in large-capacity SHPs.

(iii) **Differential Surge Tank**: It is a combination of a simple surge tank and the restricted orifice type with an internal riser pipe extending upward from the restricted orifice to near the top of the outer shell as shown in Fig. 16.5.

The internal riser is smaller in diameter than the connection with the conduit. The annular ports allow water to enter the tank. The operation of the tank depends upon the area of the ports which is proportioned to suit the operation conditions.

The water level in the internal riser rises and falls during load rejection and acceptance more quickly than in the tank. When the load is rejected water rises rapidly in the internal riser causing a differential head at the port to make water flow through the port into the tank.

During load acceptance when water is required by the turbine the water in the internal riser falls rapidly, thus creating an accelerating head rather quickly. The water from the tank flows slowly through the port to supply the turbine demand.

Thus, this type of surge tank requires less space for storing water. These are seldom used because of complicated construction.

16.2.5 Hydraulic Design of Surge Tank

(a) **Criteria for Hydraulic Design**

In view of the above discussion about the functioning, the surge tank design shall satisfy the following conditions:

1. It shall be so located that water hammer pressures of both (+) and (−) shall remain within permissible limits.

2. The tank must be stable.
3. Maximum possible upsurge (worst condition of load rejection) must be contained in the tank. A spillway or spilling arrangement is often provided to keep the upsurge within the tank.
4. The lowest down surge must be sufficiently above the conduit/penstock level so that in no case air is drawn into the penstock.
5. The range of surges should not be large enough to cause difficulty for the governor to pick up the load.

(b) **Loading Conditions**

For calculating up-surge and down-surge water levels in the tank, the worst combinations of friction, load and reservoir water level should be considered. These are given below.

For Maximum Up-Surge

- Friction loss in the system should be minimal ($n = 0.012$ for concrete-lined tunnels and $n = 0.01$ for steel pipes).
- Reservoir water level should be maximum.
- Full load rejection should be considered.

For Minimum Down-Surge

- Friction loss in the system should be maximum ($n = 0.014$ for concrete lined tunnels and $n = 0.01$ for steel pipes).
- Minimum reservoir level.
- Full load rejection followed by specific load acceptance.
- No load condition at the minimum reservoir level.
- **Surge Analysis**

The extreme surge levels in the surge tank are determined by integrating the equations of mass oscillations (refer I.S. 7396–1). The equations are non-linear. These are integrated either graphically or numerically. Several computer programmes are available.

(i) **Simple Surge Tank**

The following equations should be satisfied:

$$\frac{L}{g}\frac{dV_1}{dt} + Z \pm \beta V_1^2 + \frac{V_1^2}{2g} = 0 \quad \text{(Dynamic equation)}$$

(+) value of $\beta V^2{}_1$ to be used when flow is downstream and (-) when it is upstream.

$$A_t V_1 = A_s \frac{dz}{dt} + Q \quad \text{(Continuity equation)}$$

$$P = K_m Q\left(H + Z + \frac{V_1^2}{2g} - h_{tp}\right) \quad \text{(Power equation).}$$

The maximum up-surge (Zm above maximum reservoir level) for complete load rejection can be obtained by

$$\frac{L}{2g\phi\beta^2 V_0^2} - \frac{Z_m}{\beta V_0^2} - \frac{L}{2g\phi\beta^2 V_0^2}\left[e - \frac{2g\phi}{L}\left(Z_m + \beta V_0^2\right)\right] = 0$$

where

L = Length of head race tunnel.

ϕ = Ratio of A_s/A_t (A_s is area of surge tank and A_t area of tunnel).

V_1 = Velocity of HRT at any time upstream of surge tank (ST).

V_0 = Maximum velocity corresponding to maximum steady flow in tunnel U/S of ST.

Z = Water level in ST measured positive above reservoir level.

Q = Discharge through turbine.

h_{tp} = Total head loss in penstock.

K_m = Constant depending upon the efficiencies of turbine and generator.

The down-surge may be approximately determined from the following formula:

$$Z_{min} = h_f\left[1 + \frac{1}{2m'\left(e^{m^n}\pi - \cos^{-1} m'\right)}\right]$$

where $m' = \frac{h_f}{2V_o}\sqrt{\frac{g}{L}\left(\frac{A_s}{A_t}\right)}$

$$m^n = \frac{m}{\left(1 - m^2\right)}$$

m =damping factor

$$= 2\beta\frac{g}{L}\left(\frac{A_s}{A_t}\right)$$

$$Z_{min} = \text{minimum reservoir level in the starting} - \text{minimum surge level}$$

(ii) **Restricted Orifice Type**

Surges depend on the resistance offered by orifice (h_{or}) which is given below for an orifice area A_o.

$$h_{or} = \frac{Q_0^2}{C_d^2 \cdot A_0^2 \cdot 2g}.$$

C_d varies from 0.6 to 0.9 depending on shape, size and number of orifices. Q_o is maximum discharge through turbine. The following condition should also be satisfied to fix the orifice area.

$$\frac{Z_x}{\sqrt{2}} + \frac{1}{4}h_f \leq h_{or} \leq \frac{Z_x}{\sqrt{2}} + \frac{3}{4}h_f$$

where Z_x is surge height corresponding to change in discharge

$$Z_x = V_o\sqrt{\frac{L}{g}\frac{A_t}{A_s}}$$

h_f = total head loss in tunnel system $= \beta V_1^2$

The following equations including orifice resistance in the equations of simple surge tank should be satisfied:

$$\frac{L}{g}\frac{dV_1}{dt} + Z \pm \beta V_1^2 \pm \eta (V_1 - V_2)^2 + \frac{V_1^2}{2g} = 0$$

$$A_t V_1 = A_s\frac{dz}{dt} + Q$$

$$P = K_m\left(H + Z + V_1^2/2g - h_{tp} \pm h_{or}\right)Q = 0.$$

The maximum surge above reservoir level (Zm) can be derived from the following:

$$\frac{L}{2g\phi(\beta+\eta)^2V_0^2} - \frac{Z_m}{(\beta+\eta)V_0^2} + \left[1 - \frac{\beta}{\beta+\eta} - \frac{L}{2g\phi(\beta+\eta)^2V_0^2}\right] \times\left[e - \frac{2g\phi}{L}(\beta+\eta)\left(Z_m + \beta V_0^2\right)\right] = 0$$

where $\eta = \frac{h_{or}}{(V_1 - V_2)^2}$

V_2 = velocity in the conduit downstream of surge tank.

Z_m = can also be worked out easily by using the curve given in IS 7396 Part–I.

The downsurge can be controlled by phasing the starting of the machines, hence it is not critical.

The surge levels can be approximately worked out by following the method by Parmakian (AHEC–2.2). The maximum upsurge height (Z_{max}) above the steady state level in the surge tank for total load rejection is given by

$$Z_{max} = 1.05 b_o^{-0.89} h_f$$

where $b_0 = \frac{h_f}{Q_t}\sqrt{\frac{A_s g A_t}{L}}$

h_f = total head loss in HRT with minimum friction $\beta V_1^2/2g$.

Q_t = Discharge through the tunnel.

Minimum down surge height (Z_{min}) below steady-state level corresponding to MDDL is given by

$$Z_{min} = 0.88 b_o^{-0.91} h_f.$$

16.2.6 Structural Design Aspects

(i) **Types of Lining**

After excavating the required diameter of the surge shaft, it is lined. The type of lining will depend on the rock type. If the rock is sound but has an unfavourable jointing pattern and the shaft has a rock cover of more than twice the maximum design head, shotcreting (minimum 10 cm thick) with rock bolts would be adequate. If the rock is hard and self-supporting with a cover less than twice the head plain concrete lining may be provided. However, in case of rock of inadequate strength and stability, a reinforced concrete lining should be provided. In weak rocks with several weaknesses or where the velocities are high or cavitation is expected, a steel liner shall be provided. The liner is generally sandwiched in concrete.

(ii) **Design of Lining**

Lining shall be designed for the following loads:

(i) Internal water pressure including dynamic water level.
(ii) External water pressure due to water table or expected saturation level. It shall be considered when ST is empty.
(iii) External grout pressure—This is due to the grouting done to fill the gaps between rock and lining.
(iv) Rock pressure—Lining is not designed for any rock load. If there is any pressure due to the seepage of water into the rock, it can be withstood by the lining designed for external water and grout pressure.
(v) Seismic force—It is not considered when the lining is in contact with rock.
(vi) Live loads—Generally there are no live loads.

The thickness of the lining is designed on the basis of the design of a thick cylinder. In the case of plain concrete lining in sound rock, thickness is designed for external water pressure only. The reinforced concrete lining is designed on the assumption that the rock will share some of the internal water pressure depending on the value of the rock modulus. The sharing of water pressure can be worked out by the following formula (Refer Fig. 16.6).

$$P_r = \sigma_t \frac{E_R M_r / (M_r + 1)}{E_C M_c^2 / (M_c^2 - 1)} \left\{ \frac{(b^2 + a^2)\left((b^2 - a^2)/M_c^2\right)}{2b^2} \right\}$$

Fig. 16.6 Pressure sharing by rock

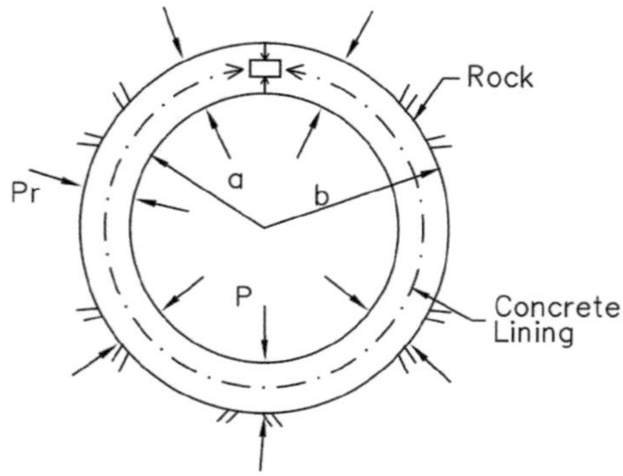

where a = Internal radius of shaft (cm).

b = External radius of shaft (cm).

E_c = Modulus of elasticity of concrete.

E_R = Modulus of elasticity of rock.

P_r = Internal pressure taken by rock (kg/cm^2).

σ_t = Permissible tensile stress in concrete (kg/cm^2).

$$M_c = \frac{1}{\mu_C} = M_r = \frac{1}{\mu_R}$$

where μ_c and μ_R are poisons ratio for concrete and rock respectively.

The portion of shaft above ground or in loose strata is designed for full internal pressure.

The thickness of the lining for external water pressure P_o (kg/cm^2) is worked out by the following formula considering it a thick-walled pipe.

$$P_o = \sigma_C \left(\frac{b^2 - a^2}{b^2 + a^2} \right)$$

where

σ_C = permissible compressive stress in concrete (kg/cm^2).

However, a minimum thickness of 0.3 m shall be provided.

Reinforcement in lining shall be provided for total hoop force due to net maximum internal pressure by the following formula.

$$A_{st} = \frac{(P - P_r)a}{\sigma_{st}}$$

where A_{st} = area of hoop steel to be provided (cm^2/cm height of shaft).

σ_{st} = permissible stress in steel (kg/cm^2).

P = total maximum internal pressure (kg/cm^2).

P_r = internal pressure taken by rock (kg/cm^2).

Hoop reinforcement should be greater of the following:

(a) 0.3% of concrete area of lining.
(b) Required for (P_r) with normal permissible stress in steel.
(c) Required for P neglecting rock support with higher stress in steel.

In case steel ribs are used for rock support during excavation, about 60% of the steel rib area may be treated as hoop reinforcement if ribs are adequately jointed and fully enclosed in lining concrete.

Nominal longitudinal reinforcement should be provided (Refer I.S. 456).

To prevent local buckling of lining it should be suitably anchored with the rock with anchors suitably spaced both in horizontal and vertical directions depending on the type of rock.

Grouting shall be done to fill gaps between concrete and rock at a pressure equal to external water pressure with a minimum of 2.5 kg/cm^2. Normally vertical spacing of holes shall be 3 m and radially these shall be at 30° spacing.

Rock around the surge shaft may be consolidated by grouting up to a depth of 6.0 m or more depending on rock condition at a pressure ranging from 2.5 to 7 kg/cm^2. Holes used for contact grouting may be used for consolidation grouting also.

16.3 Forebay

A forebay at the end of a free-flow water conductor is a structure with a small storage capacity to perform the same function as a surge tank at the end of a pressure tunnel. The difference is that the water during load rejection in the case of a forebay is wasted.

16.3.1 Design Aspects

(a) **Location**

It is located as near to the power house as possible to reduce the length of the penstock. It should be located on firm ground so that the bottom and the foundation of the side walls may rest on a firm base.

(b) **Layout**

The storage in the forebay is provided by widening and deepening the channel. The width should also be adequate to provide the intakes for the required number of penstocks. The depth should be adequate to provide sufficient water seal above the penstock opening in order to eliminate the possibility of entering air into the penstocks. The side walls should be high enough with sufficient freeboard above the canal FSL. To discharge channel flow during complete load rejection, an opening in a side wall above FSL is provided. The head required to pass canal discharge should be accommodated in the freeboard of the side walls and the channel upstream of the forebay. In many cases, the bye pass opening is fitted with automatic gates and in that case, the crest level of the opening may be below FSL. This reduces the freeboard requirement of the channel and the side walls. The bypass channel should take the canal discharge either to available natural drainage or to the tail race channel downstream of power house.

The Intake structure for the penstocks at the downstream end of forebay should accommodate a trash rack, stop log and regulation gates for the penstocks and the required transition from rectangular gate opening to circular penstock. The openings between trash rack bars should be less than the minimum opening in the speed ring vanes.

The power channel upstream of the forebay should preferably be straight in a length of about 10 times the channel width to ensure uniform flow to the penstock intake structure. For the typical layout of a forebay Fig. 2.14 may be referred.

(c) **Storage Capacity**

There is no specific criterion for deciding the storage capacity of the forebay. But the normal practice is to provide a capacity sufficient to store about 2 min of flow. This means that $L \times B \times D$ is equal to $Q \times 2 \times 60$ cum.

(d) **Hydraulic Design**

The criteria to design the components of the penstock intake structure such as inlet opening size, transitions, size of air vent and the height of water seal are the same as described in the chapter on Intake structures. The crest length of the bye pass outlet can be designed by the weir formula.

(d) **Structural Aspects**

The floor of the forebay is made of concrete panels and their joints are provided with water stops to prevent seepage. The side walls and the penstock intake are generally made as concrete gravity structures. These should satisfy the stability criterion of a gravity structure. If the height is large the side walls may be made of RCC.

References

1. Guidelines for Small Hydro Development, Standard No. 2.2 and 2.3 ‘Hydraulic and Structural Design. AHEC, IIT Roorkee (2013)
2. I.S. 7396 (Part 1).: Criteria for hydraulic design of surge tanks. BIS, New Delhi, (1985)
3. I.S. 7357.: Code of practice for structural design of surge tanks. B.I.S. New Delhi (1974)

Chapter 17
Penstocks and Pressure Shafts

17.1 General

Water is conveyed to the power station either from the dam in a storage project or from a forebay or surge shaft in case of a run-of-river project through a closed conduit that runs fully under pressure. Such a conduit is called a penstock when laid at or near the natural ground surface. Sometimes because of topographical and geological site conditions, the penstock is laid underground in a shaft and then it is called a pressure shaft. It may be a vertical or an inclined shaft. Penstocks are subjected to high internal pressure and maximum pressure is at the downstream end of the penstock, where it enters the powerhouse. So generally, these are made of steel pipe. These are also made of concrete and wood for low-head power stations. Concrete pipes (prefabricated hume pipes) are used up to a diameter of 1 m for a head up to 30 m but bigger diameter concrete pipes up to 3 m diameter for such a head are cast-in-situ. Concrete pipes are generally laid below the ground. Joints are packed carefully. Joints are normally provided at a spacing of 15 m c/c. The section of the pipes is normally circular.

For small head and discharge SHPs pipes made of wood are sometimes used in the region where wood is available locally in abundance. Wood staves are butt jointed to form a circular section. These are held together with steel bands. Leakage may be a problem from joints so their use is restricted to 30–40 m head. The thickness of staves ‘t’ (cm) is given by

$$t = \frac{D \cdot H}{21}$$

where, D is diameter in meter and H is head in meter and D.H should normally be less than 100

B. N. Asthana, *Planning and Design of Small Hydro Power Projects*,
https://doi.org/10.1007/978-3-031-96271-4_17

17.2 Layout of Penstocks

The layout and alignment of penstocks depend on economic considerations which in turn are governed by topography and geology of the site and the type of development. Where power stations are placed at the toe of a dam or near it the penstocks are laid inside the dam body with the upstream end starting from the intake structure normally placed on the upstream face of the dam with proper transition and control structure. In this case, penstocks are embedded in dam concrete. Where power stations are away from the water source, a water conductor carries water to the powerhouse. The water conductor may be either a channel or a tunnel. If the water conductor is long, the channel ends in a forebay and the tunnel ends in a surge tank which is placed close to the power plant. The water from the fore bay or surge tank is conveyed to the power plant through penstocks. These are laid either on the hill slope or slightly below the surface. If this is not possible then penstock is laid through a vertical or inclined shaft made in the rock mass. Such a penstock is called a pressure shaft. These alignments are shown in Fig. 17.1.

The layout of a penstock normally requires bends. A minimum number of bends should be provided because each bend is associated with some head loss. Normally a bend is required at the downstream of the intake structure at the upstream end of the penstock and the other at the entry to the powerhouse. Other bends are required to negotiate the topographical features of the site. The bends should normally be provided with a radius more than 5 times the diameter of penstock. Each bend is also

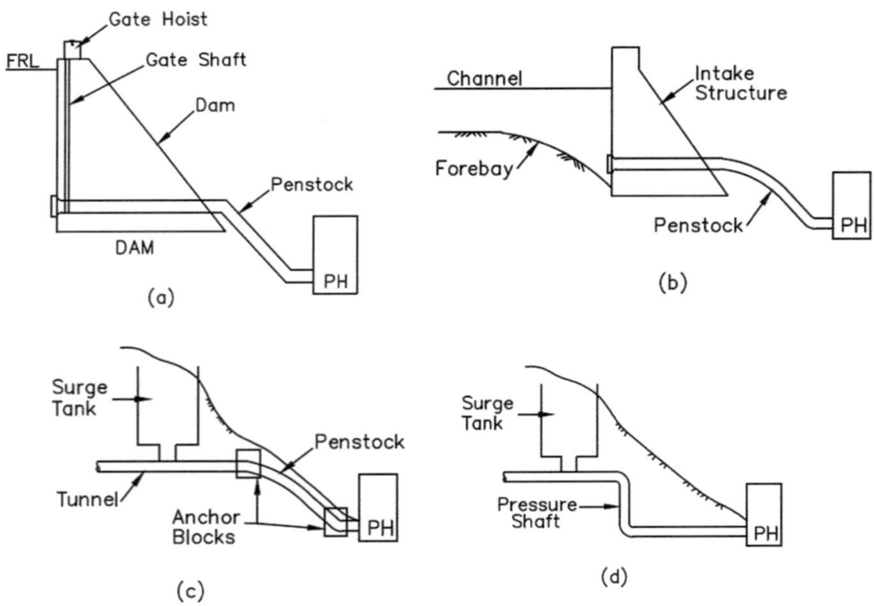

Fig. 17.1 Penstock layout

provided with an anchor block. The alignment of penstocks before entering the power house or meeting the spiral- casing is governed by the following specifications.

(i) A length of 5 to 10 times the diameter should be horizontal from the centre line of the unit, so as to have a uniform distribution of water in a spiral case.
(ii) The radius of curvature in the penstock profile should be at least five times the diameter of the penstock at the bends.

17.3 Types of Penstocks

(A) According to Placement

1. **Exposed/surface penstocks**

These are installed above natural ground level and are supported on piers or saddle-type supports. Such penstocks are not in contact with the ground surface. The gap between the ground and the bottom of the penstock is kept equal to a minimum of 600 mm for maintenance.

The advantages of this type of penstock are:
(i) Easy and regular inspection is possible.
(ii) Safety against sliding is ensured by properly designed anchorages.
(iii) Maintenance and repairs are easy.

2. **Buried Penstocks**

Buried penstocks are supported continuously on a concrete pad at the bottom of a trench which is backfilled with excavated soil. About 1.25 to 1.5 m thick back fill should be given over the penstock pipes.

The advantages of this type of penstock are the following:
(i) It is not affected by temperature changes because of soil cover. Hence expansion joints are not required.
(ii) It is not affected by the freezing of water in cold climates.
(iii) Landscape is not affected in this type of penstock.
(iv) It is not affected by hill slides, avalanches and falling trees.
(v) It is not subjected to damages from other sources.
(vi) It is not affected by earthquakes.

The disadvantages are
(i) Not accessible for inspection and fault location is difficult.
(ii) Installation may be costly if the diameter is large and the soil is hard.
(iii) It tends to slide on steep hill slopes.
(iv) Needs a special coating for protection against the corrosive action of salts present in soil.
(v) Maintenance and repairs are difficult.

(B) **According to Rigidity**

In order to reduce longitudinal and axial stresses due to temperature variations and other reasons like taking care of movements of rigid joints between pipe sections, joints are at some locations substituted by elastic joints. The occurrence and extent of pipe movement depend besides the spacing and type of expansion joints, on the alignment of penstock as well. Considering the aspect of movement following three types of penstocks are provided.

(i) Rigid pipeline with anchors.
(ii) Flexible pipeline by using expansion joints or flexible coupling.
(iii) Semi rigid pipelines—with expansion joints between each adjacent section either continuously supported or supported at a certain distance.

Generally, the third type i.e. semi rigid type of penstock is used. Such a penstock has anchors at the bends and an expansion joint downstream of each anchor. The pipe between the two anchors is supported on saddle supports.

17.4 Diameter of Penstock

Diameter of a penstock depends on the discharge and the velocity permitted in the penstock pipe ($Q = \frac{\pi}{4}D^2V$). In concrete and steel pipes a velocity of 5 to 6 m/sec is normally permitted. A smaller velocity of up to 3 m/sec has been permitted in many cases on the consideration of sediment in the flow. Higher velocity means more friction loss resulting in higher power loss. Hence, considering sediment erosion and friction loss a velocity of 4 to 5 m/sec is normally recommended. On the above considerations the following diameter for penstocks for SHPs are recommended (CBI. Pub 175).

Discharge (cumec)	Diameter (mm).
Up to 1.0	500.
1 To 2.5	700
2.5 To 5.0	1000
5 To 7.5	1200
7.5 To 10	1400
10 To 12.5	1600

17.4.1 Economic Diameter

For large discharge penstocks, economic considerations should also be taken into account for determining the diameter of a penstock. The economic diameter shall also be checked to see that the acceptable velocity is not exceeded.

The large diameter of penstock costs more but the head loss is less and the vice—versa is also true. Therefore, economical diameter is worked out by equating the cost of pipe and the cost of loss of energy.

$$\frac{dc_1}{dD} \leq \frac{dc_2}{dD}$$

where, c_1 is annual cost of penstock and c_2 is the annual cost of energy corresponding to head loss and D is diameter of penstock. On the above basis an expression for economic diameter is given below (Refer P.S. Nigam).

$$D^7 \frac{3.25}{2960} \cdot \frac{k_2}{k_1} \frac{Q^3}{H} \cdot T$$

where,

λ = friction coefficient (taken as 0.02 for preliminary estimates).

σ = Permissible stress in steel in kg/cm^2.

k1 = Annual cost of steel/penstock per kg (Rs).

k2 = value of one KWh of energy (Rs).

Q = Discharge in cumec.

H = Head in meters.

T = Hours for which energy is produced in a year.

'D' the diameter can be worked out from the above expression by trial and error.

The following formula for the economical diameter of an underground steel-lined penstock is recommended in IS 11625:

$$D^{22/3} = \frac{2.36 \times 10^6 \times Q^3 n^2 \times e \cdot p_t \times C_p}{\left[1.39C_e + 0.6C_c + \frac{121H \cdot C_s(1+i)}{\sigma_a e_j}\right] \times p}$$

where,

C_c = unit cost of concrete lining in Rs/m^3.

C_e = unit cost of excavation in Rs/m^3.

C_p = cost of 1 kWh of energy in Rs.

C_s = cost of steel in Rs/kg.

D = diameter of the penstock.

e = overall efficiency of plant.

e_j = joint efficiency of penstock.

H = head on penstock including water hammer in m.

i = percentage by which steel in penstock is overweighted due to the provision of stiffeners, corrosion allowance, etc.

n = Rugosity coefficient in Manning's formula.

p = ratio of annual fixed operation and maintenance charges to construction cost of penstock.

p_t = annual load factor.

Q = discharge through penstock in m^3/s.

σ_a = allowable tensile stress in steel in kg/cm^2.

17.5 Number of Penstocks

The general practice is to use one penstock for each turbine in the power house. This is particularly common in the case of low head power house and where the length of penstock is short. In the case of high-head plants with long penstocks, the practice is to connect several units (2 or 3) with one penstock. Such alignments are shown in Fig. 17.2

The number of penstocks for any particular installation depends on a number of factors. Economically one big size penstock costs less than several smaller penstocks. If the thickness of plate and size of one penstock becomes large, the number of penstocks can be increased to conform to the available facilities for fabrication, transportation, erection and handling. Operational convenience should also be considered. One penstock for one unit is operationally very convenient whereas one penstock for several units has several issues such as satisfactory design and fabrication of Y piece or trifurcation, and differential loading when one unit is closed. Any problem in one penstock serving several units will close all the units. On the issue of the economy, Mosonyi has quoted that if the cost of one penstock is 1 then the cost of 2, 3 and 4 equivalent penstocks will be 1.16, 1.19 and 1.22, respectively.

When one penstock is bifurcated into two or more or one main penstock is replaced by two or more equivalent pipes, the question of determining equivalent diameter arises. This is normally found by keeping the head loss the same.

Head loss in a pipe is directly proportional to $\frac{V^2}{D}$ or $\frac{Q^2}{A^2D}$ or $\frac{Q^2}{D^5}$.

So, let head loss in one pipe $hf_1 = a_1\frac{Q^2}{D^5}$.

If the pipe is replaced by 'n' pipes, the discharge in each pipe will be $\frac{Q}{n}$ and if the diameter of that pipe is D_n then head loss in the pipe will be $hf_1 = a_1\frac{Q^2}{D^5}$.

When head losses are equated then we get

$$a_1\frac{Q^2}{D^5} = a^1\frac{Q^2}{n^2D_n^5}$$

$$\therefore\ D_n = \frac{D}{n^{2/5}}$$

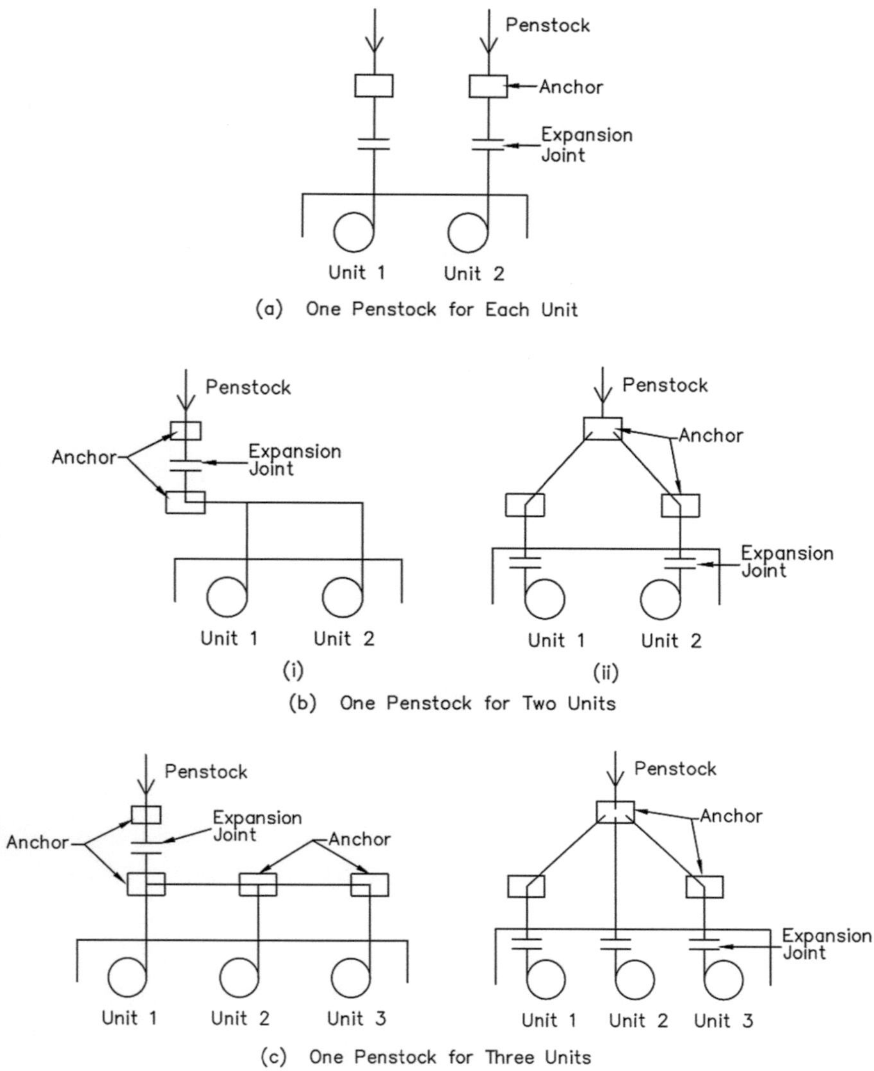

Fig. 17.2 Penstock alignments

By putting the value of 'n' as 2 or 3 we get the equivalent diameter of pipes in terms of the diameter of one pipe. So when one pipe is bifurcated into two with equal discharge then the bifurcated pipe diameter $D_2 = \frac{D}{2^{0.4}} = \frac{D}{1.32}$. Similarly when one pipe is divided in 3 pipes then $D_2 = \frac{D}{3^{0.4}} = \frac{D}{1.55}$ and when divided into 4 branches.

The thickness of pipe 't' is proportional to the diameter so $t = a_2 D$ and t_n (thickness of bifurcated pipe) $= a_2 \frac{D}{n^{2/5}}$.

$$\therefore \quad t_n = \frac{t}{n^{2/5}}$$

Then cost can be compared by determining the volume of steel. We get that cost of nth pipe will be $n^{1/5}$ times the cost of one pipe.

17.6 Hydraulic Losses in Penstock

All the losses are expressed in terms of $K\frac{v^2}{2g}$.

(i) Head loss at trash rack

The trash rack loss may be expressed as

$$h_t = k_t \frac{v^2}{2g}$$

for value of k_t refer chapter on Intake structure or IS 4880 (Part 3)

(ii) Head loss at entrance

It depends on the shape of the intake mouth. The losses are given by

$$h_e = k_e \frac{v^2}{2g}$$

K_e for normally adopted bell month shape is taken as 0.05. For other shapes refer to Table 17.1. Refer to AHEC manual No. 2.2 or IS 0.4880 (Part–3) for details.

Table. 17.1 Loss coefficient for entrance of various shapes

S. no	Type of entrance	Loss coefficient for entrance,		
		Maximum	Minimum	Average
(1)	(2)	(3)	(4)	(5)
i	Gate in thin wall-unsuppressed contraction	1.80	1.00	1.50
ii	Gate in thin wall-bottom and sides suppressed	1.20	0.50	1.00
iii	Gate in thin wall-corners rounded	1.00	0.10	0.50
iv	Square-cornered entrances	0.70	0.40	0.50
v	Slightly rounded entrances	0.60	0.18	0.25
vi	Fully rounded entrances $\frac{r}{D} \geq 0.15$	0.27	0.08	0.10
vii	Circular bell mouth entrances	0.10	0.04	0.05
viii	Square bell mouth entrances	0.20	0.07	0.16
ix	Inward projecting entrances	0.93	0.56	0.80

(iii) Friction losses

These are the main losses in the penstock. These are normally determined by the following Manning's formula:

$$h_f = \frac{v^2 n^2 L}{R^{4/3}}$$

where h_f = friction head loss (m).

V = Velocity in penstock (m/sec).

L = Length of penstock (m).

R = Hydraulic radius $\left(\frac{\text{Area}}{\text{Wetted Perimeter}}\right)$ (m).

n = Rugosity coefficient (for concrete pipe it varies from 0.012 to 0.014 and for steel pipe varies from 0.008 to 0.012).

(iv) Losses due to contraction

For gradual contraction the head loss is expressed as

$$h_c = K_c\left[\frac{V_2^2 - V_1^2}{2g}\right]$$

where h_c = head loss due to contraction (m).

V_1 = velocity at entry (m/sec).

V_2 = velocity at exit (m/sec) i.e. at the contracted section.

K_c = Contraction coefficient which is 0.1 for a flare of 10° and for 90° it is 0.5. For other flare angles a linear variation between 0.1 and 0.5 may be assumed (Refer I.S. 4880 part-3).

(v) Losses due to expansion

The flare angle for expansion should not be more than 10°. The head loss can be worked out by

$$h_e = k_e\left(\frac{V_1^2 - V_2^2}{2g}\right)$$

where h_e = head loss due to expansion (m).

V_1 = velocity at the entry end or upstream end (m/sec).

V_2 = velocity at downstream end (m/sec).

k_e = 0.2.

For other cone angles refer IS 2951 (Part-2).

(vi) Bend losses

The bend loss excluding friction loss for a circular conduit depends on the shape of bend, deflection angle and ratio of radius of bend to diameter of pipe. It is given by

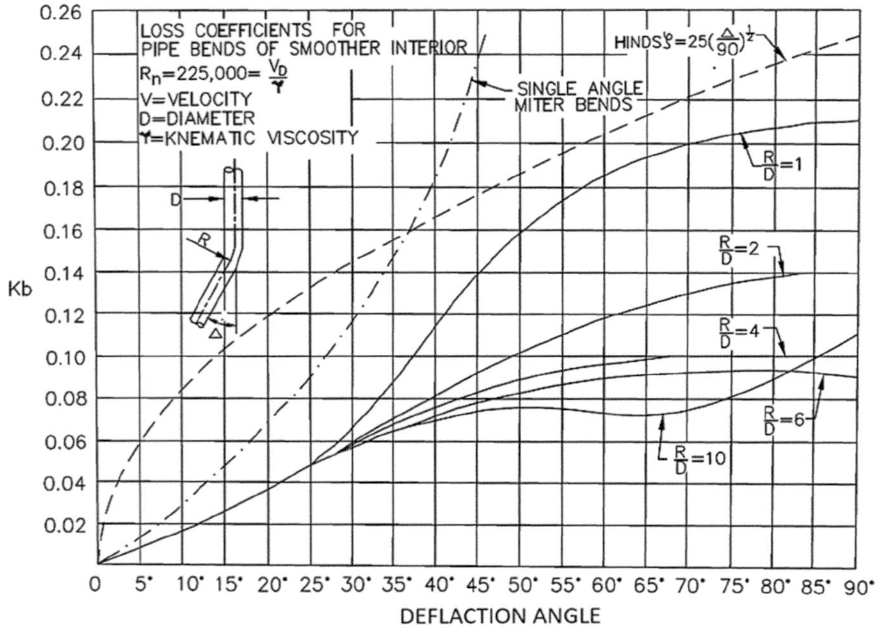

Fig. 17.3 Loss coefficients for pipe bends of smooth interior

$$h_b = k_b \frac{V^2}{2g}$$

K_b = bend loss coefficient. It can be obtained from the curves given in Fig. 17.3 for various values of R/D and deflection angles.

(vii) Valve losses

Valves are placed at two places—one at the upstream end and the other at the downstream end just before the turbine. Generally, butterfly and spherical valves are used. These are operated either in fully closed or fully open condition. Spherical valves offer very little loss in fully open condition. The head loss in butterfly valve in given by

$$h_v = k_v \frac{V^2}{2g}$$

K_v can be obtained from the curves given in Fig 17.4

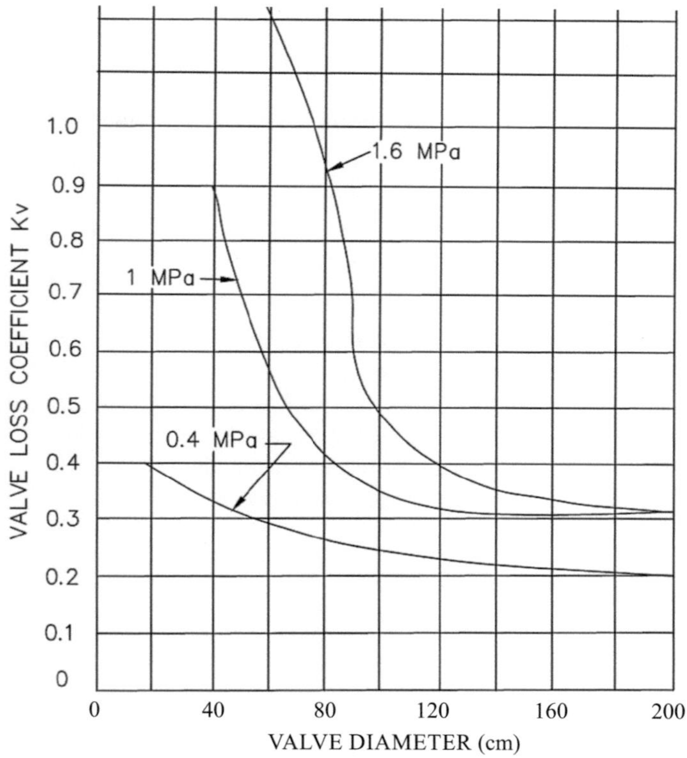

Fig. 17.4 Valve loss coefficient for butterfly valves. (*Source* IS: 2951 (Part II)-1965 (Reaffirmed 2003))

(viii) Gate losses

The head loss at the gates is given by

$$h_g = k_g \frac{V^2}{2g}$$

$K_g = 0.23$ for fully open condition
$= 0.4$ for half open.
$= 1.00$ for one fourth open.

(ix) Losses in Penstock Branches

A penstock bifurcation is termed a wye and when there are more than two branches it is called a manifold. The head losses are governed by the angle of bifurcation, ratio of cross-sectional area, type and shape of a bifurcation. It is given by

$$h_y = k_y \frac{V^2}{2g}$$

where, h_y = loss of head in Y piece (m).
V = velocity before Y piece junction (m/sec).
k_y = loss coefficient as given in Table 17.2.
q, v, D and q_b, v_b, D_b are discharge, average velocity and diameter of original and branch pipes respectively.

17.7 Water Hammer in Penstocks

Under normal flow, the penstock is mainly subjected to only internal hydrostatic pressure. But when turbine gates at the downstream end of the penstock are closed suddenly, there is a sudden pressure rise, known as a water hammer, due to the momentum of flowing water being destroyed. Ideally, there would be a series of pressure waves travelling back and forth over the length of the pipe. The celerity of these pressure waves can be calculated by the formula.

$$a = \sqrt{\frac{E_B}{p\{1 + \frac{E_B}{B}\left(\frac{D}{t}\right)\}}}$$

where,
a = celerity of wave in m/sec (It is of the order of velocity of sound in water).
E_B = Bulk modulus of fluid in t/m^2.
B = Modulus of elasticity of pipe wall in t/m^2.
p = mass density of fluid.
D = Internal diameter of penstock in m.
t = Thickness of penstock liner in m.
For steel penstocks, the equation reduces to

$$a = \frac{1425}{1 + \left(\frac{D}{100\,t}\right)}$$

If L is the length of pipe, and if gate closure time is less than 2L/a, the pressure rise may be calculated by the formula:

$$H = \frac{a \cdot V}{g}$$

where,
H = Pressure rise in m of water.
V = normal velocity of flow in m/sec.
As can be seen the pressure rise due to sudden or instantaneous closure is too high so such closure is avoided and closing time is increased beyond 2 L/a through the governor and it is termed as a slow rate. Any pressure rise due to a slow rate of closure

Table. 17.2 Loss coefficient in penstock bifurcation

	$q_b/q = 0.3$		$q_b/q = 0.3$		$q_b/q = 0.3$	
Angle or divergence β in degrees	Sharp edged	Rounded $r = 0.1\,D_b$	Sharp edged	Rounded $r = 0.1\,D_b$	Sharp edged	Rounded $r = 0.1\,D_b$
(1)	(2)	(3)	(4)	(5)	(6)	(7)
90	$\begin{cases} D_b = D \\ V_b = 0.3V \\ K_y = 0.85 \end{cases}$	$\begin{cases} D_b = D \\ V_b = 0.3V \\ K_y = 0.76 \end{cases}$	$\begin{cases} D_b = D \\ V_b = 0.5V \\ K_y = 0.87 \end{cases}$	$\begin{cases} D_b = D \\ V_b = 0.7V \\ K_y = 1.00 \end{cases}$	$\begin{cases} D_b = D \\ V_b = 0.7V \\ K_y = 0.80 \end{cases}$	$\begin{cases} D_b = D \\ V_b = 0.7V \\ K_y = 0.80 \end{cases}$
60	$\begin{cases} D_b = D \\ V_b = 0.3V \\ K_y = 0.7 \end{cases}$	$\begin{cases} D_b = 0.61D \\ V_b = 0.8V \\ K_y = 0.59 \end{cases}$	$\begin{cases} D_b = D \\ V_b = 0.5V \\ K_y = 0.59 \end{cases}$	$\begin{cases} D_b = 0.79D \\ V_b = 0.8V \\ K_y = 0.54 \end{cases}$	$\begin{cases} D_b = D \\ V_b = 0.7V \\ K_y = 0.57 \end{cases}$	$\begin{cases} D_b = D \\ V_b = 0.7V \\ K_y = 0.52 \end{cases}$
45	$\begin{cases} D_b = 0.58D \\ V_b = 0.9V \\ K_y = 0.43 \end{cases}$	$\begin{cases} D_b = 0.58D \\ V_b = 0.9V \\ K_y = 0.35 \end{cases}$	$\begin{cases} D_b = D \\ V_b = 0.5V \\ K_y = 0.42 \end{cases}$	$\begin{cases} D_b = 0.75D \\ V_b = 0.9V \\ K_y = 0.32 \end{cases}$	$\begin{cases} D_b = D \\ V_b = 0.7V \\ K_y = 0.34 \end{cases}$	$\begin{cases} D_b = D \\ V_b = 0.7V \\ K_y = 0.3 \end{cases}$

Source: IS 2951 (Part II) – 1965 (Reaffirmed 2003)
Note: These values are based on the experiments conducted at the Hydraulic Laboratory of the Technical University of Munich, Germany, for the most efficient case

is followed by a depression of some magnitude and thus highest pressure head at the lower end of penstock is given by the sum of the static head plus pressure rise. The pressure rise in rapid rate closing governors (about 2 sec closing time) may be too high for an economical design of penstock. If such fast governors are used then pressure relief valves should be used to limit the pressure rise. Thus, the governor and turbine opening are adjusted to limit the pressure rise. Mosonyi recommends the permissible pressure rise for turbines working under different heads in terms of the percentage of heads.

Head	Percentage of head
Up to 50 m	50
Up to 150 m	25
Up to 250 m	15

Another recommendation to limit pressure rise as a percentage of head depending on the type of turbine is given below.

Type of turbine	Percentage of head
Pelton wheel	15 to 30
Francis turbine	30 to 35
Kaplan turbine	30 to 50

The pressure rise is maximum at the turbine end of the penstock and linearly varies to zero at the intake end at either the fore bay or surge tank.

In case of a sudden opening of the gate, there is a drop in pressure at the end of the penstock. The gradient line from the down surge level to the maximum pressure drop at the end of the penstock is important for the layout of the penstock. No portion of penstock should normally be laid above this gradient line. If any portion of the penstock is laid above this gradient line it may be subjected to cavitation. To avoid this situation slow opening of the units is adopted as this limits the drop in pressure.

17.8 Structural Design of Exposed/Surface Penstock Liner

The surface penstocks are generally laid on natural ground. These are semi-rigid types. Anchors are provided at all the bends and at intermediate points where the distance between two bends is more than 150 m or so. Expansion joints are provided just at the downstream side of an anchor block to account for the temperature changes. Saddle supports are provided in between anchor blocks at such an interval so that excessive longitudinal stresses in the penstock shell are not induced due to bending. Normally saddle supports are spaced from 5 to 8 m centre to centre depending on the size of the penstock. Refer Fig. 17.5 (Profile of a penstock).

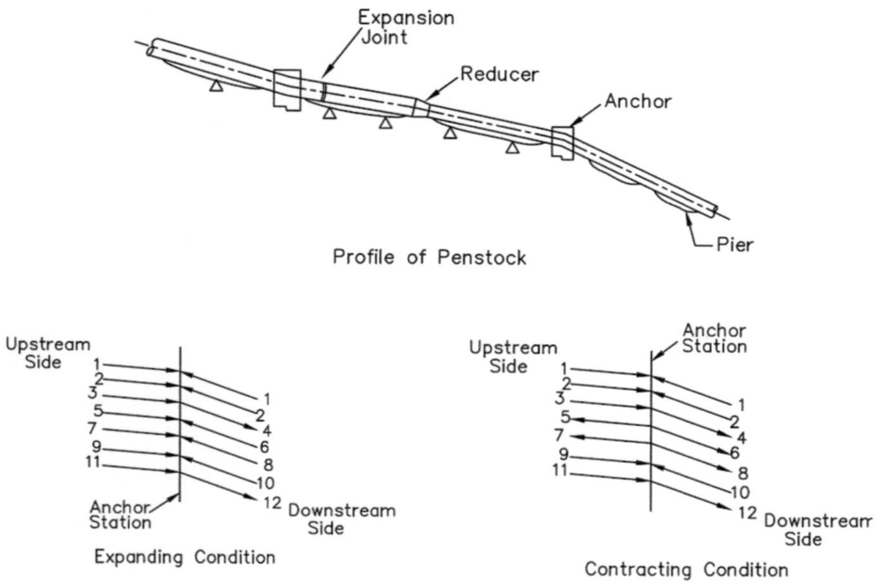

Fig. 17.5 Action of loads/forces on anchor block

The structural design of the shell thickness of penstock is primarily carried out for maximum internal pressure which induces hoop stress in the shell. It may be checked for longitudinal stresses due to various loads such as the weight of penstock and water between supports, temperature variation, friction at supports, etc. if the size of the penstock is large. The method of calculation of these stresses is given in detail in I.S. 11639 (Part–1). These are discussed briefly in the following paras:

(a) **Hoop stress in shell**

The hoop stress in the shell is due to internal pressure which is maximum at the turbine end of the pipe. It is the sum of the static head (corresponding to FRL) plus the upsurge height plus the water hammer (Refer Fig. 16.1). The stress is given by

$$f_h = \frac{PD}{2.t}$$

where, P is internal pressure, D is diameter of penstock and 't' is thickness of penstock shell.

For welded pipes, a joint factor J is taken into account. The pipe thickness is worked out by

$$t = \frac{0.1H.D}{2J.f_h}$$

where, H is the design head (m), D is diameter of pipe (m), and J is joint factor and is taken as 0.85 to 1.0 depending on X-ray/testing of joints, f_h is permissible stress (kg/cm^2), and 't' is shell thickness (m).

The allowable stress is based on a factor of safety of 3 to 3.5 on the ultimate tensile strength of the shell material. The computed thickness is increased by 1 to 3 mm for corrosion allowance. For reasonable strength required for transportation and erection a minimum thickness of 6 mm for small diameter pipes and 10 mm for bigger ones is recommended by Mosonyi. USBR has recommended a minimum thickness as below.

$$t(\text{cm}) = \frac{D + 50}{400}$$

where, D—Diameter of pipe in cm.

If the penstock length is large and it is to be designed for a high head requiring a thick shell, then it is advisable to divide the penstock into segments and to design each segment for the maximum head which may occur at the lowest point of the segment. Change in thickness in the penstock shell should be made preferably at the anchor points.

(b) **Longitudinal stresses**

1. Due to beam action

The longitudinal stress which is developed due to self-weight and weight of water in the pipe, which spans between the supports, because of beam action is calculated by

$$f_{L_1} = \frac{M}{S}\left(\text{kg/cm}^2\right)$$

where, f_{L_1} is longitudinal stress, M is bending moment considering that the pipe has continuous support, S is section modulus of pipe

$$M = \frac{(G_P + G_W)\cos\beta \times L^2}{12}$$

where, G_p and G_w are weights of pipe and water respectively in unit length of pipe (kg/cm), L is span between supports (cm), β is angle of deflection of pipe with horizontal.

$$S = \frac{\pi}{4} tD^2\left(\text{cm}^3\right)$$

$$f_{L1} = \frac{\left(G_p + G_w\right)\cos\beta \cdot L^2}{3\pi \cdot t \cdot D^2}$$

Maximum stresses will be at top and bottom. Therefore, longitudinal joints in welded pipes should not be located along the top and bottom of the pipe.

The longitudinal stresses developed due to seismic forces in the pipe spanning over the supports due to beam action are equal to seismic coefficient times the stresses computed as above. The two stresses are additive.

2 Due to axial force Pa due to the component of pipe weight in the axial direction

$$P_a = \Sigma G_P \cdot \sin \beta$$

The longitudinal stress due to water pressure acting on the valve or on a blank flange during the hydraulic test is equal to about half the hoop stress.where, ΣG_P is the total weight of all sections of the penstock from the expansion joint to the anchor block. P_a increases gradually and is maximum at the anchor block and is resisted by the anchor block.

Compressive stress in f_{L_2} the shell is given by

$$f_{L_2} = \frac{P_a}{\pi \cdot D \cdot t}$$

In addition to the above, there are other axial forces such as friction on supports, temperature variation, circumferential stretching, etc., which are generally not large enough to cause significant longitudinal stresses in the penstock of SHPs and so these are not described here. If the consideration of these forces is considered necessary, then for details I.S. 11,639 (Part-1) may be referred.

(c) **Equivalent stresses**

If hoop stress is f_h and total longitudinal stresses are f_L, these shall be combined to obtain equivalent stresses by the formula

$$f_x, f_y = \frac{f_h + f_L}{2} \pm \sqrt{\frac{(f_h - f_L)^2}{2} + q^2}$$

where, q is shear stress

The liner thickness of an exposed penstock shall withstand the equivalent stress which shall not exceed 0.5 times the minimum yield strength of the shell material. Normally, longitudinal stresses will be a small fraction of hoop stresses and so equivalent stresses will be marginally (10 to 15%) above hoop stresses. So, in SHPs generally, longitudinal stresses are not worked out but if the diameter and head are large it will be necessary to have a check on longitudinal stresses to ensure safety and economy.

17.9 Design of Anchors and Supports for Exposed Penstocks

Anchors are normally provided at the bends and supports are provided at a spacing of 5 to 8 m c/c between the anchors. The spacing of supports depends on the diameter of the penstock. So less spacing of supports is advisable for bigger-sized penstocks. The anchors and supports may be designed as per I.S. 5330. The design of anchors and supports are briefly described below.

(a) **Design of Anchors**

Forces acting at anchor blocks: The force acting on an anchor at the convex bend of a penstock due to adjacent penstock sections are shown in Fig. 17.5.

1. Hydrostatic pressure acting along the axis of the pipe on each side

$$= \mathrm{wAH}.$$

2. Dynamic force at the bend acting outside

$$= \frac{\mathrm{QwV}}{\mathrm{g}}.$$

3. Axial component of dead weight of pipe uphill from anchor up to expansion joint

$$= \Sigma \mathrm{Gp}_1 \cdot \sin \beta_1.$$

4. Dead weight of pipe from anchor downhill to expansion joint

$$= +\Sigma \mathrm{Gp}_2 \cdot \mathrm{Sin}\beta_2.$$

5. Sliding friction of pipe on piers uphill from anchor due to expansion or contraction

$$= \pm\mu\,\Sigma\left(\mathrm{Gp}_1 + \mathrm{Gw}_1\right)\cos \beta_1.$$

6. Sliding friction of pipe on piers downhill from anchor due to expansion or contraction

$$= \pm\mu\,\Sigma\left(\mathrm{Gp}_2 + \mathrm{Gw}_2\right)\cos \beta_2.$$

7. Sliding friction force at Uphill expansion joint

$= \pm\mu_1 \pi \mathrm{D}_1 \mathrm{wH}_1.\mathrm{x}$ where, x is width of packing.

8. Sliding friction force at downhill expansion joint

$= \pm\mu_2 \pi D_2 w H_2.x$ where, x is width of packing.

9. Hydrostatic pressure on the exposed end of pipe at uphill expansion joint

$$= wA_1H_1.$$

10. Hydrostatic pressure on exposed end of pipe in downhill expansion joint

$$= wA_2H_2.$$

11. Longitudinal force due to reducer above or at anchor

$$= wH(A_1 - A).$$

12. Longitudinal force due to reducer below or at anchor

$$= wH(A - A_2).$$

Suffix 1 and 2, respectively, refer to upstream and downstream portions of penstock,

where

w = unit weight of water.

V = velocity of water in penstock.

A = Sectional area of penstock.

Q = Discharge in penstock.

H = Head of water including water hammer.

Gp and Gw are the weight of penstock and contained water respectively.

Anchor blocks are designed as gravity structures for the forces as shown in Fig. 17.6.

Let F_u and F_D be total forces acting on the upstream and downstream face of the block, respectively. These can be resolved into vertical and horizontal components as shown. So, the block at a convex bend is acted upon by a horizontal force F_{UZ}–F_{DZ}

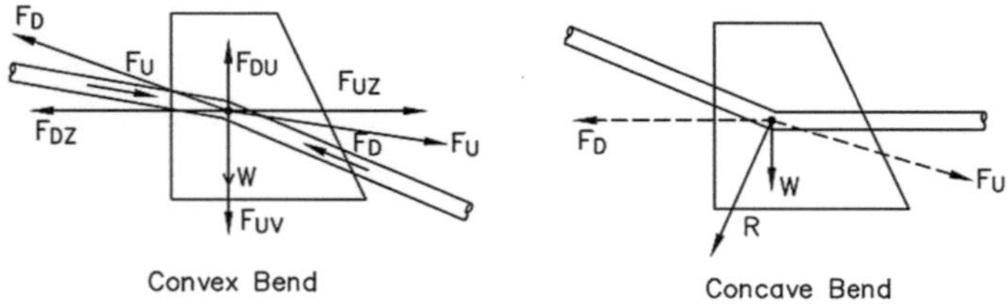

Fig. 17.6 Forces on anchors at bends

and a downward vertical force of $W + F_{UV} – F_{DV}$. For these forces the anchor block is analyzed for base pressures, overturning and sliding like a gravity structure.

In the case of a concave bend the resultant R of F_D and F_u can be resolved into horizontal and vertical components and the block can be designed as a gravity structure. In this case, the vertical component of R is additive to the weight of the block and gives stability to the block. So in this case a block of lesser self-weight is required.

In Fig. 17.6, two types of bends are shown and these two different methods of resolving the forces are shown. Any of the methods may be adopted for calculation and analyzing the block for stability.

Anchor blocks are so dimensioned that the base remains in compression i.e. the resultant of horizontal and vertical forces shall pass through the middle third of the base width. Sometimes stepped base is provided to add stability against sliding. The factor of safety against sliding should be not less than 1.5. The anchor block should be placed on firm and hard strata whose permissible bearing capacity should be more than the stresses transferred by the anchor block.

Anchor blocks are provided with nominal reinforcement on all the faces in both directions and the penstock is held with the anchor block through reinforcement in the form of U bends.

(b) **Design of supports**

Normally the supports of the penstock between the anchor blocks are shaped as saddle at a spacing of about 5 to 8 m c/c. The supports are designed for the dead weight of pipe and water and the friction at the support. Friction should be kept as low as possible. Lower frictional resistance is favourable in reducing stresses in penstock and the overturning moments at the support. Generally, concrete saddle supports as shown in Fig. 17.7 are used. The saddle angle is normally 120°. To reduce friction, either a steel plate or a plate of cast iron is placed on the saddle between the penstock pipe and concrete. Sometimes a layer of asphalt paper or graphite sheet is placed between the penstock and the steel plate to further reduce the friction. Unlined concrete saddles are used only for small diameter pipes say less than 0.8 m. The saddle support shall

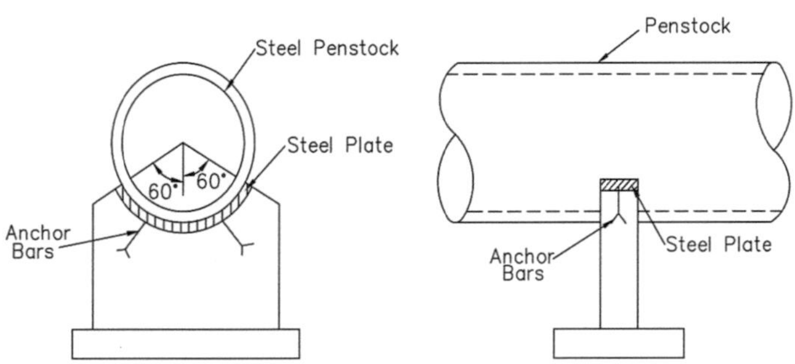

Fig. 17.7 Concrete saddle support

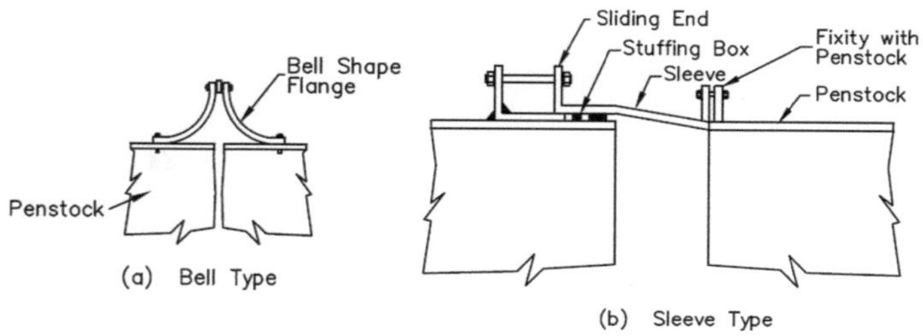

Fig. 17.8 Expansion joints

be designed as a column for vertical loads and moments. Roller-mounted bearings and rocker type supports are used for large-diameter penstocks.

17.10 Design of Expansion Joints in Exposed Penstocks

Expansion joints are provided in exposed penstocks to take care of temperature variations. Generally, two types of expansion joints, bellow type and sleeve type are used. Bellow type is normally used for small diameter (less than 1 m) pipes with a water head of up to 25 m. In this type of expansion joint, the flexibility of bell-shaped flanges permits joint movement. It is shown in Fig. 17.8a. Sleeve-type expansion joints are the most common type in use. This joint can take up to 2.5 cm expansion and contraction. It consists of a sleeve between two penstock pieces. One end of the sleeve is connected with one piece of penstock and its other end slides over the end of the second penstock piece. A conceptual stretch of the sleeve which has stuffing boxes is shown in Fig. 17.8b. Expansion joints may be designed as per I.S. 11,639 (Part-3).

17.11 Design of Buried Penstock

These are penstocks that are placed a few meters below the natural ground level in a trench. These have continuous support in the form of a concrete saddle as shown in Fig. 17.9. The advantages and disadvantages of buried penstock have been described in Para 17.3 These do not need expansion joints and intermediate supports. The pipe shell thickness is designed for hoop stresses due to internal water pressure plus water hammer pressure as described above for exposed penstocks. The shell thickness may be checked for external water pressure due to the difference between the ground water level vertically above the penstock and the penstock invert level and the backfill weight when the penstock is empty. This external pressure is generally small and so this check is generally not carried out in small SHPs. For structural design of buried penstock, I.S. 11,639 (Part 2) may be followed.

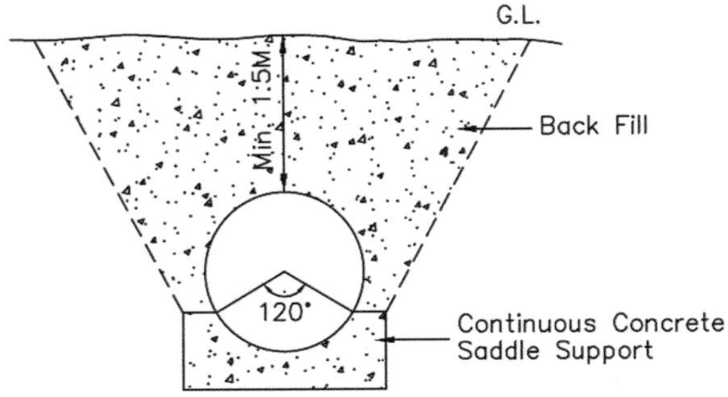

Fig. 17.9 Buried penstock

17.12 Design of Pressure Shaft

In a pressure shaft, the penstock liner is placed inside a shaft that is excavated in a rock mass either vertical or inclined. The space between the shaft and the liner is filled with concrete. The thickness of concrete is normally kept at about 400 mm. This type of penstock is normally not used in SHPs.

Generally, for designing the liner in a pressure shaft rock participation of the internal pressure is considered for which adequate cover around the shaft should be available. Rock cover is considered inadequate when it is less than the internal pressure in the vertical direction and 150 percent in the horizontal direction. The internal pressure is taken as the sum of the static head and the pressure rise. If the rock cover is inadequate and the rock is badly fractured, the liner is designed for full internal pressure. Hoop stress in this condition is computed by

$$f_{st} = \frac{P \cdot R}{t}$$

where, f_{st} is hoop tensile stress.

P is internal pressure.

R is radius of liner.

t is thickness of shell.

When the cover is adequate and the rock is competent to share the load then the sharing of pressure by liner can be worked out by the procedure given in 1S: 11639 (Part 2).

The proportioning of internal pressure between the liner and the rock can also be worked out by the formula given below

$$\lambda = \frac{\frac{b^2}{E \cdot e} - \frac{t}{p}}{\frac{b^2}{E \cdot e} + \frac{b}{E_R} \cdot \frac{m+1}{2c \cdot E_C} + \frac{b}{2c \cdot d} \cdot \frac{d^2 - c^2}{E_R}}$$

where, E = Young's modulus for steel in t/m^2.

E_c = Young's modulus for concrete in t/m^2.

E_R = Young's modulus for rock in t/m^2.
b = Inside radius of liner in m.
c = radius to outside of surrounding concrete in m.
d = radius to end of radial fissures in rock in m.
e = Thickness of steel liner in m.
t = Gap between liner and surrounding material in m.
1/m = Poisson's ratio for rock.
p = Pressure in conduit in t/m^2 (equal to head h).
λ = Proportion of pressure taken by liner and rock.

The liner thickness can be worked out after getting the share of internal pressure taken by it. Normally rock is not assumed to share more than 45 to 50% of internal pressure.

The liner is then checked for external pressure which is water pressure from outside when the shaft is empty. It shall also be checked for grouting pressure during construction. The procedure given in IS 11639 (Part 2) may be followed.

When rock participation is considered in design the stresses in steel liner under normal loading conditions without rock participation should also be checked and should not exceed 90% of the minimum yield stresses or 2/3rd of minimum ultimate tensile strength, whichever is less. In emergency conditions, it should not exceed minimum yield stress.

The thickness of the liner required from design considerations should be increased by 1.5 mm for corrosion allowance. The minimum thickness required from handling considerations is

$$t_0 = \frac{R + 0.25}{200}$$

where t_0 = minimum handling thickness in m.
R = radius of liner in m.

17.13 Fabrication and Testing of Penstocks

Small diameter pipes up to 600 m are manufactured seamlessly. They are referred to as seamless pipes. These are usually made of mild steel. Wall thickness varies from 8 to 30 mm. The length of pipes is normally 6 to 8 m.

Medium diameter (2 to 3 m) pipes are usually made of plate steel with shop-welded longitudinal joints. The plates are of medium-strength steel with tensile strength varying from 34 to 45 kg/mm^2. The welded sections of up to 3 m in diameter are fabricated in a shop in lengths of 4 to 8 m depending on transport facilities. The ends of sections are prepared and shaped for welding at the workplace. These welded joints are called transverse joints.

Large diameter pipes (more than 3 m) are delivered to the work site in segments which may be half, one-third or quarter sections. These pipe segments are assembled at the work site by welding both longitudinal and transverse joints.

All the welded joints are radiographed. All radiographs are reviewed by the welding inspector to identify the defects. The defected welds are cleaned and rewelded and again radio-graphed.

Individual sections of the penstocks are put to hydrostatic tests after they have been radio-graphed and stress-relieved. The hydrostatic test is performed at a pressure of one and a half times the design internal pressure. The defects disclosed during the pressure test are repaired by rewelding the section.

References

1. Guidelines for Small Hydro Development, Standard No. 2.2 and 2.3 'Hydraulic and Structural Design. AHEC, IIT Roorkee (2013)
2. I.S. 11625.: Criteria for Hydraulic Design of Penstock. BIS, New Delhi (2005)
3. I.S. 4880 (Part 3).: Code of Practice for Design of Tunnels Conveying Water Hydraulic Design', BIS, New Delhi (2005)
4. I.S. 11639 (Part-1).: Criteria for structural Design of Penstock: Surface Penstocks', BIS, New Delhi (2005)
5. I.S. 11639 (Part–2).: Criteria for Structural Design of Penstocks Burried/Embedded Penstocks. BIS. New Delhi (2005)
6. I.S. 2951 (Part-2).: Recommendations for Estimation of Flow Liquids in Closed Conduits', BIS, New Delhi (2003)
7. I.S. 11639 (Part-3).: Criteria for Design of Expansion Joints' BIS. New Delhi (2005)
8. Manual on Small Hydro Stations Standardization. CBI&P Pub No. 175, New Delhi (2009)
9. Mosonyi, E.: Water Power Development vol. 1, Budapest (1963)
10. Nigam, P.S.: Hand Book of Hydro Electric Engineering. Nem Chand and Co. (1985)

Chapter 18
Design of Power House

18.1 General

A power house is a structure which houses generating equipment. It includes a turbine, generator, main inlet valve, auxiliary equipment required to operate and control the main equipment and a suitable outlet (draft tube) to discharge water to the tail race. In India, power houses are mostly indoor type i.e. the generating equipment has a permanent roof over them so that the work may continue even during extreme weather conditions. An indoor type power house has a super-structure consisting of columns, cross–beams, panel walls and a roof (generally a truss supported on columns). The columns are known as gantry columns because they support an overhead travelling crane to handle heavy generating equipment. The intermediate structure supports the load of generating equipment and transfers the loads to a structure which is called a substructure. The substructure houses the water outlet structure called the draft tube. The total structure of the power house is normally a reinforced concrete structure.

The turbine used is either an impulse type (Pelton wheel) or a reaction type (Francis or propeller). It is connected with the generator either through a horizontal shaft or through a vertical shaft. The vertical shaft arrangement is generally used with 5 MW and bigger size units. The structural requirements and the layout and dimensions for the two types of arrangements of turbine and generator are different. The layout of these two types of arrangements has been discussed in detail in Chap. 4.

The portion of the power house which houses the generating equipment is generally known as the machine hall. On the one side of the machine hall, an extra area is provided for the erection of generating equipment and this area is called the erection bay. The super- structure of the machine hall is continued to the erection bay area so that the same crane can handle the equipment in both the erection bay and machine hall.

An erection bay is provided on the side of the approach road to the power house. A wide entrance is provided in the erection area so that the carrier of the equipment may enter the erection bay where the crane may unload the equipment.

B. N. Asthana, *Planning and Design of Small Hydro Power Projects*,
https://doi.org/10.1007/978-3-031-96271-4_18

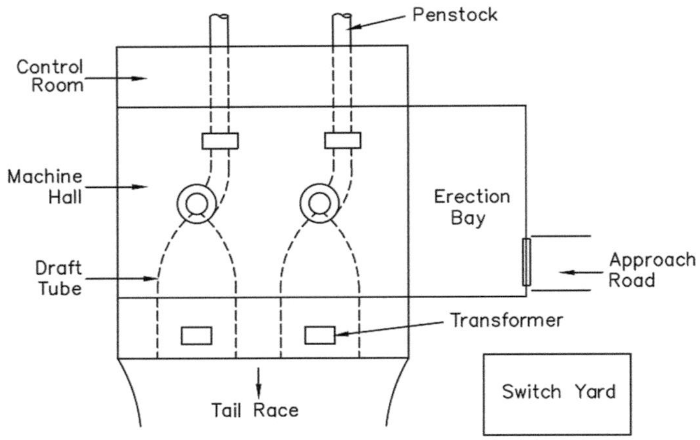

Fig. 18.1 Typical plan of a power house

The other components of the power house complex i.e. control room and main transformers are placed close to the main power house building. A typical plan of placing the components of the power house complex is shown in Fig. 18.1

The typical sections of a power house showing vertical and horizontal shaft arrangements are given in Chap. 4. It can be seen that the vertical shaft arrangement is more complex. In horizontal shaft arrangement all the equipment are placed separately on the substructure directly whereas in vertical shaft arrangement the generator load comes on the concrete encasing scroll case which disperses it to the substructure along with its own load. Since the structure of vertical shaft arrangement is more complex, its design procedure is first discussed in this chapter.

18.2 Data Required for Design

The data required to be collected for the design of a power house is listed below. For details IS 4247 (Part I) 1993 may be referred.

(a) Location of power house along with the L-Section of the water conductor system showing water levels and the stage-discharge curve at the tail race end.
(b) Foundation investigations and test results.

  (i) Bore logs up to a depth of 1.5 times the width of power house to get an idea of the strata of the foundation.
  (ii) Permeability test results.
  (iii) Allowable bearing capacity at the foundation level.
  (iv) Shear parameters of foundation strata (C & Ø).
  (v) Nature and extent of overburden if strata is rocky.
  (vi) Shear zones and bedding planes in rocky foundations.

(vii) Geophysical test data in rocky foundations.
(viii) Friction coefficient between concrete and soil/rock.
(ix) Water table and its fluctuation.

(c) Meteorological data.

(i) Daily water temperature.
(ii) Daily air temperature.
(iii) Wind velocity and direction.
(iv) Monthly Rainfall data.

(d) Layout details and drawings.

The layout shall be finalised with the equipment supplier and the following information and drawings shall be supplied by the supplier to the designer.

(i) Contour plan showing the location of power house.
(ii) General layout plan showing number and size of units, erection bay, location of the control room, tail and head race channel/tunnel, transformer space, cable tunnel, switch-yard, approach road, office and workshop space, etc.
(iii) Plan of power house at machine level (generator level and turbine floor in case of vertical setting) and at centre line of distributor.
(iv) Plan and sections showing galleries, cavities, pits, sump, and other openings.
(v) Transverse and longitudinal sections of the power house at the centre line of the machine, and longitudinal sections through upstream and downstream columns and the erection bay and control room, etc.
(vi) Details of EOT crane.
(vii) Shape and profile of outlet (draft tube) arrangement.
(viii) Shape and profile of scroll case (in case of reaction turbine).
(ix) Details of generator supports.
(x) Drainage and dewatering arrangements.
(xi) Placement of auxiliary equipment.
(xii) Sequence of erection.

18.3 Loads

Various loads such as equipment loads, crane loads (which are normally supplied by the supplier), dead load of structural members (the volume of concrete × density i.e. 2.4 T/m^3), wind load and seismic forces (taken as per I.S. code) are required for the structural design of power house.

A list of loads required is given below.

(i) Generator load (both for rotor as well as stator) at the support points (vertical as well as tangential under different conditions of operation) with impact factor and short circuit torque.

(ii) Weight of turbine runner including shaft.
(iii) Turbine characteristics.

(a) Normal and run away speed.
(b) Permissible tilt in shaft.

(iv) Weight of scroll case on each support both in empty and full of water conditions.
(v) Weight of draft tube liner and its centre of gravity.
(vi) Size and weight of main and auxiliary transformers.
(vii) Thrust due to servomotors.
(viii) Jet reaction in the case of Pelton wheel.
(ix) Crane loads (hook load + trolley load), crane surge, crane striking load, impact factor, etc.
(x) Weight of auxiliary electrical and mechanical equipment (as per I.S. 4247 Part-I). It is generally taken as 1 T / m^2.
(xi) Live load on various floors (as per I.S. 4247). It is generally taken as 1000 to 500 kg/m^2.
(xii) Design load for erection bay floor (for runner/rotor). It is generally taken as 5 T/m^2.

18.3.1 Loads for Preliminary Design

For the preliminary design of power house the equipment loads are approximately assessed from available literature in the absence of data from equipment suppliers. There are various methods in the literature to estimate equipment loads (refer CBI &P manual, AHEC manual, IS. Codes, etc.). Some expressions are given below for the purpose of load estimation.

(i) Generator load $W = g\sqrt{\frac{K}{N} - 85(\text{tonnes})}$

where, g = varies from 25 to 32.
K = Generator capacity in KVA.
N = Speed in RPM.
50 to 55% of W is the weight of rotating parts.
The impact factor on rotating weight is taken from 1.2 to 1.5
Short circuit torque Ts = 5 Tm.
Tm is torque under normal conditions and is given by

$$T_m = \frac{K}{N}(\text{Tonnes meter})$$

(ii) Turbine weight.

For the Francis turbine, $W = \frac{178}{P^{1/8}H^{.525}}[kg/HP]$.
where, P = capacity of turbine in HP.
H = Head (m).
Add 28% for control mechanism and other auxiliaries.
Other relationship is
$G = 32.6\, D^{1.75}$(tonnes).
where, D = Runner diameter in meter.
G = Weight of turbine.

For Kaplan turbine (including runner, shaft, scroll case, stay vanes, governing mechanism etc.).

$$G = (1.5\, to\, 2.5)\frac{P^{4/5}}{H}(\text{Tonnes})$$

(iii) Weight of scroll case.

$$W_s = k.d.t\left(D_1 + {}^{d}/_{2}\right)[kg]$$

where, k = 100–140 (depends on method of jointing).
d = diameter at inlet (m).
t = average plate thickness (mm).
D_1 = Diameter of runner (m).

(iv) Hydraulic forces.

(a) Horizontal thrust due to penstock—It is the function of the design head at the penstock down steam end.
(b) Hydraulic thrust in the vertical direction due to water in the runner—It can be worked out by the relation.

$$T = 785e \cdot D_1^2 \cdot H_{max}(kg) \quad \text{for Francis turbine}$$

where, 'e' depends on the specific speed of turbine Ns as below

Ns	100	400	540
e	0.11	0.61	1.0

$$T = 785\, K\, D_1^2\, H_{max}\ (kg) \quad \text{for Kaplan turbine}$$

where, K varies from 0.8 to 0.95

The above loads and other forces acting on a vertical shaft power house along with their point of application are shown in Fig. 18.2. All these loads and forces do not

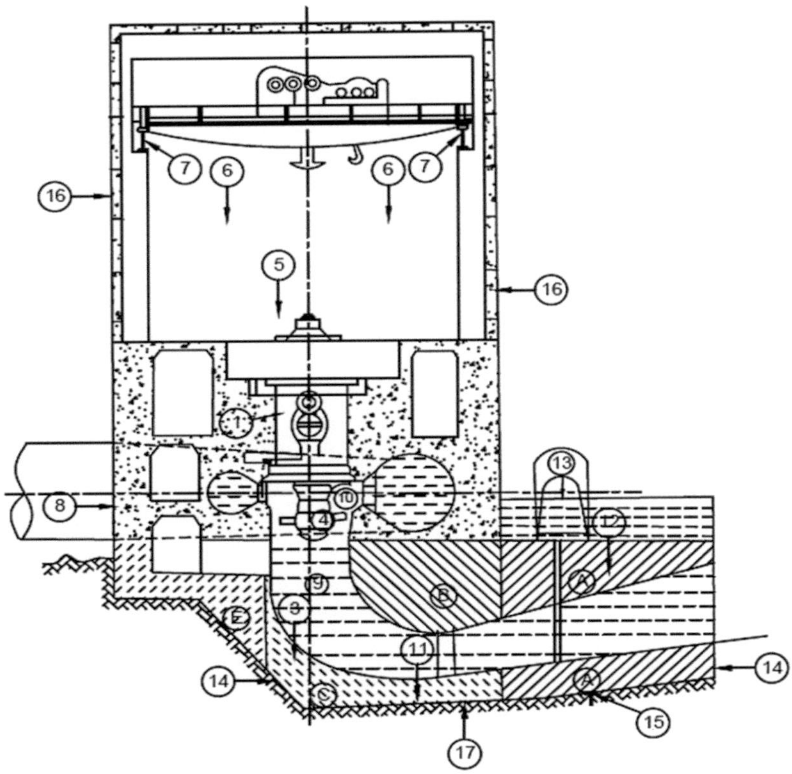

SUB–DIVISIONS OF DRAFT TUBE:

A) Cantiliver portion outside power house wall.

B) Elbow portion above draft tube.

C) Bottom slab within power house.

D) Vertical members (not shown).

E) Solid mass upstream of draft tube.

FORCES ACTING:

1) Weight of concrete in substructure etc. upto generator floor.

2) Machine load.

3) Weight of draft tube liner.

4) Weight of scroll casing liner.

5) Weight of superstructure, roof, columns, wall, etc.

6) Loaded crane.

7) Crane surge.

8) Penstock thrust.

9) Hydraulic thrust on runner.

10) Water load in scroll case.

11) water load inside draft tube.

12) Water load over draft tube.

13) Loaded draft tube gate gantry.

14) Earth pressure.

15) Water pressure due to T.W.L.

16) Wind pressure.

17) Uplift forces.

Notes:

i) Alternative loads have been shown dotted.

ii) Figures in circle denote the corresponding part load.

Fig. 18.2 Showing an indoor power house and the forces acting on it (P.S. Nigam)

act on the structure simultaneously. Hence analysis for design is carried out for load combinations applicable during different stages such as under construction stage, normal operating stage and emergency conditions due to flood, seismic or wind.

18.4 Design Concept for Vertical Shaft Power House

In the second half of last century, a large number of hydro power houses have been designed and constructed in India as well as in other countries, but no. standards for the design of power house structures are available. The design engineers consider each power station structure as an indeterminate complex structure due to its shape with a large number of cavities and openings and the forces and loads acting on it. They tackle it in their own way by breaking the structure into simple determinate elements maintaining the structural integrity of the whole structure using their experience and judgment. Hence every design organisation has its own procedure for the design of power house structure. But such two-dimensional design practices are only approximate and the concrete mass is generally heavily reinforced.

One such 2D design concept is given below which has been used for the design of several power houses using transverse and longitudinal direction analyses. This is described in detail. For illustrative example readers may refer the book by P.S. Nigam. In this approach, the vertical shaft power house structure is divided into the following three parts:

(i) Substructure which contains the draft tube.
(ii) Intermediate structure which encases scroll case and supports the generator.
(iii) Super structure which comprises gantry columns, panel walls and roof.

In the horizontal shaft power house the above mentioned part (ii) does not exist.

18.4.1 2-D Design of a Power House Structure

Before proceeding with the structural design of individual parts of the structure as described above, the power house building is checked for stability under various load combinations as a gravity structure. Safety checks for foundation stresses, overturning, sliding and floatation are made. Foundation pressures are worked out considering the substructure as a rigid foundation. The foundation pressures are assumed trapezoidal in shape.

(a) Sub Structure

The sub-structure containing the draft tube has the following structural functions:

(i) It safely supports the superimposed loads over the cavities.

(ii) It acts as a foundation structural member to transmit all the loads and forces to the foundation in such a manner that the pressures on the foundation are within the permissible bearing capacity of the foundation.

The shape of the draft tube structure, which is contained in the substructure, and the loading pattern are normally such that the structural analysis is carried out for full unit width in the transverse direction (along the flow) including the portion of the draft tube which projects outside in the downstream of power house. In the transverse direction substructure can be analysed in following ways so that it may serve the function of a foundation for the power house structure.

(i) As a flat plate or a raft.
(ii) As a continuous footing.
(iii) As a gravity structure.

The transverse analysis is normally carried out considering the substructure as a continuous footing acted upon by the trapezoidal soil reaction from the bottom and superimposed loads from the top.

The analysis in the longitudinal direction is carried out to check and provide for structural strength of the substructure so that it may support the equipment loads and other superimposed loads despite the large cavities. Longitudinal analysis is done by considering the structure as a box structure in this direction.

Another way of analyzing the substructure is by sub-dividing the structure into smaller parts as given below so that they may be analysed as simple structural elements.

1. Draft tube beyond machine hall—It is treated as a cantilever from mass concrete.
2. The concrete above the draft tube in the machine hall, which is known as elbow portion, is treated as a deep beam supported over the side walls of the draft tube.
3. Side walls are treated as short columns for loads coming over them.
4. Bottom slab of the draft tube is treated as a slab spanning between side walls and draft tube piers and is acted upon by soil reactions from below.
5. Concrete upstream of the draft tube is treated as mass concrete.

Some designers feel that no analysis and design is required for sub structure. In that case draft tube shall be treated as an opening in mass concrete and so nominal reinforcement (generally 25 bars @ 300 c/c) for tension developing around the openings shall be provided. Recent studies using 3D FEM analysis confirm this feeling of designers.

(b) Intermediate Structure

The scroll case, when it is a steel pipe, is generally fully encased in concrete. The concrete above centre line of the scroll case is treated either as an arch fixed at the outer end or hinged at the speedring or as a cantilever over the scroll case for external loads from the generator barrel and other superimposed loads. This concrete around the scroll case is generally not designed for internal pressure as some type of asphaltic membrane is placed over the steel pipe. Concrete scroll cases are designed for both external and internal loads as concrete members.

(c) Super-Structure

The super-structure generally consists of gantry columns, cross beams, panel walls and roof slab or truss and behaves as a frame structure. In the 2-D approach of design, the analysis is carried out separately in both transverse and longitudinal directions. In a transverse direction, upstream and downstream gantry columns with roof members at top form a portal for analysis. In the longitudinal direction gantry columns and cross beams form a multi—storey multi—bay frame for analysis. Gantry columns are designed in biaxial bending for the moments in two directions for the worst loading combination. The dead and live loads, combined with wind and seismic forces are considered for the analysis. Cross beams are designed for moments and shear force. A recent approach is to analyse the superstructure as a space frame for which computer programs are available (Staad-4).

18.5 Design Method for Substructure (2-D Approach)

The first step is to compute/estimate all the loads and forces acting on the power house structure which are mentioned in para 18.3. The point of application of all the forces and loads shall also be determined.

The next step is to ensure the structural stability of the power house structure under the loads and forces acting on the structure under various conditions. It is described briefly in the following paras:

18.5.1 Stability Analysis

To ensure the stability of the power house structure, the following factors are considered.

(a) Overturning.
(b) Foundation stress.
(c) Sliding.
(d) Floatation.

(a) Overturning

The factor of safety against overturning is the ratio of total stabilizing moments to the destabilizing moments. The moments of loads and forces are taken about the toe of the structure. The stabilising moments are due to vertical loads and destabilizing moments are due to horizontal forces and the uplift. The recommended factor of safety is 1.1 for the construction stage and 1.2 for the completed stage.

(b) Foundation Stresses

Since it is assumed that the substructure acts as the foundation structure (footing of the power house), all the loads and forces are transferred to the top of the substructure. The loads and forces come from super structure, generator and scroll case through the intermediate structure. The point loads are assumed to be distributed on a greater area of the substructure. The dispersal of the loads through intermediate structure concrete is assumed at 45° in all directions. So, a load is dispersed on an area equal to the base of a cone. If the available area is restricted due to cavities, then a suitable extent of truncation of the base of cone is assumed. For more details refer, to the book of P.S. Nigam.

Computations for the dispersal of loads at through concrete is very time-consuming, so some designers prefer to consider the loads acting as point loads on the substructure though this approach gives slightly more loads and reactions.

After determining the loading of all loads on the substructure, the soil reactions on the foundation are worked out on the assumption of a rigid footing by the relation

$$P = \frac{W}{A}\left(1 \pm \frac{6e}{b}\right)$$

where, W = Total vertical load.

A = Area of base.

(It is assumed that soil pressures are uniform in the longitudinal direction).

e = eccentricity (see Fig. 18.3).

b = Length of substructure in flow direction.

A typical soil reaction diagram for a loading on substructure is shown in Fig. 18.3.

The soil reactions of loads and forces are added up algebraically to work out the soil pressure for different load combinations and these are plotted to obtain the maximum soil reaction on the foundation and to check whether the entire base is under

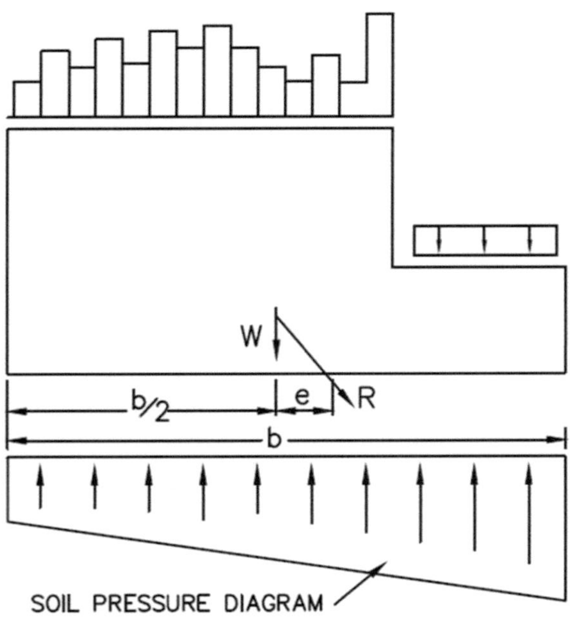

Fig. 18.3 Typical soil reaction diagram on foundation

compression i.e. maximum and minimum soil reactions at both ends are positive and the maximum soil reaction is less than the permissible bearing capacity of the foundation strata. The permissible bearing capacity for emergency load conditions i.e. with wind or earthquake.

is increased by 33 percent of the normal permissible bearing capacity. Wind and earthquake are not considered together.

The permissible bearing capacity shall be determined by in-situ and laboratory test results. In the absence of it, a suitable value may be assumed based on experience and judgment or from values available in the literature for various types of strata. The followings approximate values for permissible bearing capacity may be considered.

(1)	Soft clay or loose sand	10 t/m^2
(2)	Hard clay or compact sand gravel mix	40-50 t/m^2
(3)	Broken or shattered bed rock of any type	100 t/m^2
(4)	Slate and schist in sound condition	250 t/m^2
(5)	Massive sound rock without laminations	1000 t/m^2

If the maximum soil reaction is more than the permissible bearing capacity, one of the following measures may be adopted.

(i) Compaction of strata to improve bearing capacity.
(ii) Excavate out soft material and then backfill it with well-graded material or concrete.
(iii) Adopt pile foundation.
(iv) Increase base width by cantilevering the footing in either direction.

In alluvial strata, unequal settlement in the foundation may take place, due to the following reasons:

(i) Due to change in void ratio in soil.
(ii) Due to vibrations from generating equipment.
(iii) Due to failure of foundation in bearing or shear.

In order to avoid unequal settlement, the foundation shall be placed on one type of strata/soil and base of footing shall be horizontal. Each machine unit shall have a separate foundation. The foundation base may be stepped if power house is resting on the rock because in that case no undesirable settlement is expected.

(c) Sliding

Safety against sliding depends on the coefficient of friction between concrete and foundation strata. The criteria for checking stability against sliding is same as described for gravity dam in Chap. 12.

The experience is that problems arise in ensuring safety against sliding when the foundation is either of fine sand or clayey strata. In that situation, either one or a combination of the following measures are generally adopted.

(i) Increase the vertical load.
(ii) Reduce the uplift through drainage arrangement.
(iii) Provide shear keys.
(iv) Use piles for the foundation.

A few examples (from Mosonyi) to illustrate the use of the above measures are enclosed as Figs. 18.4, 18.5 and 18.6. A raft foundation with sheet piles has been provided for Pathri Power house which is located on Ganga Canal (UP) to utilise canal falls for power generation.

(d) Floatation

Flotation of the structure may occur due to uplift forces. The factor of safety against flotation shall be worked out by the relation

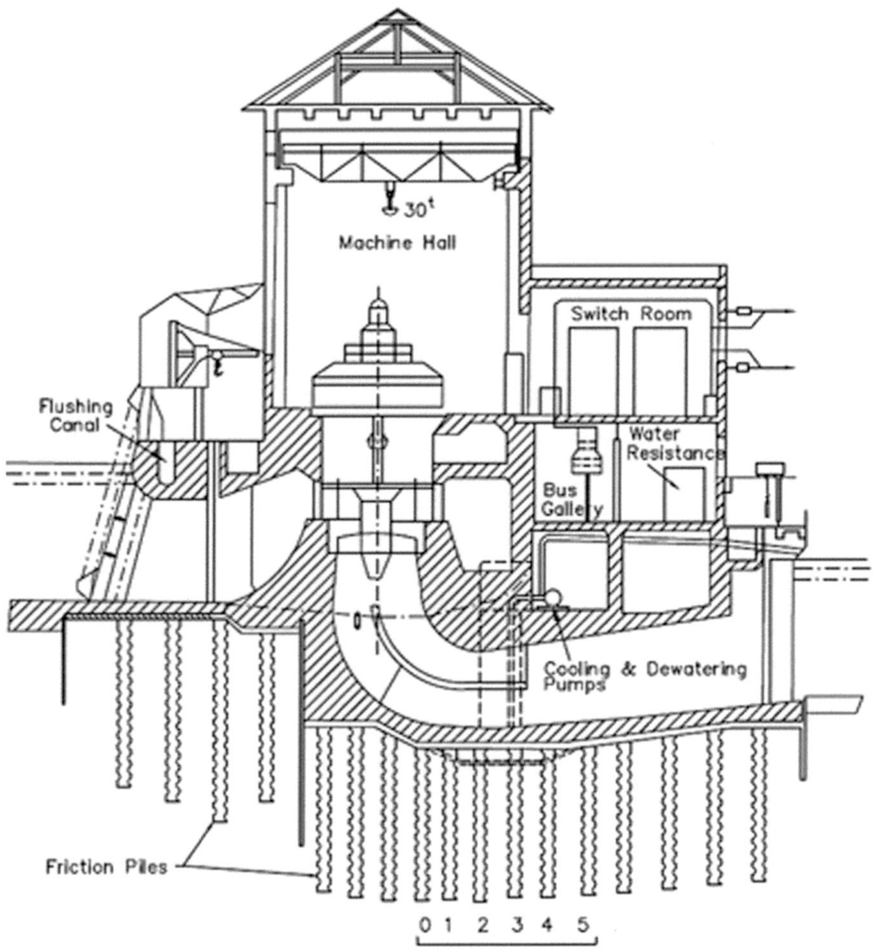

Fig. 18.4 Power house upon piles. Diettkon development, Limmat river, Switzerland, Q, 100, H = from 2.70 m to 4.65 m, N = 2720 kW, E = 19 million kWh year

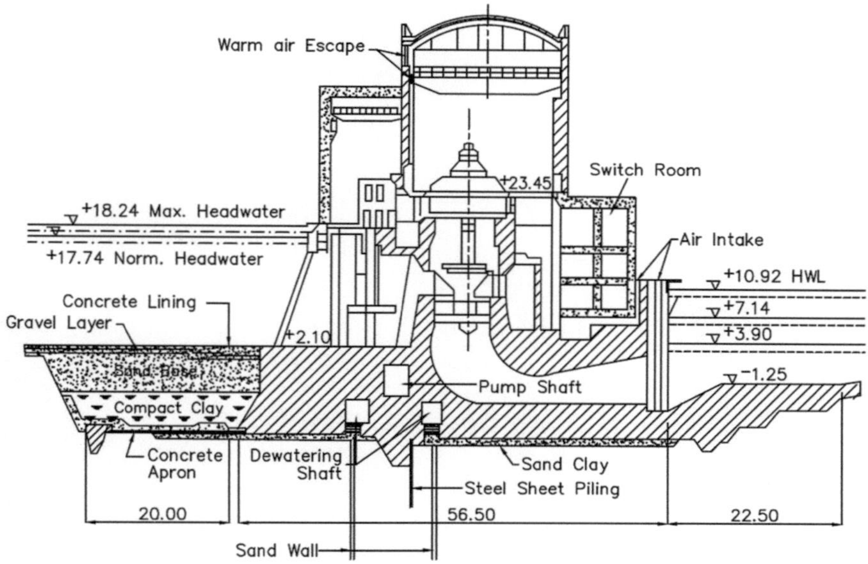

Fig. 18.5 Section through the Svir III power station, USSR (Sec also Fig. 16.66). In order to increase safety against sliding a system of screen wheels and a horizontal sand layer have been constructed

$F = \frac{\Sigma V}{U}$ where, V is vertical load and U is uplift force.

F (factor of safety) for the construction stage (with no equipment in place) and for extreme loading conditions shall be 1.1 and for the completed stage (with equipment) shall be 1.25. When it is not possible to ensure the above-recommended factor of safety, measures shall be taken to increase the weight and to reduce the uplift by provision of an apron upstream along with the drainage galleries (refer Figs. 18.5 and 18.6). In the case of power house on a rock foundation, grouted anchor bars may be provided to achieve the desired factor of safety.

18.5.2 Design of Substructure

1. Design as Continuous Beam

The structural design of the substructure can be carried out by analysing the substructure in transverse and longitudinal directions by treating the substructure as a continuous footing. This is briefly described in the following paras:

(a) Analysis in Transverse Direction

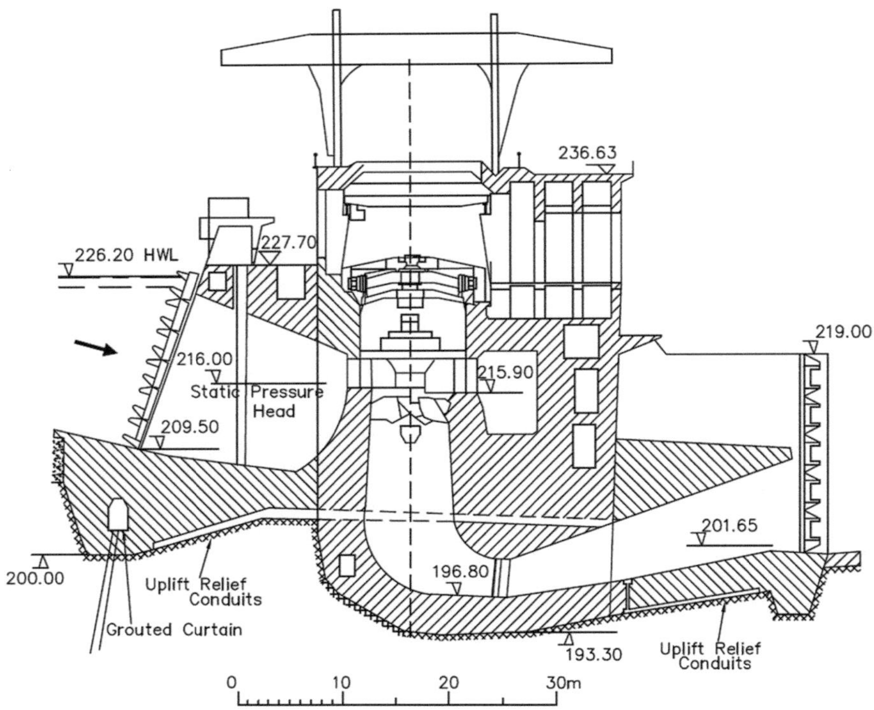

Fig. 18.6 Uplift relief conduits below the power house of Ybbs-Persenbeug plant. (After H. Kiessling)

In this analysis bending moments and shear forces are determined at various sections as shown in Fig. 18.7. These are typical sections, one in the projected portion of the draft tube and the other of the draft tube inside the power house on the transverse section. The bending moments and shear force at any of these sections of the transverse section of power house are determined considering the soil reaction and imposed loads on the substructure on one side of the section. The substructure is assumed to behave as an endless cantilever on the side of the section as is done in a continuous footing. The bending moment and shear force are worked out for each combination of loads and the maximum bending moment and shear force at any section are taken for design of that section.

The reinforcement in the transverse direction is worked out at the section for maximum BM and SF considering the section as a beam after determining the neutral axis. The permissible stresses in concrete and steel are taken as per relevant I.S. Codes. These are increased by 33 percent for emergency loading conditions. The section at the centre line of power house is considered as a U-bend. It is generally seen that no reinforcement is required at the top of the section. In that case normally 50% of bottom reinforcement is provided at the top of the section.

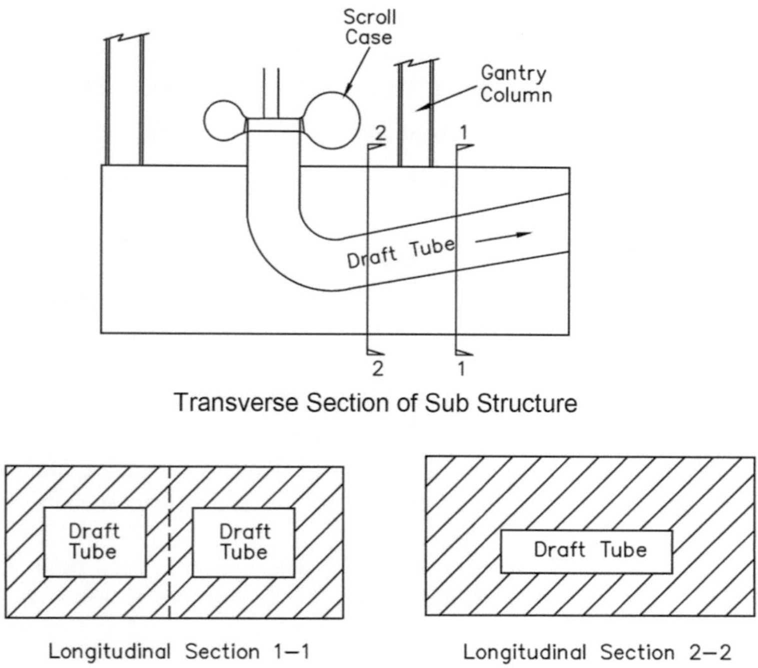

Fig. 18.7 Typical sections of the substructure (with draft tube)

No reinforcement is generally required in the sections upstream of the draft tube because of the mass concrete in that portion. However, temperature and shrinkage reinforcement as is generally required for surfaces of mass concrete is provided at the rate of 16 mm φ bars @ 250 mm. USBR recommends that nominal temperature reinforcement @ 0.2% of 0.6 m thick concrete shall be provided at the surface.

(b) Analysis in Longitudinal Direction

Box analysis

In the longitudinal direction, the structure as shown in the sections (Fig. 18.7) is analyzed and designed as a box. Unit width of the section is considered to be acted upon from the top with a uniformly superimposed loading in this direction which is determined on the top of the substructure after the distribution of point loads through intermediate-stage concrete and the soil reactions from the bottom. There are no appreciable loads from the sides. However, these may be either due to water load corresponding to the tail water level in case of portion outside the power house building or due to seismic loads. The box analysis may also be carried out for the portion of the draft tube inside the power house. The section at the centre line of power house is considered as a 'U' bend with vertical members hinged at the top (with speed ring). The boxes are analyzed by the moment distribution method to determine fixed end moments at all the joints in the members. The sections are generally made of large thicknesses so the centre line frame is analyzed in each case.

To work out the reinforcement, the moments at the face are considered if the member is more than 0.3 m thick. While detailing the reinforcement, it is ensured that the change in the area of reinforcement from one section to the other is gradual.

2. Design as Separate Structural Members

The draft tube is visualised as made up of five separate structural members as given in para 18.4.1.

- Cantilever portion outside power house
- This portion shall be analyzed and designed as a box structure as described above.
- Elbow Concrete
- There are a number of approaches to treating elbow concrete and a few are given below

    (i) It may be treated as a thick roof slab restraint from piers and side walls. Shrinkage reinforcement may be provided. Reinforcement in the first lift should be provided for the load of the second lift and supports are not removed till the second lift is in place.
    (ii) Due to the large depth, it should be designed as a deep beam for the superimposed loads from the scroll case and generator. (Mosonyi).
    (iii) Superimposed loads may be assumed to be transferred to side walls by arching action in concrete (dispersal) and the concrete spanning the draft tube opening may be designed as a lintel (U.S. Army).
    (iv) Draft tube may be treated as a cavity in mass concrete like the design of a gallery in the concrete dam.
    (v) Some designers prefer to design this portion as the top member of the box analysis (as described above).

- Bottom Slab within Power House
- In the longitudinal direction, if the slab is quite thick, temperature and shrinkage reinforcement would take care of other structural loads. If the slab is thin it may be designed, as suggested by Mosonyi, as a slab fixed on side walls.
- Vertical Members
- In the portion outside the power house wall, the vertical members are treated as the members of the box analysis. In the case of side walls inside the power house their design is carried out for loads and moments as a column fixed at both ends i.e. at the top and bottom.
- In columns, a minimum reinforcement is always provided as a percentage of area. Since the side walls are very thick this reinforcement becomes very large. So the minimum percentage should be applied to the theoretical area required for loads. I.S. Code recommends reinforcement equal to 0.8% of the theoretical area with adequate ties. US Army and USBR suggest minimum reinforcement as 1% of the theoretical area.
- Mass concrete Upstream of Draft Tube

- No analysis is recommended for this region. Only temperature and shrinkage reinforcement @ 0.2% of 0.6m concrete thickness or 16mm Ø @ 250mm is recommended.
- Several organizations recommend for the mass concrete portion a provision of dowels at horizontal and vertical construction joints at the rate of 4 to 8 sq cm/sqm of area.

18.6 Design Aspects of Intermediate Structure

18.6.1 General

Scroll case or spiral casing is that part of a reaction turbine that distributes water from the penstock uniformly and smoothly through the guide vanes to the runner. These are either of RCC or Steel. Steel spiral cases are widely used. The concrete-encasing steel scroll case is termed an intermediate structure. It supports the generator and transmits the load of the generator to the sub-structure. The two types of scroll cases are shown in Fig. 18.8. The size and shape of scroll cases are supplied by the turbine supplier.

18.6.2 Type and Selection of Spiral Casing

Spiral casings are either of RCC or Steel (Fig. 18.8).

The selection of type of spiral casing depends on the head and velocity.

(i) Limit of Gross Head

Concrete scroll case is used up to a head of 30 m and can be used up to a head of 75 m with steel lining. Steel scroll cases are generally used for heads above 30 m.

(ii) Limit of Velocity

The velocity at the inlet section may be worked out by

$$\mathrm{V} = \mathrm{K}\sqrt{\mathrm{H}}$$

where, H is rated head (m) and K is constant which is 1.0 up to 20 m head and varies from 1.0 to 0.7 up to 400 m head

Velocity in concrete casing is limited to 6 m/sec

Merits and demerits of the two types of scroll cases are as below.

Fig. 18.8 Showing spiral casings

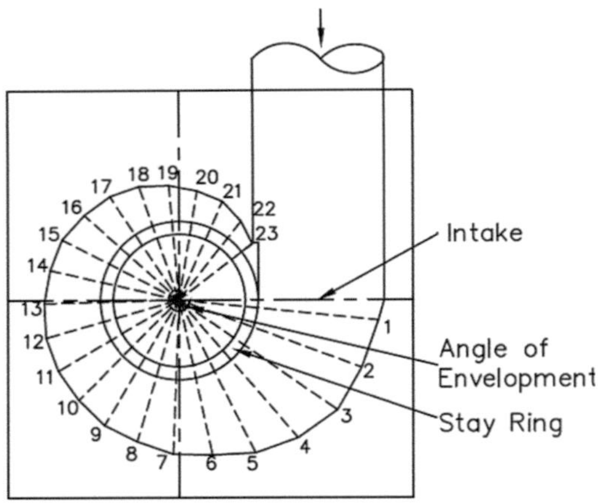

(a) Plan of Steel Spiral Case

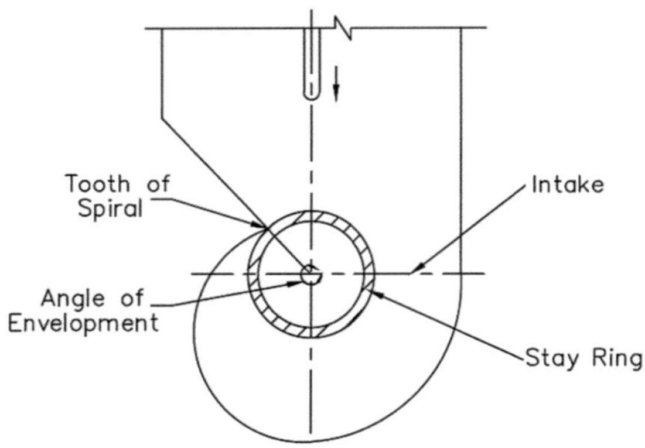

(b) Plan of Concrete Spiral Case

- Concrete casing makes the machine spacing closer because of the high ratio of height to width. For high heads (above 20 m) the detailing and placing of reinforcement becomes complicated. It needs skilled workmanship for shuttering and reinforcement placing. Special attention is required for joints, seals, anchorages, etc. for high heads.
- Steel casing can take large internal and external loads. Fabrication in the workshop reduces the time of construction but the placement of large pieces and pouring of encasing concrete need special care.

18.6.3 Design of Concrete Spiral Casing

The angle of envelopment in the concrete spiral casing is generally 180° to 225° (Fig. 18.8). The sections of concrete spiral casing are normally trapezoidal as shown in Fig. 18.9.

There is no standard practice for the design of this type of spiral casing. The main factors affecting the choice of method of analysis are the manner in which the generator is supported and the speed ring connected to the spiral case roof. Following is an outline of principles according to which a spiral casing may be designed.

The loads for which spiral casing is designed are:

(i) Internal water pressure + water hammer.
(ii) All dead and live loads coming from the top.
(iii) Load through generator foundation.
(iv) Forces from super structure.
(v) Water load due to tail water level.
(vi) Seismic forces.

The design shall be done for the following two loading conditions.

(i) Casing full of water: In this case, the loads to be considered are maximum water pressure + water hammer, dead and live loads from the roof, weight of generator and runner minus hydraulic thrust (m/c not running).
(ii) Casing empty: In this case, no internal water pressure but maximum external pressures are considered assuming that machines are not removed.

The load on the spiral case roof may either be carried by the roof itself or transmitted (at least partly) to the substructure through stay vanes.

The machine load may be carried by the spiral case roof if the span and the loads are not excessive. In this case, the structural design of the top slab may be carried out in any of the two following ways.

(i) The spiral case roof which is a reinforced concrete slab with an opening for the turbine is considered either resting upon or fixed with side walls. The slab may be considered to be structurally made up of a system of parallel beams perpendicular to the flow. Mosonyi suggests the full fixity of beams with side

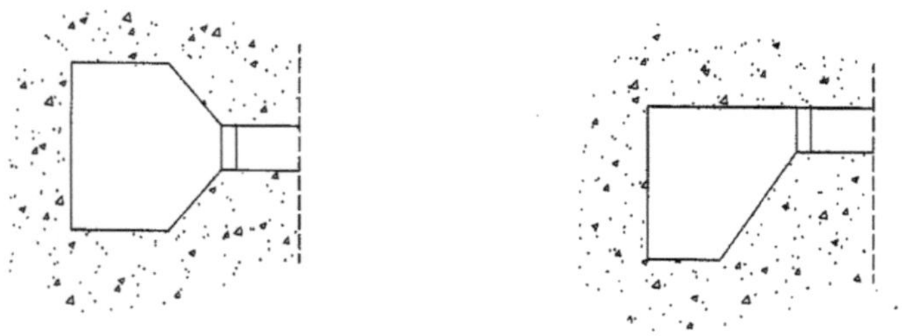

Fig. 18.9 Typical sections of concrete spiral casing

wall concrete to find the negative moments and positive moments may be taken half of negative moments. The number of beams may vary depending on size of spiral case roof.

(ii) In the other method the arrangement of beams is considered in a radial manner with one end fixed with side walls and the other end simply supported on the speed ring.

The downstream side of the spiral case is considered a cylinder and so the circular horizontal reinforcement is provided and it is fixed on the upstream side for the internal pressure. Vertical reinforcement for partial internal pressure should also be provided and it should be fixed in the roof slab.

18.6.4 Design of Steel Spiral Casing and Encasing Concrete

The angle of the envelopment is very near to 360° and the sections of steel spiral casing are circular in shape. This casing is suitable for medium and high heads as steel can resist high internal pressure.

The design, fabrication and erection of steel spiral casing is the job of the turbine supplier. The civil designer is interested in the features that affect the design of encasing concrete.

The following are the encasing concrete alternatives.

(i) Fully encased—It is the general practice to fully encase the spiral casing by the concrete (Fig. 18.10). It protects the liner, damps the vibrations, resists the unbalanced hydraulic thrust and supports the generator. But the design of concrete is indeterminate.

(ii) Half encased—It is sometimes provided. The spiral casing in this case is encased with concrete up to the centre line of the distributor. Generator has to be supported on the speed ring. It can dampen the vibrations. For the ease of construction, some manufacturers prefer it.

(iii) Un-encased—The spiral casing is placed on supports that are fixed with substructure concrete. There is no concrete around the spiral casing. The arrangement is good from cost and time considerations as well as from the design point of view. But the liner has to be made strong to resist full internal

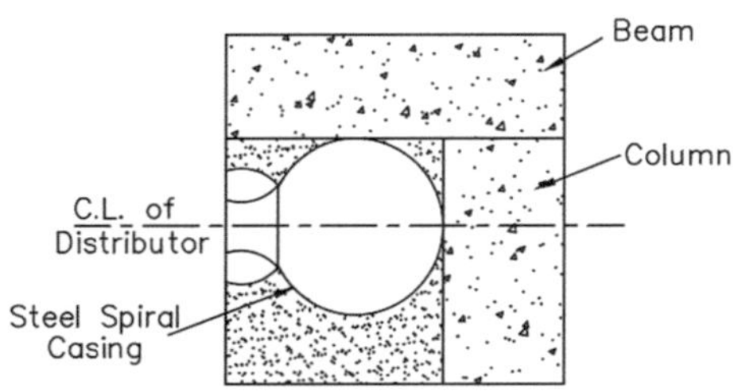

Fig. 18.10 Showing fully encased spiral casing with beam column arrangement

pressure and some structural arrangement has to be made to transfer the generator load to the substructure.

Design of Fully Encased Concrete

As stated above the general practice is to fully encase the spiral casing with concrete though the design of concrete is difficult as it is indeterminate due to

(i) Irregular shape.
(ii) Deformations is steel liner.

The external and internal loads for which encasing concrete shall be designed depend on the mode of placement of concrete over the liner. There are several practices for placing encasing concrete but the normal practice in India is as below.

Concrete is placed over the liner after placing 15 to 25 mm thick tar mastic over it and the liner is empty. In this case, during placement bracings are provided inside the scroll case to eliminate the possibility of reshaping the scroll case due to a load of green concrete. In this method of placement, the design loads are

(i) Liner is designed for full internal pressure.
(ii) Encasing concrete is designed for external loads only.

The encasing concrete over the spiral casing above the centre line of the distributor is considered for design purposes in different ways by the designers. These are as below.

(i) Beam and column arrangement—In this arrangement the concrete at the top of the casing is considered a beam. The top of beam is at generator floor level. The beam may be considered of uniform depth above the scroll case. The beam is a cantilever from the side concrete. This is a simple method and is generally adopted. The beam and column are shown in Fig. 18.10.

As an alternative, the beam may be considered made of varying depth following the profile of the scroll case with minimum depth at the crown of the scroll case. This beam may be assumed hinged at the speed ring and fixed at the other end with side concrete. The side concrete is designed as the column. The beam and column in both cases are provided with reinforcement for bending moment and shear force.

(ii) Arch arrangement—In this arrangement the top concrete is treated as an arch whose end at the speed ring is considered either fixed or hinged and the other end is fixed at the centre line of distributor. Arch is assumed of constant depth.
(iii) Ring arrangement—In this arrangement the concrete around the scroll case is considered as a ring between hinged supports at the two ends of the speed ring.

Many design organizations do not prefer any analysis for the design of this concrete and provide nominal steel for temperature and shrinkage around the opening. The external loads are assumed to be dispersed to the mass concrete on the side of scroll case.

18.7 Design Aspects of Intermediate Structure with Impulse Turbine

In a power house using an impulse turbine, there is an open pit instead of a draft tube. The scroll case is replaced by a manifold with several jets. The jets strike the walls and bottom of the pit with full kinetic energy when the load is rejected. Therefore, the pit is generally lined with highly erosion-resistant material or with steel plates. The pit is designed as a thick cylinder and is reinforced for vertical and horizontal forces. The foundation block is designed for following loads and forces.

1. All horizontal and vertical loads.
2. Dead weight of concrete.
3. Water pressure and thrust transmitted by manifold.
4. Weight and thrust transmitted through guide bearing.
5. Wind and seismic forces.

The foundation block should be checked for stability as per the procedure discussed above. Analysis should be done for the following two conditions.

1. Generator running (normal operating condition).
2. Generator stopped and rotor jacked up.

18.8 Design of Super—Structure (2 D—Design Concept)

18.8.1 General

The structure above the generator/turbine floor is called the superstructure. It consists of gantry columns, cross beams connecting the columns, panel walls between the columns supported on cross beams, roof and roof supports. These are provided in an indoor power house, which is climatically most suitable in India. Outdoor power-house does not have a superstructure. The superstructure in the 2-D design concept is analyzed and designed in two directions i.e. in transverse and longitudinal directions.

In the transverse direction gantry columns and roof support which is either a truss or a beam, if it is an RCC roof, form a portal that resists the transverse forces. The gantry columns are either of steel or RCC. Generally, RCC gantry columns are used. Sometimes steel columns are preferred to save construction time and for ease of erection and fabrication. Availability of material, economy and other site constraints are also the factors for selection of the type of gantry column material. The steel columns are normally enclosed with concrete for protection against fire. For ease of construction generally, a truss is used as a roof support.

In the longitudinal direction, the structure is a multistorey frame consisting of gantry columns and cross beams, which connect the gantry columns. These frames resist the longitudinal forces.

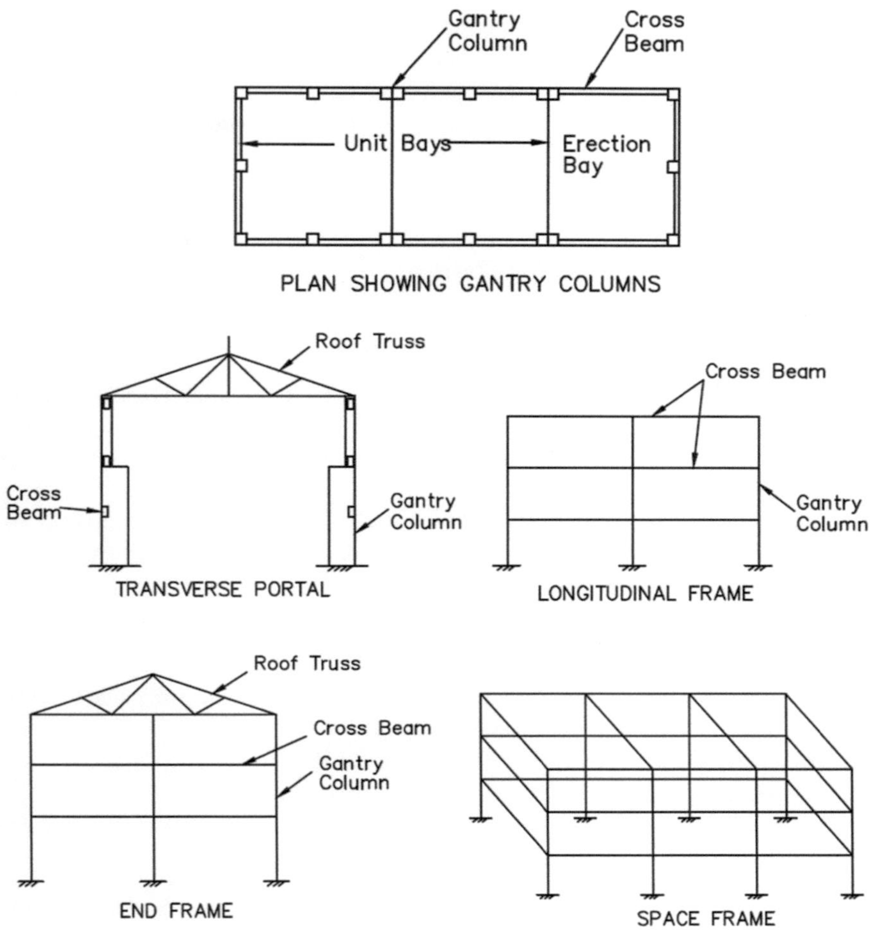

Fig. 18.11 Showing superstructure frames

The end walls are also multistorey frames that resist and transfer wind loads to the longitudinal frames. In Fig. 18.11, the above frames of the superstructure are shown.

18.8.2 Spacing of Columns

Gantry columns are generally spaced equally in longitudinal direction and upstream and downstream columns face each other so that they may form a portal in transverse direction. Generally, a column is required to be provided in the centre of the bay because normally the spacing between two columns in a longitudinal direction is about 6 to 7 m. When a joint is provided between the unit bays to avoid unequal settlement in a power house on an alluvial foundation, twin columns with an expansion joint are provided at the joint as shown in the plan in Fig. 18.11. The foundation of columns should be on mass concrete either at the generator or turbine floor. The column foundation on a beam or an opening should be avoided. In downstream a column may be founded on the draft tube pier.

18.8.3 Load and Forces for Design

The following loads and forces are considered for the design of superstructure.

(i) Dead load—It is self-load of the structure i.e. of roof, columns, beams, panel walls and the floor, if any.
(ii) Live load—It is the live load on roof and the floor. These may be taken from IS 875 and IS 4247.
(iii) Wind load—It is the force exerted by the wind on vertical panel walls and the parapet of the roof. Normally vertical component of wind force acting on a triangular truss is not considered.
(iv) Crane load—Following crane loads are considered. I.S. 4247 may be referred. Generally, these loads are supplied by the supplier.

    (a) Weight of fully loaded crane at its extreme position on either side to work out the maximum reaction on columns.
    (b) Vertical impact factor (25% if not specified).
    (c) Crane surge (10% of static load to be shared equally by the two rails).
    (d) Crane strike force depends on the velocity of the crane and is given by the supplier.

(v) Earthquake forces—Both horizontal and vertical accelerations due to earthquakes shall be considered as per I.S. 1893 depending on the zone in which the power house is located.
(vi) Temperature stress—Due to temperature variation the structure is designed to withstand force due to half of total temperature variation which is taken as 2/3rd of average maximum annual variation.

Maximum annual variation = [mean daily maximum temperature of the hottest month—that of the coldest month].

The design based on the above loads shall also be checked for erection and construction loads and crane testing load.

18.8.4 Permissible Stresses

Permissible stresses in concrete and steel should be taken as per I.S. 456 and I.S. 800 respectively. M25 grade concrete is generally used for gantry columns and cross beams. Load combinations and corresponding increases in permissible stresses are as below:

S. no	Load combinations	Percentage increase in permissible stress
1	Dead + live loads + crane loads u/s or d/s + Crane surge u/s or d/s	Nil
2	1 + wind in either direction	33
3	1 + temperature (rise or fall)	25
4	2 + maximum tail water level	50
5	1 + seismic forces	33
6	1 + fully loaded crane strike force	33

18.8.5 Principle of Frame Analysis

A two-dimensional analysis in the transverse and longitudinal direction can be carried out as below:

1. Transverse Analysis

In the transverse direction, the gantry columns are connected with the roof support for a portal. If the roof support is an RCC beam it will be fixed with the gantry columns at the top and the portal can be analyzed by the usual method of analysis i.e. method of moment distribution for determining bending moment and shear force for the loads acting on its members.

In case the roof support is a truss, it is hinged at both ends at the top of gantry columns to form a bent. The two columns in such a bent are fixed at the base and are considered hinged at the top by means of an inextensible link formed by the truss. The base of the columns may be at the same level or at different levels depending on the fixity with mass concrete. The analysis of such a bent which is shown in Fig. 18.12 is carried out as below.

The bent comprising of two columns AB and CD of height h_1 and h_2 respectively is connected by an inextensible member BD with hinged ends and is acted upon by a force H at height Z on column AB. In such a bent the deflection at B shall be equal to defection at D. This will induce a reaction R in the top member.

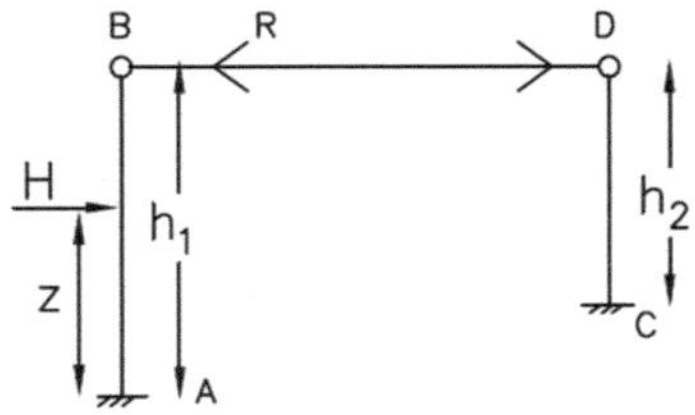

Fig. 18.12 Analysis of a bent

Let deflection at B due to load H be Δ_H
and deflection at B due to reaction R be Δ_{RB}
and deflection at D due to reaction R be Δ_{RD}.
Equating net deflections at B and D we get.

$$\Delta_H - \Delta_{RB} = \Delta_{RD} \tag{18.1}$$

Deflections are determined by considering columns as cantilevers.

The solution of expression (1) gives the value of R. Thus, moments at the base of columns will be

$$\begin{aligned} M_A &= H.Z - R.h_1 \\ M_c &= R.h_2 \end{aligned} \tag{18.2}$$

The computations of deflection in the case of gantry columns of varying sections are time-consuming. Monograms given in USBR TM N0. 633 for defections in such varying section columns are very useful.

The moments for various loads should be worked out at different levels and should be summed up algebraically to get moments for various load combinations. The columns are designed for maximum moments.

2. Longitudinal Analysis

In the longitudinal direction, the gantry columns and cross beams, which are spaced at 4 to 5 m, form multistorey multi-bay frame. It can be analyzed by the standard methods of frame analysis (moment distribution by iteration) for obtaining moments and shear force in various members due to loads acting on the frame. The moments at various levels under different loads are summed up algebraically to get moments for various load combinations. The columns in this direction are designed for maximum moments. Similarly, the design moments for X-beams are also computed.

18.8.6 Design of Frame Elements

1. Gantry Columns

The column shall be designed for the moments for the worst combination of loads. It is subjected to vertical loads and biaxial bending. In the case of steel columns, the design is simple because the stresses due to loads and moments are summed up and the sum shall be less than the permissible stresses in steel as given below-

$$\frac{P}{A} + \frac{M_X}{I_X} Y_Z + \frac{M_Z}{I_Z} y_X \leq f_s,$$

where P, M_x and M_z are vertical load, transverse moments and longitudinal moments, I_x and I_z are moment of inertia of the section and f_s is permissible stress, A is the area of steel section.

In the case of concrete columns, the design can be carried out as below.

(a) When one direction moment is less or equal to 25% of the other direction moment, the following condition shall be fulfilled.

$$\frac{\text{Compressive stress due to vertical load}}{\text{Allowable stress}} + \frac{\begin{array}{c}\text{Bending stress due to transverse moment}\\ +\text{Bending stress due to Longitudinal moments}\end{array}}{\text{Allowable bending stress}} \leq 1$$

(b) When the moment in one direction is more than 25% of the other direction, the design shall be carried out by applying load and moments simultaneously on the section. One method of designing columns in biaxial bending has been illustrated with an example in P.S. Nigam book (Chapter Generator Foundation). The following procedure has been adopted for testing the assumed column section with reinforcement.

(i) Find the resultant moment and the inclination of the axis about which it acts.
(ii) Calculate the eccentricity.
(iii) Assume an inclined neutral axis from the compression edge and draw it. Divide the compression area into strips parallel to the neutral axis and determine the maximum compressive stress and tensile stress in steel and check the assumed value of the neutral axis.
(iv) If the calculated value does not match with the assumed value of the neutral axis then assume another value and repeat the procedure and check whether the stress in concrete and steel is within permissible values.

Column Foundation

The columns are fixed with mass concrete. The foundation shall be treated as a load placed at the edge of mass concrete. It should be checked for tension in concrete and if required suitable and adequate reinforcement shall be provided at the surface of mass concrete. Column reinforcement should be taken into mass concrete up to sufficient depth to provide adequate fixity.

2. Design of X-Beams

These are the members of the longitudinal frame. These are designed for dead and live loads and the bending moment and shear force obtained from frame analysis. These may be either of steel or reinforced concrete. Steel beams are preferred for quick construction but their fixity with concrete columns needs special care. Precast RCC beams are also used.

3. Panel wall

These are of RCC, RB (reinforced Brickwork) or precast concrete blocks. These are made fixed with cross beams and designed for wind load. The openings in panel walls shall be properly reinforced. Pressure pipes, if any, should not be embedded in panel walls.

4. Roof and Roof Supports

The roof is generally made of steel truss covered by corrugated steel sheets (minimum thickness 1.25 mm) supported on a system of rafters and purlins. Normally a triangular truss is provided with height span ratio of 1/4. If the span is large then N—truss is used. RCC roofing with a minimum slope of 1 in 120 is rarely adopted because of construction difficulties. The truss is designed for dead load, live load and wind load. The truss, being hinged to the gantry columns, has to be designed for the thrust generated in it. Waterproofing of the roof shall be carried out in accordance with relevant I.S. Codes.

18.9 Design of Generator Foundation

The generator mainly consists of the stator (stationary part), which has armature windings and a rotor (rotating part), which has field coils, shaft, spider rim, etc. Stator rests directly on the generator foundation. Rotor weight is transmitted through the thrust bearing, so its weight transmission depends on the location of the thrust bearing. If this bearing is at top. The arrangement is called the suspended type. The rotor weight is transmitted to the generator foundation at a lower pedestal in a barrel-type generator foundation which is commonly used.

18.9.1 Barrel Type Generator Foundation

Barrel type of the generator foundation is shown in Fig. 18.13. The barrel is fixed at the bottom with concrete which encases the scroll case and is free at top. It has pedestals at two levels. At a higher level pedestal, the stator is supported and at a lower one the rotor is supported. It has the advantage of the lower setting of a generator. The thickness of the barrel depends on the radius of the generator housing and the turbine pit. It is structurally indeterminate due to openings and recesses provided and the discontinuity caused due to the passageway to the turbine pit.

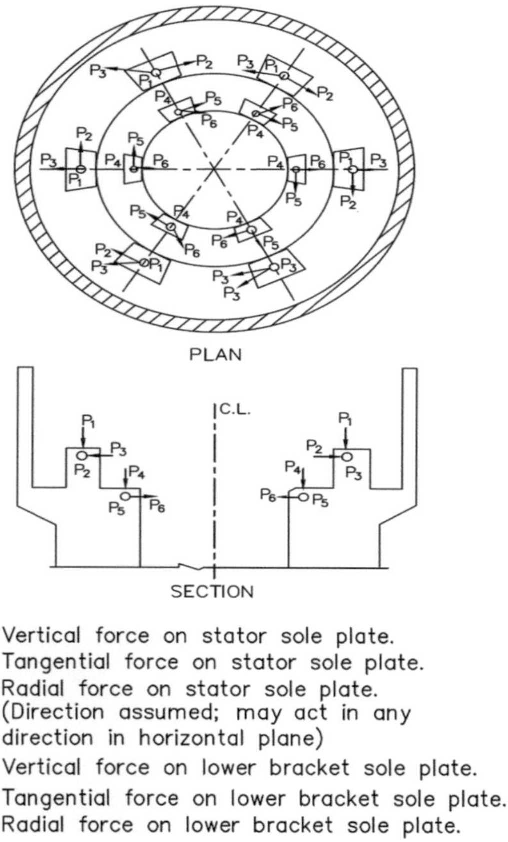

Fig. 18.13 Machine loads for design of generator foundation (P.S. Nigam)

18.9.2 Design Aspects

(a) Loads and Forces

The loads and forces for the design of barrel type generator foundation shown in Fig. 18.13 are taken in accordance with I.S. 7207. These are:

(i) Dead load.
(ii) Live load on floor, if any.
(iii) Vertical load on stator sole plate.
(iv) Vertical load on rotor sole plate.
(v) Tangential forces due to short circuit torque on the stator sole plate and braking torque on rotor plate.
(vi) Radial force acting on both the stator and rotor plates due to one-sided magnetic attraction.
(vii) Force due to eccentricity of rotating parts.
(viii) Seismic force depending on the seismic zone.

(ix) Temperature stress due to temperature variations.

Loads and forces (iii) to (vii) are given by the suppliers. The preliminary estimate can be made for generator loads and short circuit torque as given in para 18.3.

(b) Design Concepts

As stated above the generator barrel foundation is indeterminate. However, for design purposes, the barrel is analysed for loads and forces with the following assumptions.

(i) Barrel is fixed at the bottom in scroll case concrete and free at top.
(ii) Vertical load dispersal in concrete is at 45°.
(iii) Height of the barrel is usually small compared with the diameter.
(iv) Vertical loads on sole plates are point loads on the edge of concrete.
(v) Short circuit torque and seismic loads are emergency loads and are not considered to act together.

Generally, reinforcement in the barrel is required for shear stress due to short-circuit torque. Stresses due to other forces and loads are very small and require no reinforcement. The shear reinforcement is generally provided as circular hoops with vertical bars on both faces. The shear reinforcement due to short circuit torque may be calculated in a barrel by an approximate method given by Roark (refer P.S. Nigam book for details). The barrel can also be analysed using the FEM technique.

The pedestals for rotor and stator sole plates shall be checked for bearing stresses in concrete.

Generally, M 30-grade concrete is used. The minimum reinforcement used is 25 mm Ø tor steel @ 300 mm c/c for both hoop bars and vertical bars on both faces. Concrete and steel shall be according to I.S. 456 and I.S. 800 respectively.

(c) Stresses

Permissible stresses shall be reduced to 55% in steel and 40% in concrete to safeguard against fatigue failure which may be caused due to repeated loads from the turbine and generator. For emergency loading (seismic load or short circuit torque) the above-reduced stresses should be increased by 33 for both steel and concrete.

(d) Vibrations

In order to avoid resonance, the natural frequency of generator supports should be different from the frequency of impressed forces. For computation, the generator floor and supporting structure shall be assumed fixed with the substructure. The natural frequency of the structure is generally kept 20 to 40% more than the speed of the machine to avoid resonance and it is called high tuned structure. Low tuned structures are not preferred. The forcing frequency should correspond to runway speed.

Since the computation of frequency of power house structure is complicated and tedious, the following empirical criteria are followed-

(i) USBR guideline

- Volume of concrete in the substructure in cubic feet should not be less than 2 times the machine capacity in KW. In most existing plants, it is found to vary up to 30 times.
- Weight of concrete in generator support and scroll case encased concrete should be at least 2 to 2.5 times the weight of rotating parts.

(ii) CWC Guidelines

- Volume of concrete in substructure should be 4 to 6 cubic feet per KVA of rated generator capacity out of which 2 cubic feet per KVA should be in second-stage concrete.

Some designers suggest that in the case of power house on rock, the substructure should be anchored with foundation rock to provide additional safety.

18.10 Design of Horizontal Setting Power House

In a horizontal setting power house, which is normally adopted for small capacity SHPs, the shaft connecting the turbine with the generator is kept horizontal. The various layouts using different turbines have been described in Chap. 4.

In this type of power house, the structure consists of super structure and substructure. There is no intermediate structure. The machines (turbine and generator) are directly founded on the substructure. The gantry columns of the superstructure are also founded directly on the substructure and these transfer loads and moments directly to the substructure. The substructure contains the turbine discharge outlet structure (draft tube) as in the case of the vertical setting power house.

Therefore, in such a power house the structural design is required for super structure and the sub-structure only, the procedure for which has been described in the above paras.

The 2-D design concepts for these structures have also been described briefly in AHEC Manual No. 2.2 & 2.3.

18.10.1 Generator Foundation

In this case, the generator rests on the substructure and all loads are transferred to the substructure through anchor bolts. In the horizontal shaft machine, both stator and rotor rest on the common bed plate through the pedestal bearing. The arrangement is shown in Fig. 18.14. The loads acting on the foundation of a generator are also

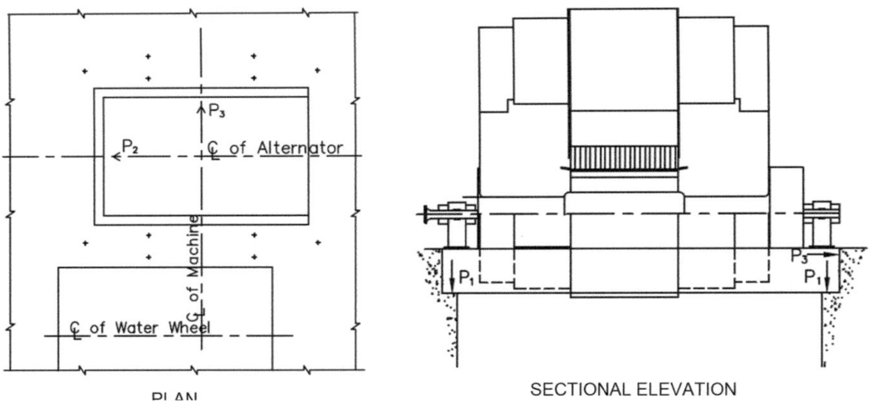

Fig. 18.14 R.C.C. Generator foundation for horizontal shaft arrangement

shown in the figure. For the design of foundations, the following loads are considered which are supplied by the manufacturers.

- Weight of stator, rotor bearing, bed plate, etc.
- Weight of over-hanging exciter and runner.
- Maximum vertical load due to short circuit torque.
- Axial hydraulic thrust.

18.11 Three Dimensional (3-D) Analysis of Power House Structure

Though a large number of power house structures have been designed and constructed on the 2-D design concept described above, the designers feel that 2-D analysis and design do not represent the true structural behaviour of the structure. Nigam first attempted to analyze the power house structure to understand the 3-D structural behaviour through experimental methods using the theory of photo elasticity. Through the photo elastic method, he analysed the sub structure and intermediate structure separately for some loads. He also analyzed the space frame of the superstructure using the computer program developed by Saini et al. These studies and their results are given in detail in the book by P.S. Nigam. Since then several 3-D analyses of the power house structure have been carried out using computer programs. STAAD-4 has been used for analyzing and designing superstructure space frame and FEM programs have been used to analyze substructure and intermediate structure separately as well as a composite structure. A 3-D analysis by Chauhan et al. (2005) investigated stresses and deflections in concrete around the spiral casing and draft-tube due to different loadings using ANSYS Software in a composite structure of intermediate and substructure of a power house along with its foundation. They also investigated the effect of the elastic modulus of foundation rock on stresses in power house concrete and concluded that stresses and deflections do not appreciably change with the change in elastic modulus of the foundation except in the elbow portion. The study also indicated that the power house structure may be assumed as fixed with the foundation if the rock is strong.

The general conclusions of all the 3-D studies are as below:-

- The intermediate and substructure should be considered as a composite structure.
- Tension in concrete requiring reinforcement occurs generally around the openings such as draft tube, turbine pit and scroll case, etc., and under the vertical loads such as supports and gantry columns.
- In the substructure, the elbow concrete is most stressed as compared to other structural elements of draft tube.
- The space frame analysis of superstructure gives 15 to 20% lesser moments in columns than 2-D analysis. It also shows some torsional moments in the cross beams.

In the case of a horizontal shaft power house, the superstructure can be analyzed and designed as a space frame and the substructure as a plate element using the STAAD-4 program.

18.12 Design of Power House Founded on Rock

On the rock foundation which normally has a high bearing capacity, the power house building does not need a spread footing as in the case of power house on an alluvial foundation. The design procedure of a power house on an alluvial foundation has been discussed in the preceding paras. In the case of a rock foundation, the piers and side walls of the draft tube may directly rest on rock with separate independent footings and the bottom slab may be a sort of lining over the rock to provide a smooth surface for flow. Weep holes and rock anchors may be provided in the bottom slab as a safeguard against uplift. The draft tube elbow and divide walls may behave as portals having machine loads on top. The scroll case encasing concrete and superstructure may be treated separately as discussed in the above paras.

In the case of a horizontal shaft power house on rock, the runner and generator may have separate footings. The gantry columns also can have separate footings directly on rock foundations.

References

1. Chauhan, Gopal, Tripathi, V., Asthana, B.N., Singh, Y.: Structural design concept of hydropower station. In: Proceedings 3rd Specialty Conference on the Conceptual Approach to Structural Design, Singapore (2005)

2. Guidelines for Small Hydro Development, Standard No. 2.2 and 2.3, 'Hydraulic and Structural Design. AHEC, IIT. Roorkee (2013)
3. I.S. 4247 (Part-1,2,3).: Code of Practice for Structural Design of Surface Hydroelectric Power Station Part-1 Data for Design, Part-2 Superstructure, Part-3 Sub structure. BIS, New Delhi (2003)
4. I.S. 7207.: Criteria for Design of Generator Foundation for Hydroelectric Power Stations. BIS, New Delhi (2003)
5. I.S. 875.: Code of Practice for Design Loads (other than Earthquake) for Building and Structures-Part 3: Wind Loads,' BIS, New Delhi (1987)
6. I.S. 1893.: Criteria for Earthquake Resistant Design of Structures. BIS, New Delhi (1998)
7. Mosonyi, E.: Water Power Development, vol. 1' Budapest (1963)
8. Nigam, P.S.: Hand Book of Hydro Electric Engineering. Publication of Nem Chand and Co., Roorkee (1985)

Chapter 19
Hydro Mechanical Equipment

19.1 General

Hydro-mechanical equipment in hydro power projects are also termed as hydraulic control equipment. Their purpose is to control and regulate the flow of water through various structural components of the hydro power project. These also facilitate the repair and maintenance of various structural components of the project. These equipment are generally called gates and valves and are used to control the flow where the head is low, medium and high. In SHPs, the heads encountered are generally low and medium. A head up to 15 m is termed as low and a head between 15 and 30 m is termed as medium. Where it is required to regulate the flow of water at very high pressures, valves are generally used. In a valve, the closing member is operated and it remains within the water passage. But in a gate, the closing member is taken out of the flow. Various types of gates and valves have been used in hydropower projects depending largely on the water head, location and the purpose to be served.

19.2 Location of Gates and Valves

In a run-of-river type SHP there are maximum number of control and regulation locations that require gates and valves. For the river diversion if the structure is a barrage the gates are required to pass the river flow and if the diversion structure is a dam then its spillway also has gates. Sometimes sluices in dams are provided with gates. An intake structure is normally part of a diversion structure and it requires regulation gates. It is also provided with emergency gates. A run-of-river scheme in the Himalayan region normally requires a desilting arrangement after the intake and it requires control gates at the inlet, outlet and escape channel depending on the type of desilting arrangement. There will be an intake with gates for the water conductor after the desilting arrangement. If the water conductor is long enough it

B. N. Asthana, *Planning and Design of Small Hydro Power Projects*,
https://doi.org/10.1007/978-3-031-96271-4_19

will be terminated in either a surge tank or a forebay depending on the type of water conductor. A penstock is provided from the forebay/surge tank to carry water to the power house. A gate is provided at the inlet of the penstock. A gate is also required at the head of the bye pass channel of the forebay. Generally, a valve is provided at the end of the penstock before the flow enters the turbine. When flow leaves the turbine through an out flow channel (draft tube) a gate is provided to segregate the power house from the tailrace during the repair and maintenance of the turbine.

In the case of the dam toe power house, normally a valve at the turbine inlet and a gate at the draft tube are required and in case of canal fall SHP gates are provided at the intake and the draft tube.

19.3 Purpose of Gates

The gates at various locations serve different purposes. The gates are classified on the basis of the purpose they serve. This classification is as below:

1. Regulation gates—These are gates to regulate the flow with operational facilities capable of operating the gate at partial opening.
2. Emergency gates—These are meant to be used in an emergency when the main gate is under repair and so these also require operational facility.
3. Bulk head/stoplog gates—These are generally used during a closure required for repair and maintenance purposes of regulation gates. An operational facility is not required for this type of gate. These are normally made in small height (1 to 1.5 m) pieces which are placed in position through a mobile crane.

19.4 Type of Gates and Valves

The following types of gates and valves are generally used in small hydro projects.

Gates

1. Vertical lift gates—The two types of vertical lift gates that are generally used are as below.
   (a) Fixed wheel gates—These are used at barrages, spillways, intakes, etc.
   (b) Slide gates—These are used in sluices and the intakes of small discharge channels.
2. Radial Gates—These are generally used on spillways where height of gate is comparatively more as compared to the length.
3. Bye pass gates—These are the gates that are provided at the head of a bye-pass where automatic operation is required.

Valves

The valves are normally used at the inlet of the turbines and are called MIV (main inlet valve). The two types of valves used are:

(a) Butterfly valve—It is normally used for low and medium-head turbines.
(b) Spherical valve—It is used where the head is high.

The above gates and valves are described in the following paras.

19.5 Fixed Wheel Vertical Lift Gate

These gates are rectangular in shape. The height of gate is normally around half of the span. The span is the clear spacing between the piers. The height is the difference between the crest and the water level (FRL in case of dams and FSL in case of intakes) plus the free board which is generally 0.3 to 0.5 m.

The three essential components of this type of gate are

(a) Closing member—It checks flow through the opening. It comprises a skin plate and horizontal girders (Fig. 19.1).
(b) Groove—It is made in the piers in which the gate moves satisfactorily.
(c) Hoisting arrangement—It is the arrangement that is capable of hoisting the gate at the specified speed. Generally, a mechanically operated rope-drum arrangement is provided for hoisting. This can be both electrically and mechanically operated arrangement.

The closing member of the gate consists of a structural frame work to which a skin plate is attached. The skin plate is normally placed on the upstream face. The

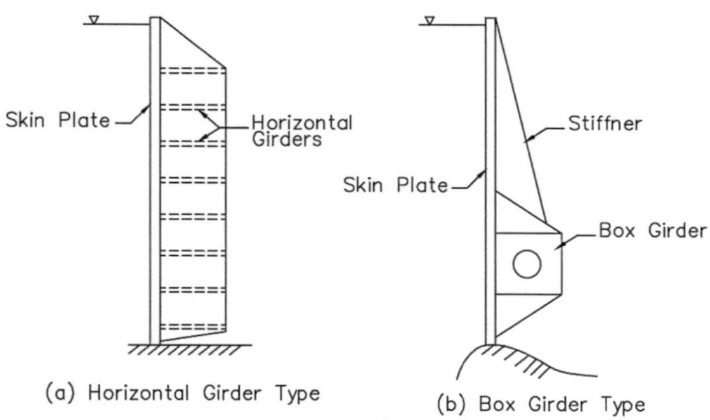

Fig. 19.1 Vertical lift gates

closing member should be strong enough to withstand the hydraulic pressure and it should also be water tight. The hydraulic pressure for which the gate is designed is the hydro static and hydro dynamic forces considering wave effects and seismic loads. The hydraulic pressure is transmitted by the structural framework to the wheels attached to it. The wheels move in the grooves and transmit the pressure to the pier.

The hoisting mechanism of the gate is normally made strong enough to withstand the weight of the gate as well as the force to overcome the friction developed due to water pressure and movement of gate. The gate is normally designed to close under its own weight. If it requires positive thrust for closing, the hoist shall be suitably designed for the purpose.

The frame work, to which the skin plate is attached, generally comprises horizontal girders (normally I-Joists are used) stiffened by vertical stiffeners. The horizontal girders are also stiffened by heavy vertical end girders. The wheels are attached to the end girders (Refer Fig. 19.2 for details). Sometimes in large span gates, the horizontal girders take the form of a box girder which transmits the water load through two or four wheels to the grooves. These types of gates are shown in Fig. 19.1b. Horizontal girder-type gates are generally used. Hence, this type of gate is described below in detail in the following paras:

19.5.1 Structural Details

The horizontal girder type fixed wheel gates are simple in design. The structural components of such a gate are as below. These may be seen in Fig. 19.2, which shows a typical vertical lift gate.

The closing member of the gate consists of

(i) Horizontal girders
(ii) Vertical stiffners
(iii) Skin plate
(iv) End girders
(v) Wheels
(vi) Seals and Guide rollers.

The groove in the piers, in which the wheels move, has the following embedments.

(i) Wheel track
(ii) Seal bases
(iii) Guides
(iv) Anchor bolts.

The hoisting arrangement consists of the following.

(i) Hoist
(ii) Lifting arrangement may be either a hook or lifting beam
(iii) Ropes
(iv) Counter-weight.

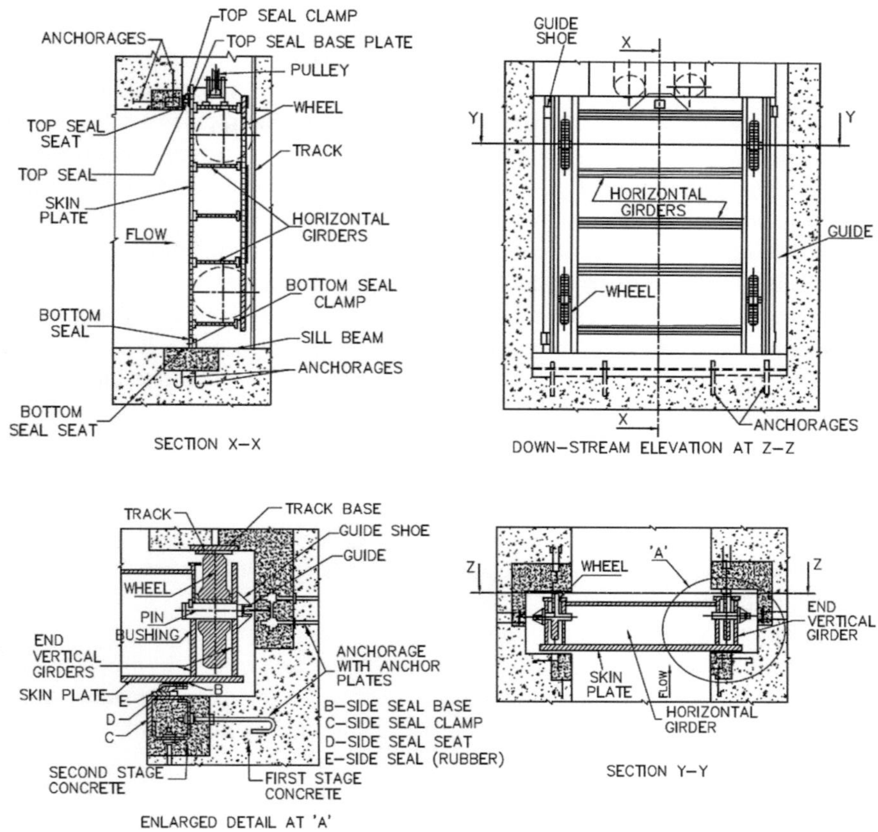

Fig. 19.2 Typical arrangement of various components of fixed wheel gates (I.S. 4622)

In addition to the above, a hoist bridge on piers is required for placing the hoisting equipment.

19.5.2 Design Requirements of Structural Components of Gate

(a) Horizontal Girders

Horizontal girders also called main girders are normally I- Joists but may also be of fabricated or rolled type sections. These are acted upon as point loads from the vertical stiffeners and are designed as beams simply supported on the end girders. The spacing of the horizontal girders shall preferably be such that all the girders shall carry almost equal loads, so that same size girders may be used in a gate.

Spacing of Girders—The method of equal load sharing may neither require a girder near the top nor a girder very near to the bottom nor close spacing between two top girders. This situation may not be desirable. Hence, the spacing of the girders is generally determined by the trial and error method. The spacing of girders on the consideration of sharing equal loads is thus adjusted from other requirements of general rigidity of frame work, ease of inspection and maintenance and the possibility of damage due to accidental causes.

The spacing of the girders on the consideration of equal sharing of water load can be determined graphically as below and can be used as a guide.

The graphical method is shown in Fig. 19.3. In the figure triangle a b c represents the water pressure applied on the gate. A semicircle is drawn with a b as diameter. Then a b is divided into equal parts aa_1, a_1a_2, a_2a_3, a_3a_4, a_4b depending on the number of girders to be used.

From points a_1, a_2, a_3 ...perpendiculars are drawn to meet the semi-circle at points b_1, b_2, b_3 ......

Through the points b_1, b_2, b_3 .... circular arcs with centre at 'a' are drawn which meet the line a b at points c_1, c_2, c_3...

Through points c_1, c_2, c_3... lines parallel to 'bc' are drawn to meet the line 'ac' at points d_1, d_2, d_3...

Thus, lines c_1 d_1, c_2 d_2, c_3 d_3... divide the area 'abc' into equal parts.

Hence water load is divided into equal parts represented by ac_1 d_1, d_1c_1 d_2c_2,

The girders placed at the centroid of these areas will carry an equal load which is given by the product of the corresponding area multiplied by the specific weight of water.

The above spacing of girders based on sharing water load equally is taken as a guide but normally a girder is provided flush with the top of the gate and another girder is placed adequately above the bottom leaving some space to properly fix the bottom seal of the gate. Hence the top and bottom girder will carry less load but generally, all girders of same size are provided. It is also the practice to provide a lighter section for the top girder flushing with the gate top and disregard it in stress

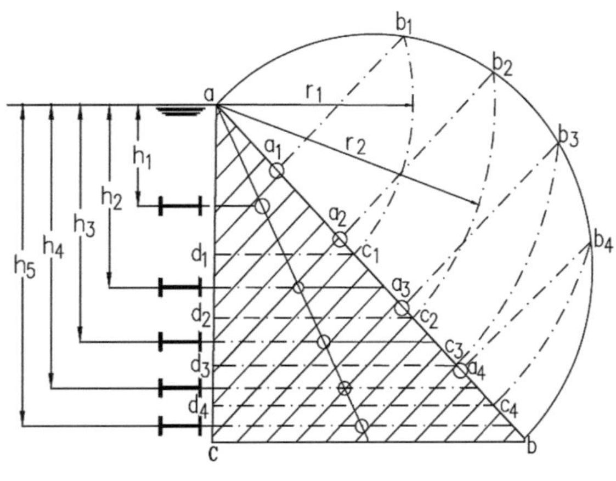

Fig. 19.3 Graphical method of determining girders (P.S. Nigam)

calculation. Thus the second girder coincides with the centroid of the first triangular strip of Fig. 19.3.

Design Aspects of Girders: The skin plate transmits water load to vertical stiffeners which transmit load to the horizontal girders. The load sharing between stiffners depends on the panel size of the skin plate and the support conditions. Since the determination of load sharing is rather indeterminate, a safe and simple assumption given below may be made for the design.

'Skin plate transmits all its water load directly to the beam (horizontal girder) disregarding the sharing of load through the stiffners'.

In the design of the beam section, a part of the skin plate shall be assumed to be acting as the flange but the estimation of the width of the skin plate acting as the flange is indeterminate. Hence, the co-acting width is normally worked out by using empirical formula given by IS 4622

$$\text{Co - acting width} = 40\text{t} + \text{b}$$

where, t = thickness of skin plate
b = width of bottom flange.

(b) Vertical Stiffners

The spacing of vertical stiffners is decided in such a manner that the skin plate panels between stiffners and horizontal girders are of such a size that the thickness of the plate is within economical limits. The vertical stiffners are treated generally as fixed with the horizontal girders. The loading on the stiffners is the water load transferred by the two adjacent panels. It depends on the type of support of the four sides of the panels. A safe simplification is to assume that the water load is transferred as a rectangular and uniform loading on the full panel between two vertical girders. The section of the vertical stiffners is a T-section and the co-acting width of the skin plate is part of the flange. Co-acting width of the skin plate may be taken the same as given above for horizontal girders.

(c) End Girders

End girders are vertical girders at the two ends. The horizontal girders are connected with the end girders and transmit all the water load on the gate and the load of the gate to the end girders as point loads. The end girder is considered supported as a continuous beam on the wheels.

The bending moments and shear force for design of end girder are, thus, worked out by the moment distribution method. The end girder is either a rolled I-joist or a fabricated section.

As stated above the end girder is supported on the wheels by the pin (axle of wheel) passing through the web of the girder. Thus, the pin is butting against the upstream portion of the web (refer Fig. 19.2). So, the web should be suitably strengthened for the reactions.

(d) Skin Plate

Skin plate is normally fixed on the upstream side of the framework. It is welded with the horizontal girder as well as with the vertical stiffners. The plate is divided into small panels and the water load on a panel is distributed in two directions.

Normally skin plate is designed for the most stressed panel and the same thickness is provided through all the panels.

The skin plate is designed co-acting with the vertical stiffners as well as with the horizontal girder. For details of design procedure refer to I.S. 4622.

The actual thickness of the skin plate to be provided shall be at least 1.5 mm more than the theoretical thickness based on stress computations, in order to take care of corrosion. The thickness of the skin plate inclusive of corrosion allowance shall not be less than 8 mm.

(e) Wheels and wheel track

(i) Wheel

The gate wheels are suitably designed to withstand the hydrostatic load which is transferred as the reaction from the end girder. The wheels move on the wheel track embedded in the gate groove. The wheels transfer the load to the pier (refer Fig. 19.2).

The wheels are made of cast steel or wrought steel and carry enormous loads. The wheels may be flanged or flangeless. The flangeless wheels are preferred. The tread of the wheel may be flat when self-aligning bearings are used between the wheel and wheel pin.

The plane of contact of the wheel and track may be taken as line or an arc of a circle. If all the wheels are in contact with the track, the load on each of the wheels can easily be determined by finding reactions at the supports of the end girder. Normally it is considered in design that due to misalignment any one wheel may be off the track. Wheel loads can then be evaluated by omitting various wheels one at a time. Due to tedious computations, this condition is normally accounted for by taking 50% extra load on the wheel.

A factor of safety of 3 and 2 may be adopted for normal and emergency loads respectively. The wheel is designed for the line contact stress between the wheel and the track. The stress is calculated by the Hertz formula as

$$\mathrm{F} = 3190\sqrt{\frac{\mathrm{P}}{\mathrm{D}}}$$

where, P is load per inch width of the roller and D is diameter of roller. The stress is limited to the ultimate tensile strength of wheel material. The diameter of the wheel can be suitably adjusted.

I.S. 4622 recommends the following formula for working out contact stress between wheel and track

$$\mathrm{f_c} = 0.418\sqrt{\frac{\mathrm{P.E}}{\mathrm{r.l}}}$$

where, $\mathrm{f_c}$ = contact stress in tonnes/cm^2

P= wheel load in tonnes

E= Modules of elasticity in tonned/cm^2

R= radius of wheel in cm
L = tread width in cm.

The permissible contact stress is given by the critical stress of the projected area divided by the factor of safety. The projected area of the tread is the product of the wheel diameter and the tread width. The critical stress is given by

$$\text{critical stress} = (\text{Brinell hardness number} \times 0.169) - 15.174\left(\text{N/mm}^2\right)$$

$$\text{Projexted area} = \frac{\text{P}}{\text{f}_\text{c}}$$

If diameter of wheel is selected on practical consideration, then the tread is given by

$$\frac{\text{Projected area}}{\text{Diameter}}.$$

A check on maximum shear stress and contact stress should also be made as per details given in I.S. 4622.

(ii) **Wheel Track**

Wheel track shall provide a smooth machined surface for the wheel to roll on it easily and to transmit load to pier concrete through the track base. The hardness of the track surface should be about 50 points more than the Brinell hardness number of the wheel tread.

The thickness of track plate for line contact is worked out by following formula:

$$\text{b} = 1.55\sqrt{\frac{\text{P}}{\text{l}} \times \frac{\text{r}}{\text{E}}}$$

where,

b = Half contact width in mm, other notations are as given above.
Z_1 = depth to the point of maximum shearing stress in mm = 0.786 b.

The thickness of wheel track shall not be less than $6.Z_1$. The track base is embedded in concrete. The stress in concrete shall be worked out by using the following formula. The bearing stress in concrete shall be taken as per I.S. 456. The concrete shall be at least of grade M20.

$$\text{p} = 0.2813\text{P}\left[\frac{\text{E}_\text{c}}{\text{E}_\text{s} \cdot \text{I} \cdot \text{w}^2}\right]^{1/3}$$

where,

p = bearing stress in concrete in N/mm^2
P = total wheel load in N.
E_c = modules of elasticity of concrete in N/mm^2
E_s = modulus of elasticity of steel in N/mm^2

I = moment of inertia of track base in mm^4
w = width of track base in contact with concrete in mm.

In case bearing stress in concrete p is more than the permissible bearing stress of concrete then either P (wheel load) shall be reduced by increasing the number of wheels or the dimensions of base plate shall be increased.

The edge distance of the track base from the groove face shall be a minimum of 150 mm to safeguard the damage to the groove face concrete due to load on the track base. The criteria given in I.S. 4622 may also be referred.

(f) **Guide Rollers and Guide Shoes**

When the gate is operated in the groove, there is a possibility of the gate to drift sideways. Gate guide rollers/shoes shall be provided on the sides of the gate to limit the lateral motion of the gate to a maximum of 6 mm in either direction. The rollers are normally flanged and these move either on a steel plate or a rail fixed with groove concrete through bolts. I.S. 4622 recommends a minimum of two guide rollers or shoes to be provided on each side of the gate. These should be structurally strong enough to withstand the load they are likely to be subjected to depending on hoisting and hydraulic conditions. These are normally provided in high head gates. Figure 19.2 shows the wheel and the embedment details for the wheel track, guide shoes and seals in a gate groove.

(g) **Guides**

The guides are fixed inside the groove. The guide for the gates fixed with guide rollers (refer Fig. 19.2) comprises of flat plate anchored into concrete by angle cleats and anchor bolts. The thickness of the plate (as recommended by I.S. 4622) shall not be less than the values given below.

Type of gates	Thickness of plate.
Low head gates	20 mm.
Medium head gates	32 mm.
High head gates	40 mm.

The guides are provided in full range of gate travel.

(h) **Seals and Seal Seats**

Seals are fixed on the sides and the bottom of the gate to prevent seepage. The practice is to provide music note type rubber seals. The seals are fixed with seal clamps and stainless steel screws to ensure a positive water pressure between the seal and the gate and to bear tightly on the seal seat to prevent leakage. For small or low head gates a hollow bulb music note type seal is provided whereas for medium and high head gates solid bulb music note type seals are recommended (I.S. 11855). Typical details of fixing seals in the bottom and sides of a gate are shown in Fig. 19.4.

The minimum width of seal seats shall be 80 mm. The minimum thickness of stainless steel plates for low head gates shall be 6 mm and 8 mm for medium head gates. The seal seats shall be finished smooth and shall be embedded in groove concrete.

Fig. 19.4 Showing seal and seal fixing arrangement

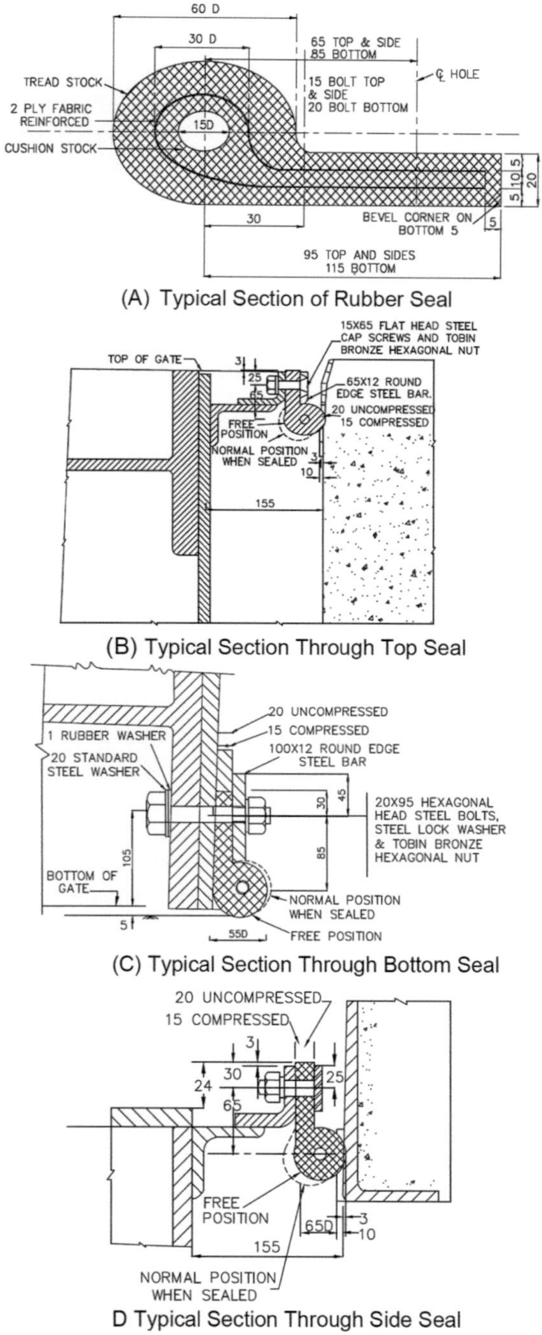

(i) **Sill Beam**

A sill beam which is usually an I-Joist is provided at the crest for proper seating of bottom seal. The surface of the sill beam is machined smooth and kept flush with the surrounding concrete. The sill beam is fixed with concrete through bolts.

(j) **Hoist**

Normally rope drum hoists are provided for low and medium head gates. These are operated both mechanically and electrically. Hydraulic hoists are used for high head gates.

The hoisting capacity comprises following forces:-

(i) Weight of gate and of all its components.
(ii) Frictional forces comprising

1. Wheel friction
2. Seal friction
3. Guide friction.

(iii) Hydrodynamic load like down pull / uplift etc.
(iv) Forces due to any other specific requirement.

The hoist capacity worked out from above shall be increased by at least 20% for reserve capacity.

The operating speed of hoist is normally 30 cm/minute. The hoisting arrangement is kept on a hoist bridge supported on the piers through the steel structure. The height of hoist bridge shall allow lifting of the gate clear of water surface when it is not in operation and lifted for maintenance.

For designing hoist, the following codes may be referred:

I.S. 6938 Code of practice for rope drum hoists.

I.S. 10,210 criteria for design of hydraulic hoists.

19.6 Slide Gates and Structural Components

As the name suggests, slide gates are those in which a gate leaf slides in a frame embedded in the groove. Since sliding friction is large as compared to the rolling friction of wheels in a fixed wheel gate, slide gates are used for small size openings in low head projects. The hoisting capacity requirement in a slide gate is more. The slide gates which are operated by screw/rope hoists are normally used for small size openings in low head projects. High head slide gates in dam sluices are operated by hydraulic hoists.

The slide gate comprises the gate leaf and the embedded parts/frame in the groove.

The design details given in the following paras are based on I.S. 5620.

(i) Gate Leaf

The gate leaf is the operating part and is made of a skin plate that is rigidly fixed with a frame structure. The supporting structure is made of suitably spaced horizontal and vertical members. These are steel structural members like I-Joists, channels or angles. In small canal heads wooden gate leaves are sometimes used but normally the practice is to use welded steel structural gate leaf.

The skin plate may be either on the upstream or on downstream side. The upstream skin plate has the advantage that trash does not accumulate in the supporting structure of skin plate and so this type is normally preferred.

The sealing cum bearing plates which are normally made of brass or bronze are screwed on the face of the gate leaf for the purpose of sealing and load transfer to the concrete. A low head slide gate with upstream skin plate and upstream sealing arrangement is shown with groove embedments in Fig. 19.5.

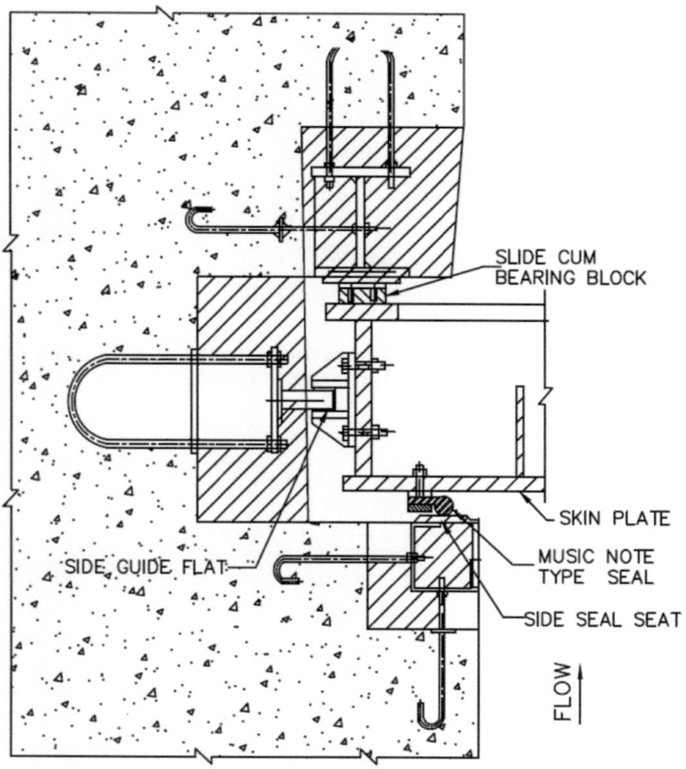

Fig. 19.5 Showing low head slide gate with upstream sealing and skin plate

(ii) **Embedded Parts/frame**

The embedded parts in the groove are normally sealing surfaces, bearing block and guides for the gate leaf and these are shown in Fig. 19.5. The embedded parts are firmly anchored with concrete. The bearing track generally extends at least one gate height above the opening and guides extend up to full travel of gate.

In low head slide gates sometimes a frame is embedded in the groove. Sometimes a full face gate frame is provided at the face of the opening. The two arrangements of the embedded frame are shown in Fig. 19.6.

(1) **Seals**

Seals are fixed with the gate leaf. Generally, rubber or metal seals are used. Occasionally for very small heads wooden seals are used which are commonly of sal, shisham or deodar. These are fixed with bolts.

Fig. 19.6 Showing embedded frames for a slide gate

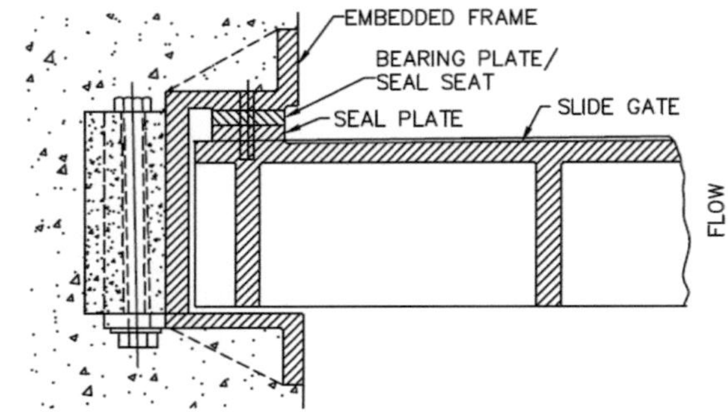

a. Typical Section Showing Low Head Slide Gate and Embedded Frame

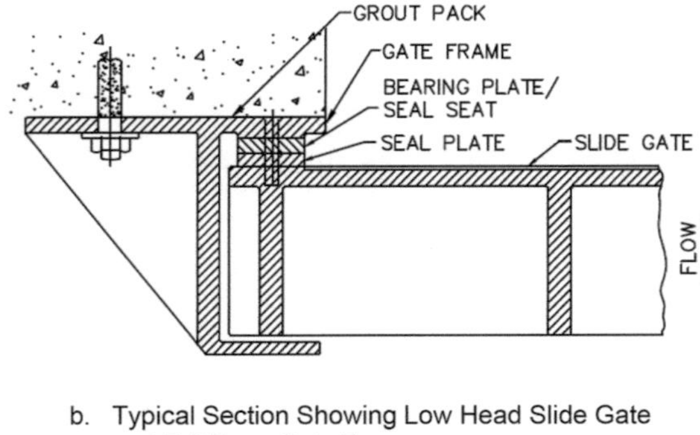

b. Typical Section Showing Low Head Slide Gate and Full Face Gate Frame

Rubber seals of music note type are used at the top, and sides of the gate leaf and wedge type are used for the bottom. These are fixed with gate leaves by seal clamps and stainless steel bolts/screws.

Metal seal plates are made of brass or bronze and fixed with the gate leaf by counter-sunk screws of stainless steel.

(2) **Bearing Plates**

These are made of either cast iron or steel or stainless steel etc. and are welded to the embedded parts. Where these are used with sealing plate the material should be of same type but harder than that of sealing plate.

(3) **Guides**

A suitable guide shall also be provided on the embedded parts to limit lateral and longitudinal movements of the gate leaf to 5 mm (refer Fig. 19.5).

For more details about slide gates I.S. 5620 may be referred.

(iii) **Hoist**

The slide gates of small size operating under low heads are generally operated by screw hoists. These are normally hand operated but electric motors can also be used to operate the gate. The medium head gate is operated by rope-drum hoists which have been discussed with fixed wheel gates. Big size high head slide gates are normally operated by hydraulic hoists.

The screw hoist is simple in construction and consists of cast iron or fabricated steel pedestal resting on a platform. A screw stem passes through the pedestal and platform and is connected to the gate by means of nuts. A set of gears reduces the requirement of power for rotation, so that the gate may be moved by hand operation. Hoists of capacity 7.5 tonnes are normally provided for manual operation. For details of the design of screw hoists refer I.S. 11228.

19.7 Radial Gates

These gates are generally used on spillways but these are also used in canal heads, conduits and tunnels. Figure 19.7 represents a typical radial gate. In this type of gate, the gate frame comprising the skin plate and horizontal girders is connected to end beams which are supported on main struts or radial arms that remain in compression due to hydrostatic forces acting on the gate skin plate. The resultant of hydrostatic forces in a simple gate design passes through the gate trunnion. The skin plate is normally curved in shape (arc of a circle) but gates with flat plates have also been used due to ease of construction. Studies have indicated that radial gates are often the cheapest type of gates. The merit lies in its mechanical simplicity. These have been used satisfactorily where height is equal or more than the span. These have been used for a maximum height of about 20 m. These have also been used where the head is small. A gate of size 20 m in span and 2.5 m in height has been used in a project in Iceland.

Fig. 19.7 Radial gate

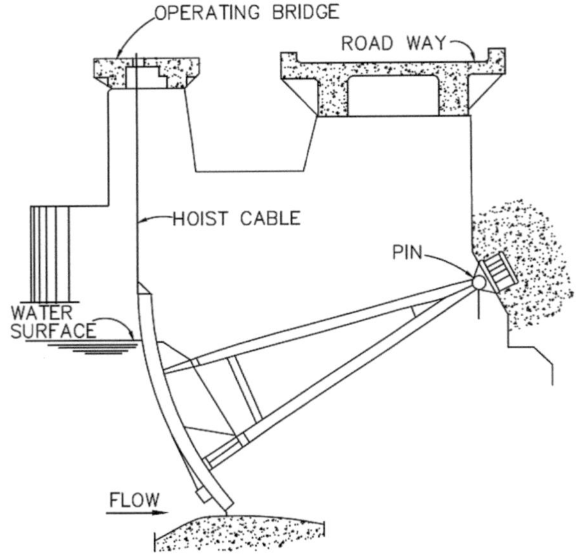

This type of gate has the following advantages over vertical lift gates.

1. Simplicity in construction
2. Easy access for maintenance
3. No grooves in pier are required. So, it offers favourable and smooth hydraulic flow conditions.
4. It does not require wheel and wheel assembly
5. Hoist capacity requirement is less.

The only difficult structural problem with this type of gate is the anchorage of the trunnion with concrete piers.

19.7.1 Hydraulic Forces on Radial Skin Plate

The hydraulic forces acting on a radial skin plate are shown in Fig. 19.8. The total hydrostatic force P is the resultant of horizontal force H which is equal to w $h^2/2$ and the vertical uplift force which is equal to the area of triangle ABC multiplied by w, the unit weight of water. The resultant P acts at 2/3 depth from the water surface i.e. 2/3 h. It passes through centre of rotation so it does not affect the hoisting capacity of the gate required for lifting.

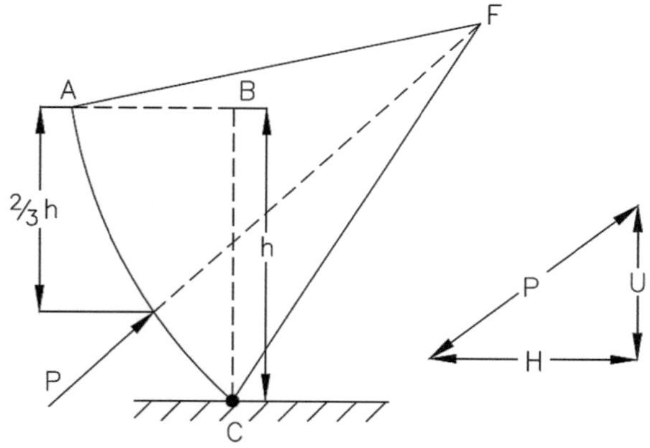

Fig. 19.8 Showing forces acting on skin plate

19.7.2 Design Aspects of Radial Gate Components

The design aspects of a radial gate involve the following:

(a) Fixing the following parameters of the gate:

1. Location of trunnion
2. Radius of gate
3. Location of sill
4. Location and type of hoists.

(b) Following components of the gate require structural design:

1. Skin plate
2. Stiffners
3. Horizontal girders
4. End arms
5. Trunnion hub, pin, bush etc.
6. Load carrying anchors
7. Anchorage girders
8. Thrust block and Trunion tie (if inclined arms are used)
9. Seals, seal seat, seal base, sill beam
10. Guide roller and
11. Anchor bolts.

The considerations in fixing the parameters and in the structural design of various components are given in following paras:

(a) **Fixing Parameters**

(i) **Location of Trunnion**

The trunnion and its assembly shall be so located that there is at least 1.5 m clearance above the water profile on the spillway crest under the condition of maximum discharge. The trunnion shall in no case be allowed to submerge under water. However, in the case of conduits/tunnels sometimes the trunnion gets submerged in water. Under such conditions, precautions shall be taken to prevent corrosion of trunnion parts.

The trunnion shall be so located that the resultant hydraulic thrust on the gate in the closed position for reservoir full condition lies as close to the horizontal as possible. This will reduce the upward or downward force which will otherwise be imposed on anchorage girder.

USBR from economic and operational considerations recommends that the height of the trunnion pin above a horizontal line from the sill shall be limited to 0.5 to 0.75 H, where H is the height of gate.

The location of the trunnion should allow the gate to be fully raised or lowered without interfering with the spillway or hoist bridge or any other part of structure.

(ii) **Radius of Gate**

The radius of the gate, that is the distance from the centre of trunnion pin to the inside face of the skin plate shall as far as possible vary between H and 1.25 H, where H is the vertical distance between top of gate and a horizontal through the sill.

(iii) **Location of Sill**

The sill of the gate is generally located slightly downstream of the crest to avoid cavitation of downstream glacis during partial gate opening.

The sill shall, as far as possible, be located at a place where a vertical plane tangent to the upstream face of the skin plate will intersect the spillway downstream of the crest. This will ensure the requirement of placing the sill in the downstream.

It shall also be ensured that the downstream shift of sill from the spillway crest is minimal to economise on the height of gate.

(iv) **Location of Hoist**

Radial gates are generally operated by rope-drum hoists. Hydraulic hoists are used to operate gates under high head in dam conduits or tunnels.

In the case of spillways and canal heads, hoists are generally installed on the roadway or on piers or on a deck below the roadway.

The hoist shall be so located that the hoisting force is applied to the gate at the largest possible radius and the hoisting angle does not change much during travel of the gate.

The hoisting is generally done from the upstream side of the gate, but it can be done from downstream in case reservoir water is corrosive. The hydraulic hoists are generally placed on downstream side of gate.

(b) **Structural Design Considerations (I.S. 4623)**

1. Skin Plate

Skin plate is rigidly fixed with stiffeners which are either horizontal or vertical or both. Horizontal stiffners are normally supported by suitably spaced vertical stiffners which are connected together by horizontal girders. The hydraulic force is transmitted through horizontal girders to the two end vertical girders. The end arms butt on the vertical girders.

The skin plate is designed as panels bending across the stiffners for maximum load. The design procedure is the same as discussed with vertical lift gates. The minimum thickness of plate is 8 mm. The panel size should be adjusted such that the plate thickness required for loads and that when increased by 1.5 mm for corrosion is around 8 mm.

2. **Vertical Stiffners**

The spacing of vertical stiffners depends on the design of skin plate and the spacing of horizontal girders. The spacing of vertical stiffners is normally 40 to 60 cm. These are designed as beams spanning between two horizontal girders. In design, the co-acting width of skin plate may be considered. These are made either of standard rolled sections or of built up sections.

3. **Horizontal Girders**

- Number and spacing of Horizontal Girders

Normally two horizontal girders are used for the gate height up to 8.5 m and three girders for gate height of 8.5 m to 12 m. Four and more girders are used for gates of a height of more than 12 m. To simplify the erection of the gate minimum number of girders shall be used.

The spacing of horizontal girders shall be so adjusted that the maximum positive and negative moments in the vertical stiffeners are equal.

- Design of Girders

The horizontal girders are designed for the loads transferred through vertical stiffners as point loads. These can also be for simplicity, designed for uniform loading which is transferred through the two panels of skin plate connected to it. The girders are designed taking into consideration the fixity at arms support. The girders are wide flange beams or built up girders depending on size of gate. The girders are checked for shear at the points where they are supported by end arms. The horizontal girders are suitably braced to ensure rigidity.

4. **End Arms**

The end arms transfer water load from the horizontal girders to the trunnion which transfers load to the pier. The arms are designed as columns for the axial load and bending moments transmitted by the horizontal girders. The compressive stress should remain within permissible limits for l/r ratio of the arm where l is effective length and r is least radius of gyration.

The arms are suitably braced by bracings in between the arms. These shall be so spaced that the l/r ratio of the arm in both the longitudinal and transverse directions are nearly equal.

The arms are made straight or parallel along the pier face. Inclined arms may be used to economise on horizontal girders. Inclined arms may be fixed to the horizontal girders at about 1/5th of the gate span from each end of the girders for the economic design of girders. Inclined arms develop thrust on the pier. So, to reduce the load on the pier a moderate slope for inclined arms is preferred.

5. **Trunnion Hub**

The trunnion hub is a thick circular cylinder sufficiently long to accommodate the arm of the gate which is rigidly fixed to it to transfer the load to the pin. The trunnion hub rotates about the pin. It is made of cast steel. It can be made of web and flanges also. The thickness of steel hub may be 0.3 d where d is diameter of pin subject to a minimum of 135 mm.

The inside of the trunnion hub is smooth finished and a bronze bushing is fixed in it. Side-type bushings are force fit in the trunnion hub and a running fit on the trunnion pin. Sometimes roller type bearings are used to reduce hoisting capacity.

Trunnion pins are normally supported at both ends on the trunnion bracket which is either fixed to the anchorage or support girder by bolts or welding. The trunnion bracket shall transfer total load from trunnion to the anchorage. Where convenient the pin may be a cantilever from the anchorage box embedded in the piers. The trunnion pin is designed for bending, shear and bearing for the total load transferred through the trunnion hub. The load is taken uniformly distributed over the length of pin. The trunnion pin is a medium fit in the bearing lugs of the support bracket and is suitably locked against rotation. The pin is made of corrosion-resistant steel. The trunnion pin along with the girder and anchorage is shown in Fig. 19.9.

6. **Trunnion Girder**

The trunnion girder (Fig. 19.9) transfers load from the pin to the concrete pier. In an intermediate pier, the trunnion girder is an overhanging beam. The overhanging portions of the girder are anchored into pier concrete through anchor rods. The trunnion girder is designed to withstand bending moment, shear force and torque. The design is checked for the condition when both gates are closed and also for the condition when one gate is closed and the other is open. The girder is normally made of a box section with web and flange plates. Suitable diaphragm plates are provided inside the trunnion girder to resist torque. The width of the girder should also be checked against the bearing stress of concrete.

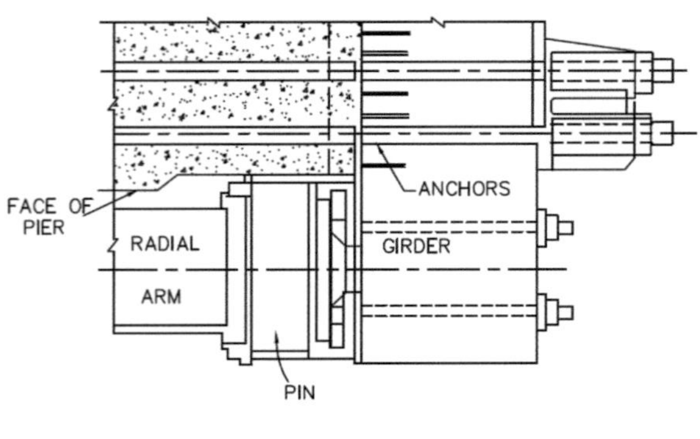

Fig. 19.9 Showing trunnion and anchorage at pier

In case of small gates used in canals, etc. the elaborate arrangement of trunnion girder and its anchorage may be replaced by supporting the bearing pedestal of the trunnion on a concrete projection from the pier which can be designed as a cantilever for the loads.

7. **Load Carrying Anchors**

Anchorage of the girder with the pier can be done in different ways. One anchorage arrangement consists of an embedded girder in pier concrete connected with a trunnion girder through tie bars. The other arrangement is to transfer water load from the trunion girder into the pier through a series of anchor rods as shown in Fig. 19.9 in which the load is transferred as bond stress between the anchors and the concrete. In the case of an embedded girder, the load is transferred as bearing stress between the concrete and the embedded girder at the upstream end of the tie bars.

The load carrying anchors are either flats placed vertically or rods. In the case of prestressed anchors, these are either rods or cables. The load carrying anchors are either welded to the trunnion girder or fixed by nuts.

The length of embedment of anchors for bonded type anchorage shall be based on permissible bond stress of concrete subject to a minimum of two-thirds the radius of the gate leaf. The anchors are hooked at ends or provided with anchor plates.

To avoid cracking of face concrete the bonded anchors shall have a minimum cover of 500 mm from the concrete face.

The length of anchors in case of upstream embedded girder where the load is transferred in bearing shall be such as to limit the shear stress in the 45° planes at the embedded girder to a safe permissible value.

All the load carrying anchors shall be suitably pretensioned to the same value to ensure that the anchors are of the same length and carry effectively their own portion of total load on the trunnion.

8. **Thrust Block or Trunnion Tie**

The thrust block or trunnion tie is required only if inclined arms are used with the gate. Figure 19.10 shows the inclined arm and the location of the thrust block. These are used to resist horizontal forces from the trunnion to the girder behind the trunnion. The thrust block is fixed to the trunnion girder/yoke girder and is designed to withstand the bending and shear force caused by the side thrust on the trunnion due to the inclined arm. A thrust washer should be used between the trunnion hub and the trunnion bracket to transfer the thrust.

To allow for the elongation of load-carrying anchors, bronze or stainless steel clad sliding surfaces shall be provided on the face of the thrust block and the matching face of the trunnion bracket. The trunnion tie shall span from one trunnion hub to the other and shall be rigidly fixed with the trunnion hub either by welding or bolts.

A steel pipe with flanges at ends for fixing or standard rolled structural sections like I-Joist may be used for the trunnion tie. The deflection in the tie for self-load shall remain within permissible limits.

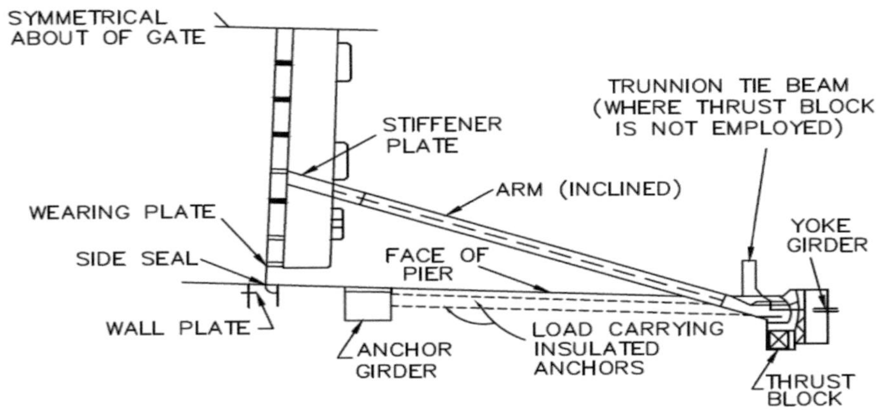

Fig. 19.10 Arrangement of radial gate with respect to face of pier

9. **Seals**

The seals shall be fixed on all water sides of the gate in such a way that they bear tightly on seal seats to prevent leakage in fully closed position of gate. For reducing seal friction metal clad seals may be used. The cladding may be of brass, bronze, stainless steel or corrosion-resisting steel to ensure water-tight mating of gate leaf with embedded frame. Generally, rubber seals are used. These shall be as per I.S. 11855.

10. **Seal Seat, Seal Seat Base and Sill Beam**

The seal seat of stainless steel shall be either welded or screwed to the seal seat base. The screws shall be of stainless steel or of corrosion-resistant steel. The seal seat shall be smooth-finished. The minimum thickness of the seal seat plate shall be 6 mm for low head gates and 8 mm for high head gates after machining.

The seal seat base on which seal seat is fixed may be of any structural section. Angles or flats are used to fix the seal seat base with embedded anchor bolts. After fixing, the seal seat shall preferably flush with the surrounding concrete.

The sill beam shall be made of stainless-steel flats welded or screwed by stainless steel screws with the rolled steel beam. The surface of the steel flat shall be machine finished and its minimum thickness after finish shall be 6 mm. The top of the sill beam shall flush with the surrounding concrete. The sill beam shall be anchored with concrete.

11. **Guide Roller**

Guide rollers shall be provided on the sides of the gate to limit the lateral motion or side sway of the gate to not more than 6 mm in either direction. The two guide rollers shall be provided on each side of the gate. These shall remain within the wall plate area when the gate is fully raised. Wall plates may be of structural steel. The rollers shall be adjustable and removable.

12. **Anchor Bolts**

To ensure proper alignment of embedded parts, anchorages are provided in two stages. Anchor bolts shall be provided in the first stage concrete with suitable block out openings. The seal base and seal seat assemblies are placed in second-stage concrete. For proper bonding of first stage and second stage concrete suitable dowels should be provided and the interface surface should be properly prepared. The anchor bolts shall be provided with double nuts and washers. It is preferable to embed anchor plates in first stage concrete and to weld anchor bolts with plates subsequently. The minimum size of anchor bolts is suggested in I.S. 4623 as 16 mm dia.

19.8 Bye Pass Gate

In case of a forebay, an automatic gate is generally provided at the crest of bye-pass head. The gate should operate automatically to release excess water as soon as the water head in the forebay rises due to sudden closer or reduction in generation of power at the machine in the power house. Among several types of such gates (CBI&P, Pub 305), the widely used gate is float operated fish belly gate. It is shown in Fig. 19.11. In this type, the lower plate of the gate simulates the belly of a fish. In a lowered position, the gate sits in a groove in the crest, and the upper skin plate of the gate follows the nappe of the flow. These gates have been used for heights up to 6 m.

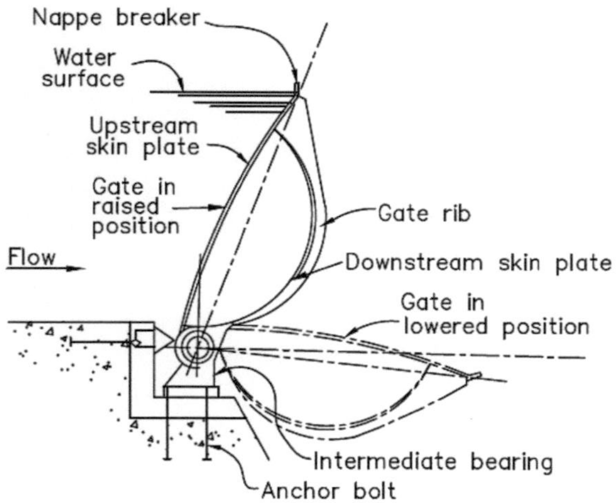

Fig. 19.11 Fish belly type hinged gate

19.9 Valves

Generally, valves are placed at the end of the penstock. These act as the main inlet valve in the power house. The valve is placed at the entry of the scroll case of the turbine. Butterfly valve is normally used for heads up to 100 m and for higher heads spherical valve is used. These two valves are briefly described below.

19.9.1 Butterfly Valve

A butterfly valve consists essentially of a cylindrical body with a circular leaf mounted on a transverse shaft with two bearings diametrically opposite to each other in the valve body. Details are shown in Fig. 19.12. The operating mechanism which is outside the valve rotates the leaf/disc to 90° to move it from a fully closed to a fully open position.

For small-sized valves, operating at low or medium heads, a gear-reducing unit driven by an electric motor may be used as an operating mechanism. For high heads and big valves, the operating mechanism is made of hydraulic cylinders or rotors. Normally the leaf driving unit is placed on the valve body to make a self-contained unit.

The valve body is shaped to avoid abrupt changes in velocity. Between the valve and scroll case a reducer may be used to give a constant increase in velocity. Butter fly valves up to a size of 6 m diameter have been used.

In order to eliminate leakage in a closed position in high head installation operating with sediment laden water some form of adjustable sealing device is used.

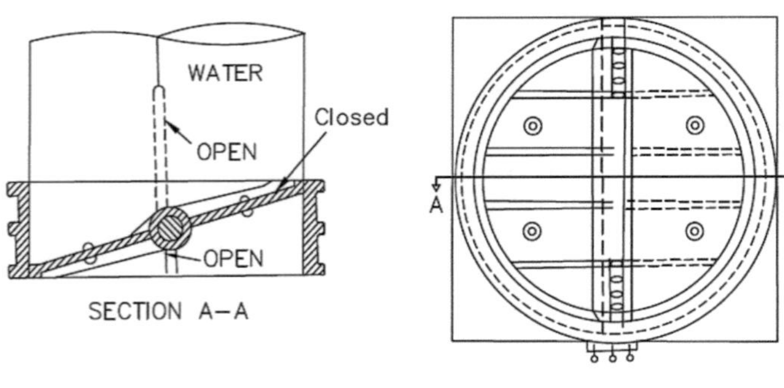

Fig. 19.12 Butterfly valve (P.S. Nigam)

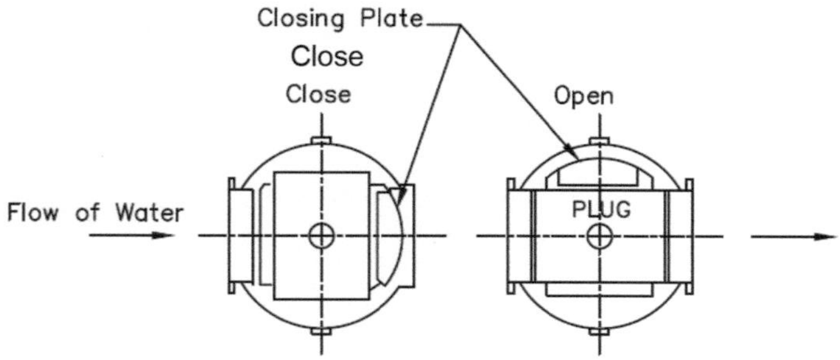

Fig. 19.13 Showing spherical valve. (P.S. Nigam)

19.9.2 Spherical Valve

The shape of this type of valve is spherical and so it is commonly known as spherical valve. The closing member is a plug shaped pipe so it is also called a plug valve. This valve is used for high heads beyond the limit for butterfly valves. It is shown in Fig. 19.13.

The valve consists of a plug which is a portion of pipe carrying two bearing trunnions supported in the spherical body mounted in the penstock. The plug has a closing plate at one end. Normally valve is of small size and has high velocity and so a diverging-shaped section is introduced between the valve and the scroll case to restore turbine approach velocity.

This type of valve has the following advantages:

- The spherical shape can withstand high pressure
- Hydraulic losses are less
- Leakage is less because the closing plate is not affected by flow
- Size required for the valve is small so the space required is less.

Figure 19.13 explains the operation of the valve. When the valve is closed the plug cylinder is perpendicular to the axis of the pipe and the closing plate mounted on the cylinder closes the opening. While opening the plug the cylinder is rotated and the cylinder comes in line with the axis of pipe. The flow passes through the plug cylinder. Before opening the valve, the closing plate is raised from the seat by relieving high pressure on the plate through a bye pass mechanism.

References

1. CBI&P.: Manual on Development of Small Hydroelectric Projects. Pub. 305, New Delhi (2009)
2. I.S. 4622.: Recommendations for Structural Design of Fixed Wheel Gates. B.I.S., New Delhi (2003)
3. I.S. 5630.: Recommendations for Structural Design Criteria for Low Head Slide Gates. B.I.S., New Delhi (2020)

4. I.S. 11228.: Recommendation for Design of Screw Hoists for Hydraulic Gates', B.I.S. New Delhi (1985)
5. I.S. 4623.: Recommendations for Structural Design of Radial Gates. B.I.S, New Delhi (2000)
6. I.S. 5620.: Recommendation for Structural Design of Low Head Slide Gates. BIS. New Delhi (2020)
7. Nigam, P.S.: Handbook of Hydro Electric Engineering. New Chand and Co., Roorkee (1985)

MIX
Papier aus verantwortungsvollen Quellen
Paper from responsible sources
FSC® C105338

If you have any concerns about our products,
you can contact us on
ProductSafety@springernature.com

In case Publisher is established outside the EU,
the EU authorized representative is:
Springer Nature Customer Service Center GmbH
Europaplatz 3, 69115 Heidelberg, Germany

Printed by Libri Plureos GmbH
in Hamburg, Germany